# SOLUTIONS MANUAL

for the

# MECHANICAL ENGINEERING REFERENCE MANUAL

Ninth Edition

Michael R. Lindeburg, P.E.

PROFESSIONAL PUBLICATIONS, INC.
Belmont, CA 94002

**In the ENGINEERING REFERENCE MANUAL SERIES**

Engineer-In-Training Reference Manual
  Engineering Fundamentals Quick Reference Cards
  Engineer-In-Training Sample Examinations
  Mini-Exams for the E-I-T Exam
  1001 Solved Engineering Fundamentals Problems
  E-I-T Review: A Study Guide
  Diagnostic F.E. Exam for the Macintosh
  Fundamentals of Engineering: Thermodynamics
Civil Engineering Reference Manual
  Civil Engineering Quick Reference Cards
  Civil Engineering Sample Examination
  Civil Engineering Review Course on Cassettes
  Seismic Design of Building Structures
  Seismic Design Fast
  Timber Design for the Civil P.E. Examination
  Fundamentals of Reinforced Masonry Design
  246 Solved Structural Engineering Problems
Mechanical Engineering Reference Manual
  Mechanical Engineering Quick Reference Cards
  Mechanical Engineering Sample Examination
  101 Solved Mechanical Engineering Problems
  Mechanical Engineering Review Course on Cassettes
  Consolidated Gas Dynamics Tables
  Fire and Explosion Protection Systems
Electrical Engineering Reference Manual
  Electrical Engineering Quick Reference Cards
  Electrical Engineering Sample Examination
Chemical Engineering Reference Manual
  Chemical Engineering Quick Reference Cards
  Chemical Engineering Practice Exam Set
Land Surveyor Reference Manual
  1001 Solved Surveying Fundamentals Problems
Engineering Economic Analysis
Engineering Law, Design Liability, and Professional Ethics
Engineering Unit Conversions

**In the ENGINEERING CAREER ADVANCEMENT SERIES**

How to Become a Professional Engineer
The Expert Witness Handbook—A Guide for Engineers
Getting Started as a Consulting Engineer
Intellectual Property Protection—A Guide for Engineers
E-I-T/P.E. Course Coordinator's Handbook
Becoming a Professional Engineer
Engineering Your Start-Up
High-Technology Degree Alternatives

**SOLUTIONS MANUAL for the MECHANICAL ENGINEERING REFERENCE MANUAL**
Ninth Edition



Printed in the United States of America

ISBN: 0-912045-73-6

Professional Publications, Inc.
1250 Fifth Avenue, Belmont, CA 94002
(415) 593-9119

Current printing of this edition: 1

# TABLE OF CONTENTS

# Notice to Examinees

Do not copy, memorize, or distribute problems from the Principles and Practice of Engineering (P&P) Examination. These acts are considered to be exam subversion.

The P&P examination is copyrighted by the National Council of Examiners for Engineering and Surveying. Copying and reproducing P&P exam problems for commercial purposes is a violation of federal copyright law. Reporting examination problems to other examinees invalidates the examination process and threatens the health and welfare of the public.

# Mathematics

WARM-UPS

1 $\sum_{J=1}^{5}(J+1)^2-1 = (1+1)^2-1 + (2+1)^2-1 + (3+1)^2-1 + (4+1)^2-1 + (5+1)^2-1$

$= 2^2+3^2+4^2+5^2+6^2-5 = 85$

2 THE ACTUAL VALUE IS

$y(2.7) = 3(2.7)^{.93} + 4.2 = 11.756$

$y(2) = 3(2)^{.93} + 4.2 = 9.916$

$y(3) = 3(3)^{.93} + 4.2 = 12.534$

THE ESTIMATED VALUE IS

$9.916 + .7(12.534 - 9.916) = 11.749$

THE ERROR IS

$\frac{11.756 - 11.749}{11.756} = .0006$ OR $.06\%$

3 LET d BE THE DIAMETER

$V_{SPHERE} = \frac{4}{3}\pi r^3 = \frac{4}{3}\pi\left(\frac{d}{2}\right)^3 = .524d^3$

$V_{CONE} = \frac{\pi}{3}r^2h = \frac{\pi}{3}\left(\frac{d}{2}\right)^2 h = .262d^2h$

BUT $.524d^3 = .262d^2h$

$h = 2.00d$

4

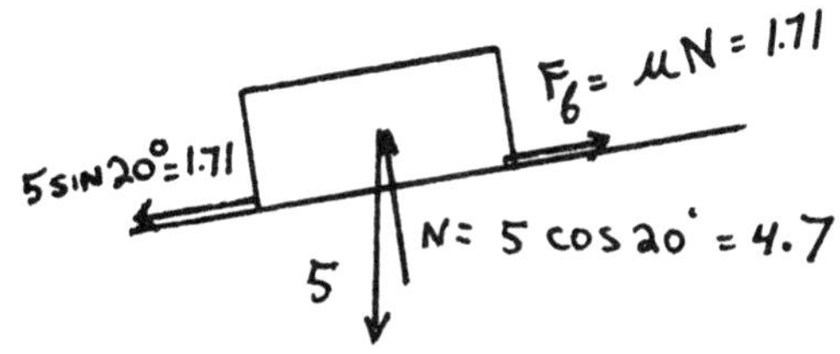


5 EXPAND BY 2ND COLUMN

$-2\begin{vmatrix} 4 & 3 \\ 9 & 5 \end{vmatrix} = -2(20-27) = 14$

6 FROM EQN 6.3

$250° + 460° = 710°R$

$\frac{5}{9}(250-32) = 121.1°C$

7 FROM PAGE 1-42,

$K = 1.71\text{ EE-}9 \frac{BTU}{FT^2\text{-}HR\text{-}°R^4}$

$$\frac{\left(1.71\text{ EE-}9 \frac{BTU}{FT^2\text{-}HR\text{-}°R^4}\right)\left(17.57 \frac{WATT\text{-}MIN}{BTU}\right)\left(\frac{1}{60} \frac{HR}{MIN}\right)}{\left(0.3048 \frac{m}{ft}\right)^2 \left(\frac{5}{9} \frac{K}{°R}\right)^4}$$

$= 5.66\text{ EE-}8 \frac{WATTS}{M^2\text{-}K^4}$

8 $y = 6 + .75(2-6) = 3.0$

9 THE SLOPE IS $\frac{9.5-3.4}{8.3-1.7} = .924$

USING THE FIRST POINT,

$(y-3.4) = .924(x-1.7)$

10 LET X BE THE NUMBER OF ELAPSED PERIODS OF .1 SECOND. LET $Y_x$ BE THE AMOUNT PRESENT AFTER X PERIODS

$Y_1 = 1.001\,Y_0$

$Y_2 = (1.001)^2 Y_0$

$Y_N = (1.001)^N Y_0$

NOW $\frac{Y_x}{Y_0} = 2 = (1.001)^N$

$\log(2) = N\log(1.001)$

$N = 693.5$ PERIODS

$t = 69.35$ SECONDS

CONCENTRATES

1 FIRST, REARRANGE

$x + y = -4$

$x + z = 1$

$3x - y + 2z = 4$

NOW, USE CRAMER'S RULE (PAGE 1-6)

$$\begin{vmatrix} 1 & 1 & 0 \\ 1 & 0 & 1 \\ 3 & -1 & 2 \end{vmatrix} = 1\begin{vmatrix} 0 & 1 \\ -1 & 2 \end{vmatrix} - 1\begin{vmatrix} 1 & 1 \\ 3 & 2 \end{vmatrix} = (0+1) - (2-3) = 1+1 = 2$$

$$\begin{vmatrix} -4 & 1 & 0 \\ 1 & 0 & 1 \\ 4 & -1 & 2 \end{vmatrix} = -2 \quad \begin{vmatrix} 1 & -4 & 0 \\ 1 & 1 & 1 \\ 3 & 4 & 2 \end{vmatrix} = -6 \quad \begin{vmatrix} 1 & 1 & -4 \\ 1 & 0 & 1 \\ 3 & -1 & 4 \end{vmatrix} = 4$$

$x^* = \frac{-2}{2} = -1 \quad y^* = \frac{-6}{2} = -3 \quad z^* = \frac{4}{2} = 2$

**2** ALWAYS GRAPH THE DATA FIRST TO SEE IF IT IS A STRAIGHT LINE. IN THIS CASE, IT IS.

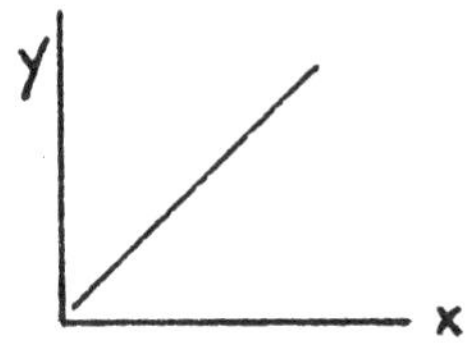


USE LINEAR REGRESSION (PAGE 1-13)

$N = 7$

$\Sigma x = 12550$ $\qquad \Sigma y = 12300$

$\bar{x} = 1792.9$ $\qquad \bar{y} = 1757.1$

$\Sigma x^2 = 3.117 \text{ EE7}$ $\qquad \Sigma y^2 = 3.017 \text{ EE7}$

$(\Sigma x)^2 = 1.575 \text{ EE8}$ $\qquad (\Sigma y)^2 = 1.513 \text{ EE8}$

$\Sigma xy = 3.067 \text{ EE7}$

$$m = \frac{7(3.067\text{ EE7}) - (12550)(12300)}{7(3.117\text{ EE7}) - (12550)^2} = .994$$

$$b = 1757.1 - .994(1792.9) = -25.0$$

so $y = .994x - 25.0$

THE CORRELATION COEFFICIENT IS

$$r = \frac{7(3.067\text{ EE7}) - (12550)(12300)}{\sqrt{\left[(7)(2.117\text{ EE7}) - (12550)^2\right]\left[(7)(3.017\text{ EE7}) - (12300)^2\right]}}$$

$$= \sim 1.00$$

**3** UPON GRAPHING THE DATA, WE SEE THAT IT IS NOT A STRAIGHT LINE

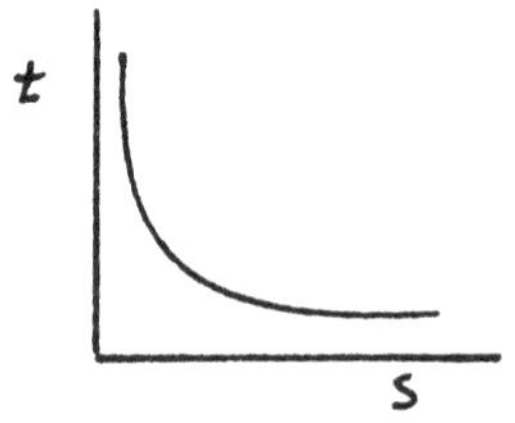


IT LOOKS LIKE AN EXPONENTIAL WITH FORM

$$t = b e^{ms}$$

OR PERHAPS

$$\log t = b + ms$$

TRY MAKING THE VARIABLE TRANSFORMATION $R = \log t$

| S | R |
|---|---|
| 20 | 1.633 |
| 18 | 2.149 |
| 16 | 2.585 |
| 14 | 3.041 |

$N = 4$

$\Sigma S = 68$ $\qquad \Sigma R = 9.408$

$\bar{S} = 17$ $\qquad \bar{R} = 2.352$

$\Sigma S^2 = 1176$ $\qquad \Sigma R^2 = 23.215$

$(\Sigma S)^2 = 4624$ $\qquad (\Sigma R)^2 = 88.51$

$\Sigma SR = 155.28$

$$m = \frac{4(155.28) - (68)(9.408)}{4(1176) - (68)^2} = -.2328$$

$$b = 2.352 + .2328(17) = 6.3096$$

so $R = 6.3096 - .2328S$

or $\log t = 6.3096 - .2328S$

**4** THIS IS A FIRST-ORDER LINEAR DIFFERENTIAL EQUATION. (PAGE 1-31)

$$\mu = \exp\left[\int -1\, dx\right] = e^{-x}$$

$$y = e^{x}\left[2\int e^{-x} x e^{2x} dx + c\right]$$

$$= e^{x}\left[2xe^{x} - 2e^{x} + c\right]$$

BUT $y = 1$ WHEN $x = 0$

$$1 = 1\left[0 - 2 + c\right] \quad \text{or } c = 3$$

so $y = 2e^{2x}(x-1) + 3e^{x}$

**5** SOLVE THE CHARACTERISTIC QUADRATIC EQUATION:

$$R^2 - 4R - 12 = 0$$

$$R = 6, -2$$

so $y = a_1 e^{6x} + a_2 e^{-2x}$

**6** LET $X_t$ = POUNDS OF SALT IN TANK AT TIME $t$

$X_0 = 60$

$X'$ = RATE AT WHICH SALT CONTENT CHANGES

2 = POUNDS OF SALT ENTERING EACH MINUTE

3 = GALLONS LEAVING EACH MINUTE.

THE SALT LEAVING EACH MINUTE IS

$$3\left(\begin{matrix}\text{CONCENTRATION}\\ \text{IN LB/GAL}\end{matrix}\right) = 3\left(\frac{\text{SALT CONTENT}}{\text{VOLUME}}\right) = 3\left(\frac{X}{100-t}\right)$$

$$X' = 2 - 3\left(\frac{X}{100-t}\right)$$

OR $$X' + \frac{3X}{100-t} = 2$$

THIS IS FIRST ORDER LINEAR (PAGE 1-31)

$$\mu = \text{EXP}\left[3\int\frac{dt}{100-t}\right] = (100-t)^{-3}$$

$$X = (100-t)^3\left[2\int\frac{dt}{(100-t)^3} + K\right]$$

$$= 100 - t + K(100-t)^3$$

BUT $X = 60$ AT $t = 0$

SO $K = -.00004$

$$X = 100 - t - .00004(100-t)^3$$

$$X_{60} = 37.44 \text{ POUNDS}$$

**7** IF $C$ IS POSITIVE, THEN $N(\infty) = \infty$, WHICH IS CONTRARY TO THE GIVEN DATA. SO $C \le 0$.

IF $C = 0$, THEN $N(\infty) = \frac{a}{1+b} = 100$ WHICH IS POSSIBLE DEPENDING ON $a, b$

IF $C = 0$, THEN $N(0) = \frac{a}{1+b} = 10$, WHICH CONFLICTS WITH THE PREVIOUS STEP.

SO $C < 0$. THEN $N(\infty) = a$, SO $\underline{\underline{a = 100}}$

NOW $N(0) = \frac{a}{1+b} = \frac{100}{1+b} = 10$, SO $\underline{\underline{b = 9}}$

$$\frac{dN}{dt} = -100(1+9e^{ct})^{-2}(9)e^{ct}(c)$$

IF $t = 0$, THEN $C = \underline{\underline{-.0556}}$

**8** $$\frac{dy}{dx} = 3x^2 - 18x$$

$3x^2 - 18x = 0$ AT ALL EXTREME POINTS

$x^2 - 6x = 0$ AT $x = 0$, $x = 6$

$$\frac{d^2y}{dx^2} = 6x - 18$$

$6x - 18 = 0$ AT INFLECTION POINTS

$x = 3$ IS AN INFLECTION POINT

$6(0) - 18 = -18$, SO $x = 0$ IS A MAXIMUM

$6(6) - 18 = 18$, SO $x = 6$ IS A MINIMUM

**9** THE ENERGY CONTAINED IN ONE GRAM OF ANY SUBSTANCE IS

$$E = MC^2 = (.001)\text{ KG}\,(3\text{ EE}8)^2(\text{M/S})^2$$

$$= 9\text{ EE}13 \text{ JOULES}$$

$$(9\text{ EE}13)\text{ J}\left(\frac{1}{1000}\right)\frac{\text{KJ}}{\text{J}}(.9478)\frac{\text{BTU}}{\text{KJ}}$$

$$= 8.53\text{ EE}10 \text{ BTU}$$

$$\#\text{ TONS} = \frac{8.53\text{ EE}10\text{ BTU}}{(13{,}000)\frac{\text{BTU}}{\text{LB}}(2000)\frac{\text{LB}}{\text{TON}}}$$

$$= 3281 \text{ TONS}$$

**10**

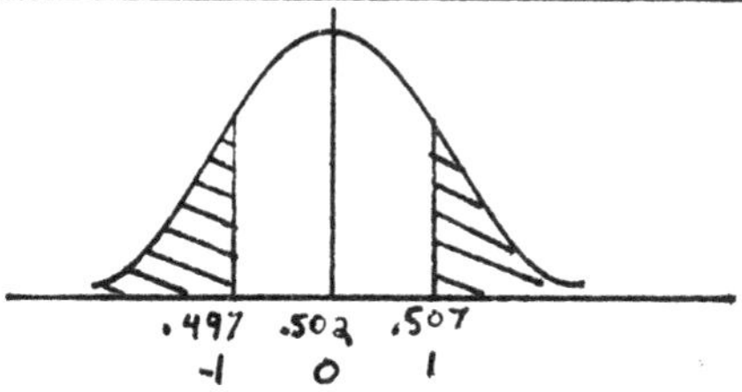


THE STANDARD NORMAL VARIABLES ARE

$$z_1 = \frac{.502 - .497}{.005} = 1$$

$$z_2 = \frac{.507 - .502}{.005} = 1$$

a) $p\{\text{DEFECTIVE}\} = 2[.5 - .3413] = .3174$

b) $$p\{3,2\} = \frac{3!}{(3-2)!\,2!}(.3174)^2(1-.3174)^1$$
$$= .2063$$

c) $(8)(200)(.3174) = 507.8$

**11**

THE RANGE OF SPEEDS IS (48 - 20) = 28. SINCE THERE ARE NOT A LOT OF OBSERVATIONS, 10 CELLS WOULD BE BEST. CHOOSE THE CELL WIDTH AS $\left(\frac{28}{10}\right) \approx 3$

| INTERVAL | MID-POINT | FREQ. | CUM FREQ. | CUM % |
|---|---|---|---|---|
| 20-22 | 21 | 1 | 1 | .03 |
| 23-25 | 24 | 3 | 4 | .10 |
| 26-28 | 27 | 5 | 9 | .23 |
| 29-31 | 30 | 8 | 17 | .43 |
| 32-34 | 33 | 3 | 20 | .50 |
| 35-37 | 36 | 4 | 24 | .60 |
| 38-40 | 39 | 3 | 27 | .68 |
| 41-43 | 42 | 8 | 35 | .88 |
| 44-46 | 45 | 3 | 38 | .95 |
| 47-49 | 48 | 2 | 40 | 1.00 |

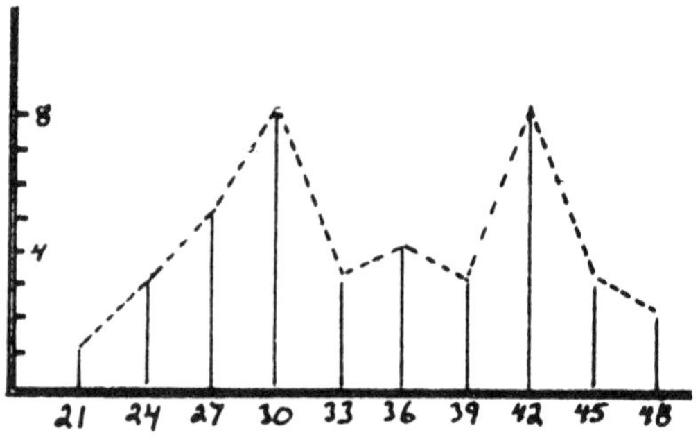


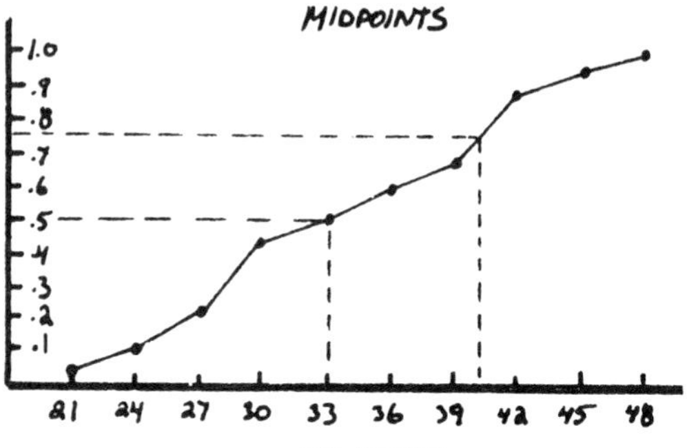


f) USE THE CUMULATIVE DISTRIBUTION GRAPH. FOR 75% (.75), THE CELL MID-POINT IS APPROXIMATELY 40

g) USE THE CUMULATIVE GRAPH TO FIND THE MID-POINT FOR 50%. THIS OCCURS AT APPROXIMATELY 33.

$\Sigma x_i = 1390$, SO MEAN $= \frac{1390}{40} = 34.75$

h) $\Sigma x^2 = 50496$

$$\sigma = \sqrt{\left(\frac{50496}{40}\right) - \left(\frac{1390}{40}\right)^2} = 7.405$$

i) $$s = \sqrt{\frac{N}{N-1}}(\sigma) = \sqrt{\frac{40}{39}}(7.405) = 7.500$$

j) $s^2 = 56.24$

**12** NO CONTRACT DEADLINE WAS GIVEN, SO ASSUME 36 AS A SCHEDULED TIME. LOOK FOR A PATH WHERE (LS - ES) = 0 EVERYWHERE

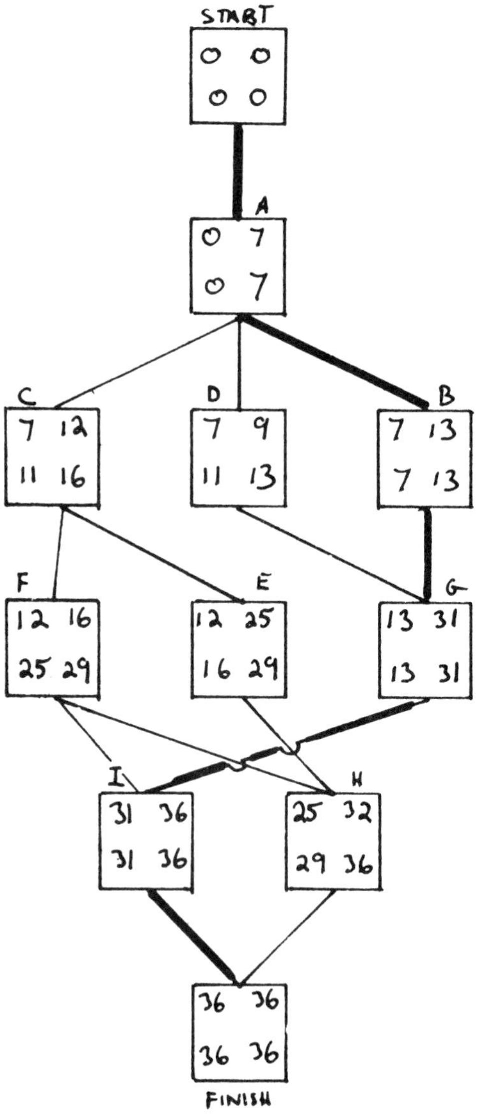


c) 36

d) 36

e) 0

f) FLOAT IS SAME AS SLACK = 0

**13** TO SOLVE THIS AS A REGULAR CPM PROBLEM, IT IS NECESSARY TO CALCULATE $t_{mean}$ AND $\sigma$ FOR EACH ACTIVITY. FOR ACTIVITY A,

$$t_{MEAN} = \frac{1}{6}\left[1 + (4)(2) + 5\right] = 2.33$$

$$\sigma_A = \frac{1}{6}(5-1) = .67$$

THE FOLLOWING TABLE IS GENERATED IN THE SAME MANNER.

| ACTIVITY | $t_{mean}$ | $\sigma$ |
|---|---|---|
| START | 0 | 0 |
| A | 2.33 | .67 |
| B | 10.5 | 2.17 |
| C | 11.83 | 2.17 |
| D | 4.17 | .83 |
| FINISH | 0 | 0 |
| | 28.83 | |

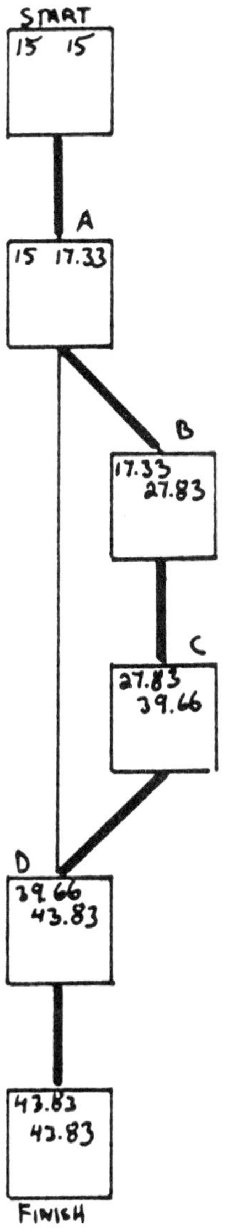


BY OBSERVATION, THE CRITICAL PATH IS START-A-B-C-D-FINISH, THE PROJECT VARIANCE IS

$$\sigma^2 = (.67)^2 + (2.17)^2 + (2.17)^2 + (.83)^2 = 10.56$$

THE PROJECT STANDARD DEVIATION IS

$$\sigma = \sqrt{10.56} = 3.25$$

SO, WE ASSUME THE COMPLETION TIMES ARE NORMALLY DISTRIBUTED WITH A MEAN OF 28.83 AND A STANDARD DEVIATION OF 3.25

THE STANDARD NORMAL VARIABLE IS

$$z = \frac{28.83 + 15 - 42}{3.25} = .56$$

AREA UNDER TAIL FOR $z = .56$ IS .2123

SO $.5 - .2123 = .2877$ (28.77%)

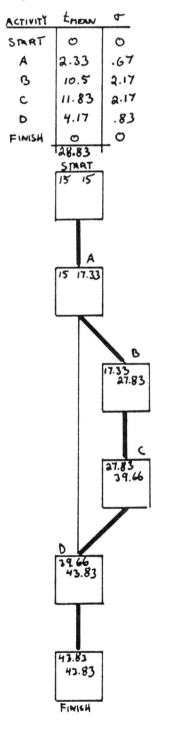

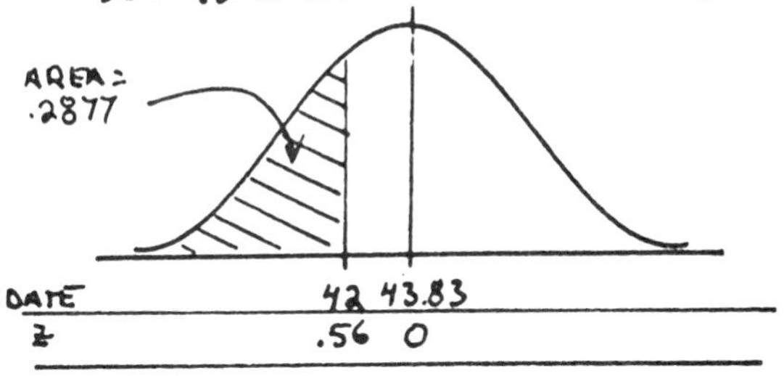


**14** $\lambda = 20$

FROM PAGE 1-22,

(a) $p\{x=17\} = f(17) = \dfrac{e^{-20}(20)^{17}}{17!} = .076$

(b) $p\{x \leq 3\} = f(0) + f(1) + f(2) + f(3)$

$$= \frac{e^{-20}(20)^0}{0!} + \frac{e^{-20}(20)^1}{1!} + \frac{e^{-20}(20)^2}{2!} + \frac{e^{-20}(20)^3}{3!}$$

= 2 EE-9 + 4.12 EE-8 + 4.12 EE-7 + 2.75 EE-6

= 3.2 EE-6

**15** FROM PAGE 1-22,

$\mu = 1/23$

$$p\{x > 25\} = 1 - F(25) = e^{-\left(\frac{1}{23}\right)(25)}$$

$$= .337$$

**16** FIRST, PLOT THE DATA TO SEE IF IT IS LINEAR

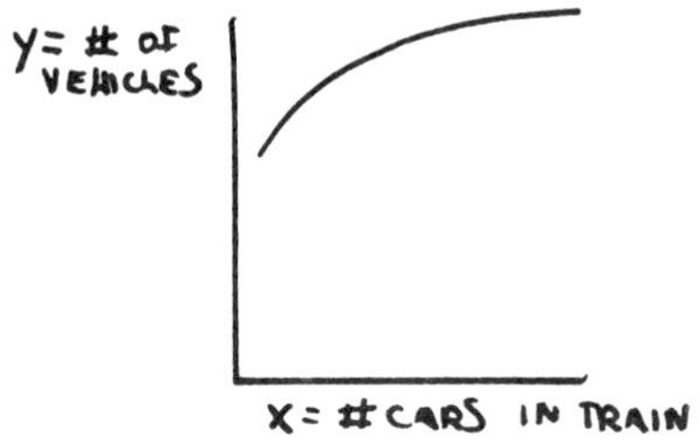


IT DOESN'T LOOK LINEAR, SO TRY THE FORM

$y = a + bz$

WHERE $z = \log_{10} x$

| $z$ | $Y$ |
|---|---|
| .3 | 14.8 |
| .7 | 18.0 |
| .9 | 20.4 |
| 1.08 | 23.0 |
| 1.43 | 29.9 |

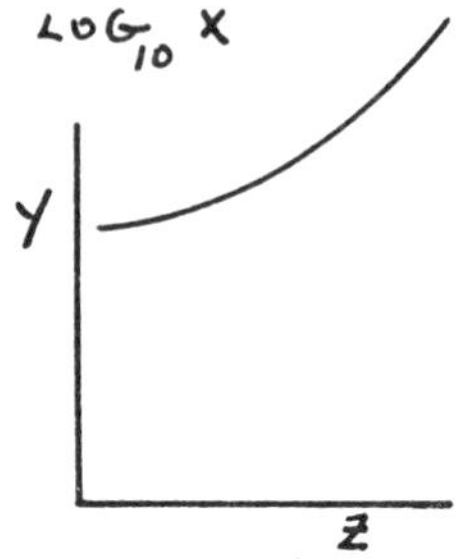


THAT ISN'T LINEAR EITHER, WE'VE 'OVERCOMPENSATED' THE CURVE, TRY THE FORM

$y = a + bw$

WHERE $w = z^2 = (\log_{10} x)^2$

| $W$ | $Y$ |
|---|---|
| .09 | 14.8 |
| .49 | 18.0 |
| .81 | 20.4 |
| 1.17 | 23.0 |
| 2.04 | 29.9 |

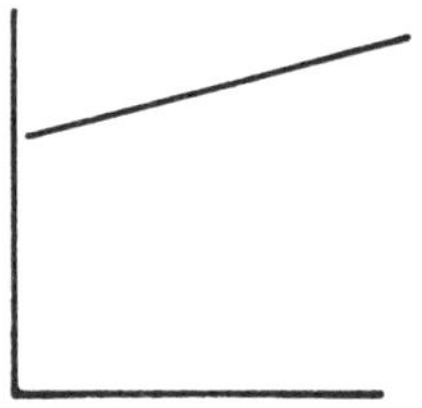

FROM PAGE 1-13

$\Sigma w_i = 4.6$ $\Sigma y_i = 106.1$

$(\Sigma w_i)^2 = 21.16$ $(\Sigma y_i)^2 = 11257.2$

$\Sigma w_i^2 = 6.43$ $\Sigma y_i^2 = 2382.2$

$\bar{w} = \frac{4.6}{5} = .92$ $\bar{y} = \frac{106.1}{5} = 21.22$

$\Sigma w_i y_i = 114.58$

$$m = \frac{(5)(114.58) - (4.6)(106.1)}{(5)(6.43) - 21.16} = \frac{84.84}{10.99}$$

$$= 7.72$$

$b = 21.22 - (7.72)(.92) = 14.12$

SO,

$$y = 14.12 + 7.72\,w$$
$$= 14.12 + 7.72\,z^2$$
$$= 14.12 + 7.72\,(\log_{10} x)^2$$

THE CORRELATION COEFFICIENT IS

$$r = \frac{(5)(114.58) - (4.6)(106.1)}{\sqrt{[(5)(6.43) - 21.16][(5)(2382.2) - 11257.2]}} \approx 0.999$$

{THE TRANSFORMATION $y = a + b\sqrt{x}$ YIELDS $y = 9.11 + 4\sqrt{x}$ AND $r = 0.999$, EQUALLY GOOD.}

**17**

(a) USE THE 'CHARACTERISTIC EQUATION' METHOD TO SOLVE THE HOMOGENEOUS CASE. (IT IS MUCH QUICKER TO USE LAPLACE TRANSFORMS, HOWEVER)

$x'' + 2x' + 2x = 0$ DIFF. EQ.

$R^2 + 2R + 2 = 0$ CHARACTERISTIC EQ.

COMPLETE THE SQUARE TO FIND $R$

$R^2 + 2R = -2$

$(R+1)^2 = -2 + 1$

$R + 1 = \pm\sqrt{-1}$

$R = -1 \pm i$

SO,

$x(t) = A_1 e^{-t}\cos t + A_2 e^{-t}\sin t$

NOW, USE THE INITIAL CONDITIONS TO FIND $A_1$ AND $A_2$.

$x(0) = 0$

$0 = A_1(1)(1) + A_2(1)(0)$

SO, $A_1 = 0$

DIFFERENTIATING THE SOLUTION,

$x'(t) = A_2[e^{-x}\cos x - \sin x\, e^{-x}]$

USING $x'(0) = 1$

$1 = A_2[(1)(1) - (0)(1)]$

SO, $A_2 = 1$

AND, THE SOLUTION IS $x(t) = e^{-t}\sin t$

(b) WITH NO DAMPING, THE DIFFERENTIAL EQUATION WOULD BE

$x'' + 2x = 0$

THIS HAS A SOLUTION OF $x = \sin\sqrt{2}\,t$, SO $\omega_{nat} = \sqrt{2}$

(c) $x(t) = e^{-t}\sin t$

$x'(t) = e^{-t}\cos t - \sin t\, e^{-t}$

$= e^{-t}(\cos t - \sin t)$

FOR $x$ TO BE MAXIMUM, $x'(t) = 0$. SINCE $e^{-t}$ IS NOT ZERO UNLESS $t$ IS VERY LARGE, $\cos t - \sin t$ MUST BE ZERO. THIS OCCURS AT $t = .785$ RADIANS, SO

$x(.785) = e^{-.785}\sin(.785)$

$= .322$

(d) USE THE LAPLACE TRANSFORM METHOD

$x'' + 2x' + 2x = \sin(t)$

$\mathcal{L}(x'') + 2\mathcal{L}(x') + 2\mathcal{L}(x) = \mathcal{L}(\sin(t))$

$s^2\mathcal{L}(x) - 1 + 2s\mathcal{L}(x) + 2\mathcal{L}(x) = \frac{1}{s^2+1}$

$\mathcal{L}(x)[s^2 + 2s + 2] - 1 = \frac{1}{s^2+1}$

$$\mathcal{L}(x) = \frac{1}{s^2+2s+2} + \frac{1}{(s^2+1)(s^2+2s+2)}$$

$$= \frac{1}{(s+1)^2+1} + \frac{1}{(s^2+1)(s^2+2s+2)}$$

USE PARTIAL FRACTIONS TO EXPAND THE SECOND TERM:

$$\frac{1}{(s^2+1)(s^2+2s+2)} = \frac{A_1+B_1 s}{s^2+1} + \frac{A_2+B_2 s}{s^2+2s+2}$$

CROSS MULTIPLYING,

$$= \frac{A_1 s^2+2A_1 s+2A_1+B_1 s^3+2B_1 s^2+2B_1 s+A_2 s^2+A_2+B_2 s^3+B_2 s}{(s^2+1)(s^2+2s+2)}$$

$$= \frac{s^3[B_1+B_2]+s^2[A_1+A_2+2B_1]+s[2A_1+2B_1+B_2]+2A_1+A_2}{(s^2+1)(s^2+2s+2)}$$

COMPARING NUMERATORS, WE OBTAIN THE FOLLOWING 4 SIMULTANEOUS EQUATIONS:

$$\begin{aligned} B_1 + B_2 &= 0 \\ A_1 + A_2 + 2B_1 &= 0 \\ 2A_1 + 2B_1 + B_2 &= 0 \\ 2A_1 + A_2 &= 1 \end{aligned}$$

USE CRAMER'S RULE TO FIND $A_1$

$$A_1 = \frac{\begin{vmatrix} 0 & 0 & 1 & 1 \\ 0 & 1 & 2 & 0 \\ 0 & 0 & 2 & 1 \\ 1 & 1 & 0 & 0 \end{vmatrix}}{\begin{vmatrix} 0 & 0 & 1 & 1 \\ 1 & 1 & 2 & 0 \\ 2 & 0 & 2 & 1 \\ 2 & 1 & 0 & 0 \end{vmatrix}} = \frac{-1}{-5} = 1/5$$

SO, THE REST OF THE COEFFICIENTS FOLLOW EASILY FROM THE EQUATIONS:

$$A_1 = 1/5$$
$$A_2 = 3/5$$
$$B_1 = -2/5$$
$$B_2 = 2/5$$

THEN,

$$\mathcal{L}(x) = \frac{1}{(s+1)^2+1} + \frac{1/5}{s^2+1} + \frac{-2/5\,s}{s^2+1} + \frac{3/5}{s^2+2s+2} + \frac{2/5\,s}{s^2+2s+2}$$

TAKING THE INVERSE TRANSFORM,

$$x = \mathcal{L}^{-1}\{\mathcal{L}(x)\} = e^{-t}\sin t + \tfrac{1}{5}\sin t - \tfrac{2}{5}\cos t + \tfrac{3}{5}e^{-t}\sin t + \tfrac{2}{5}\left[e^{-t}\cos t - e^{-t}\sin t\right]$$

$$= \tfrac{6}{5}e^{-t}\sin t + \tfrac{2}{5}e^{-t}\cos t + \tfrac{1}{5}\sin t - \tfrac{2}{5}\cos t$$

## TIMED

1. This is a typical hypothesis test of 2 population means. The two populations are the original population from which the manufacturer got his 1600 hour average life value and the new population from which the sample was taken. We know the mean ($x$ = 1520 hours) of the sample and its standard deviation ($s$ = 120 hours), but we do not know the mean and standard deviation of a population of average lifetimes. Therefore, we assume that

a) the average lifetime population mean and the sample mean are identical ($x = u_{\bar{x}}$ = 1520 hours.)

b) the standard deviation of the average lifetime population is

$$\sigma_{\bar{x}} = \frac{s}{\sqrt{n}} = \frac{120}{\sqrt{100}} = 12 \qquad \text{(Eqn. 1.220)}$$

The manufacturer can be reasonably sure that his claim of a 1600 hour average life is justified if the average test life is near 1600 hours. 'Reasonably sure' must be evaluated based on an acceptable probability of being incorrect. If he is willing to be wrong with a 5% probability, then a 95% confidence level is required.

Since the direction of bias is known, a one-tailed test is required. We want to know if the mean has shifted downward. We test this by seeing if 1600 hours is within the 95% limits of a distribution with a mean of 1520 hours and a standard deviation of 12 hours. From page 1-29, 5% of a standard normal population is outside of $z = 1.645$. The 95% confidence limit is, therefore,

$$1520 + 1.645(12) = 1540$$

The manufacturer can be 95% certain that the average lifetime of his bearings is less than 1600 since 1600 is not between 1520 and 1540.

If the manufacturer is willing to be wrong with a probability of only 1%, then a 99% confidence limit is required. From page 1-24, $z = 2.33$ and the 99% confidence limit is

$$1520 + 2.33(12) = 1548$$

The manufacturer can be 99% certain that the average bearing life is less than 1600 hours.

2

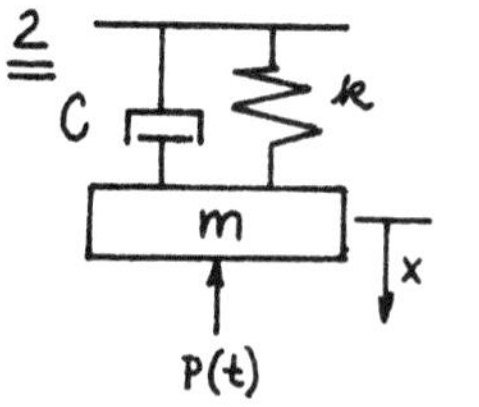


$$m = \frac{8.0 \text{ lbm}}{32.2 \frac{\text{ft}}{\text{sec}^2}} = 0.25 \text{ slugs}$$

$$C = 0.50 \text{ lbf-sec/ft.}$$

$$k = \frac{8.0 \text{ lbf}}{5.9 \text{ in}\left(\frac{\text{ft}}{12 \text{ in}}\right)} = 16.27 \frac{\text{lbf}}{\text{ft.}}$$

The differential equation (using Eq. 16.107) is

$$\frac{m}{g_c}\left(\frac{d^2x}{dt^2}\right) = -kx - C\left(\frac{dx}{dt}\right) + P(t)$$

$$0.25\,x'' + 0.50\,x' + 16.27\,x = 4\cos(2t)$$

$$x'' + 2x' + 65x = 16\cos(2t)$$

Initial conditions are:

$$x_0 = 0 \qquad x_0' = 0$$

TIMED #2 CONTINUED

Taking the Laplace transform of both sides,

$$\mathcal{L}(x'') + \mathcal{L}(2x') + \mathcal{L}(65x) = \mathcal{L}(16\cos(2t))$$

$$s^2\mathcal{L}(x) - sx_0 - x_0' + 2s\mathcal{L}(x) - 2x_0 + 65\mathcal{L}(x) = 16\left[\frac{s}{s^2+4}\right]$$

$$\mathcal{L}(x) = \frac{16s}{(s^2+4)(s^2+2s+65)}$$

Use partial fractions:

$$\mathcal{L}(x) = \frac{16s}{(s^2+4)(s^2+2s+65)} = \frac{As+B}{s^2+4} + \frac{Cs+D}{s^2+2s+65}$$

$$16s = As^3 + Bs^2 + 2As^2 + 2Bs + 65As + 65B + Cs^3 + Ds^2 + 4Cs + 4D$$

Then

$A + C = 0$ $\qquad C = -A = -\frac{61}{8}B$

$B + 2A + D = 0$ $\qquad B + 2A - \frac{65}{4}B = 0 \Rightarrow A = \frac{61}{8}B$

$2B + 65A + 4C = 16$

$65B + 4D = 0$ $\qquad D = -\frac{65}{4}B$

$$2B + 65\left(\frac{61}{8}B\right) + 4\left(-\frac{61}{8}B\right) = 16$$

$B = 0.0342521$

$A = 0.2611721$

$C = -0.2611721$

$D = -0.5565962$

Now

$$\mathcal{L}(x) = \frac{0.2611721s + 0.0342521}{s^2+4} - \frac{0.2611721s + 0.5566}{(s+1)^2 + 8^2}$$

$$= 0.26\left(\frac{s}{s^2+2^2}\right) + 0.017\left(\frac{2}{s^2+2^2}\right) - 0.26\left[\frac{s-(-1)}{(s-(-1))^2+8^2} + \frac{1.1311472}{(s-(-1))^2+8^2}\right]$$

Using Transform Tables in Appendix A, take the inverse transforms, giving

$$\boxed{x(t) = 0.26\cos(2t) + 0.017\sin(2t) - 0.26\left[e^{-t}\cos(8t) + 0.14e^{-t}\sin(8t)\right]}$$

3 Using Eq. 1.210,

(a) $$\bar{x} = \frac{\Sigma x_i}{n} = \frac{1249.529 + 1249.494 + 1249.384 + 1249.348}{4} = \frac{4997.755}{4} = 1249.4388$$

Small sample ($n < 50$), so use sample standard deviation, given by Eq. 1.215,

$$s = \sqrt{\frac{1249.529^2 + 1249.494^2 + 1249.384^2 + 1249.348^2 - \frac{4997.755^2}{4}}{3}}$$

$s = 0.0862168$

From Table 1.5, 90% fall within $1.645s$ of $\bar{x}$, or

$$1249.4388 \pm 1.645(0.0862168) = 1249.4388 \pm 0.1418 = 1249.5806,\ 1249.2970$$

By observation, all points fall within this range, so $\boxed{\text{all}}$ are acceptable

(b) none

(c) readings must fall within 1.645 standard deviations of mean (from Table 1.5, 90% corresponds to 0.45 from $\bar{x}$)

(d) $\bar{x} = 1249.4388$

(e) at 90%, error $= 1.645s = 0.1422$

(f) If a surveying crew places a marker, measures a distance x and places a second marker, and then measures the same distance x back to the original marker, the ending point should coincide with the original marker. If, due to measurement errors, the ending and starting points do not coincide, the difference is the "closure error".

In this example, the survey crew moves around the four sides of a square, so there are two measurements in the x-direction and two measurements in the y-direction. If the errors, $E_1$ and $E_2$, are known for two measurements $x_1$ and $x_2$, the error associated with the sum or difference $x_1 \pm x_2$ is

$$E_{x_1 \pm x_2} = \sqrt{E_1^2 + E_2^2}$$

In this case, the error in the x-direction, $E_x$, is

$$E_x = \sqrt{(0.1422)^2 + (0.1422)^2} = 0.2011$$

The error in the y-direction, $E_y$, is calculated the same way, and is 0.2011.

$E_x$ and $E_y$ are combined by the Pythagorean theorem to yield

$$E_{\text{closure}} = \sqrt{(0.2011)^2 + (0.2011)^2} = \boxed{0.2844}$$

(g) In surveying work, error may be expressed as a fraction of one or more legs of the traverse. Assume that the total of all four legs is to be used as the basis. Then $\frac{0.2844}{(4)(1249)} = \frac{1}{17{,}567}$

TIMED #3 CONTINUED

(h) In surveying, a second-order error is smaller than one part in 10,000. $1/17{,}567$ is smaller than a second-order error, so the answer is **YES**, this error is within the second order of accuracy.

(i) accuracy: how close the result is to being true -- how large the experimental error is.

precision: repeatability of results. One may have the same results repeatedly, all of which are in large error.

(j) systematic error: an error which is always present and unchanged in size and direction. An example is a steel tape that is 0.2 feet short -- it always reads 0.2 ft short.

4

THIS IS A FLUID MIXING PROBLEM.

THE MASS OF WATER IN THE LAGOON IS

$$m_i = \left(\frac{\pi}{4}\right)(120\ \text{FT})^2(10\ \text{FT})\left(62.4\ \frac{\text{lbm}}{\text{FT}^3}\right)$$

$$= 7.06 \times 10^6\ \text{LBM}$$

THE FLOW RATE IS

$$\phi(t) = \left(30\ \frac{\text{GAL}}{\text{MIN}}\right)\left(8.345\ \frac{\text{LBM}}{\text{GAL}}\right) = 250.4\ \text{LBM/MIN}$$

THE MASS OF CHEMICALS IN THE LAGOON WHEN C = 1 PPB IS

$$m_f = (1 \times 10^{-9})(7.06 \times 10^6\ \text{LBM}) = 7.06 \times 10^{-3}\ \text{LBM}$$

$$S_{in}(t) = 0$$

$$S_{out}(t) = C(t)\,\phi(t)$$

$$= \frac{m(t)(250.4\ \text{LBM/MIN})}{7.06 \times 10^6\ \text{LBM}} = 3.55 \times 10^{-5}\, m(t)$$

THE DIFFERENTIAL EQUATION IS

$$m'(t) = S_{in}(t) - S_{out}(t)$$

$$= -3.55 \times 10^{-5}\, m(t)$$

$$m'(t) + 3.55 \times 10^{-5}\, m(t) = 0$$

THE CHARACTERISTIC EQUATION IS

$$R + 3.55 \times 10^{-5} = 0 \quad \text{so} \quad R = -3.55 \times 10^{-5}$$

THE SOLUTION TO THE DIFFERENTIAL EQUATION IS

$$m(t) = A\, e^{-3.55 \times 10^{-5} t}$$

SINCE $m = 90$ LBM WHEN $t = 0$, $A = 90$ LBM,

$$m(t) = 90\, e^{-3.55 \times 10^{-5} t}$$

SOLVE FOR $t$ WHEN $m(t) = 7.06 \times 10^{-3}$

$$7.06 \times 10^{-3} = 90\, e^{-3.55 \times 10^{-5} t}$$

TAKING THE NATURAL LOG,

$$-9.45 = -3.55 \times 10^{-5}\, t$$

$$t = 2.66 \times 10^5\ \text{MIN} = \boxed{185\ \text{days}}$$

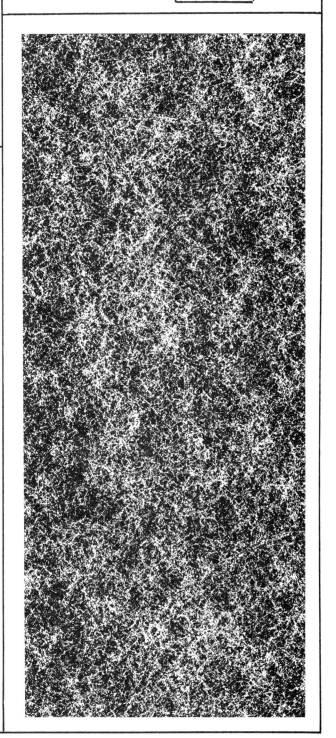

# Engineering Economics

WARM-UPS

1 $F = 1000\,(F/P, 6\%, 10)$

$= 1000\,(1.7908) = 1790.80$

2 $P = 2000\,(P/F, 6\%, 4)$

$= 2000\,(.7921) = 1584.20$

3 $P = 2000\,(P/F, 6\%, 20)$

$= 2000\,(.3118) = 623.60$

4 $500 = A\,(P/A, 6\%, 7)$

$= A\,(5.5824)$

$A = \dfrac{500}{5.5824} = 89.57$

5 $F = 50\,(F/A, 6\%, 10)$

$= 50\,(13.1808) = 659.04$

6 EACH YEAR IS INDEPENDENT.

$\dfrac{200}{1.06} = 188.68$

7 $2000 = A\,(F/A, 6\%, 5)$

$= A\,(5.6371)$

$A = \dfrac{2000}{5.6371} = 354.79$

ALTERNATE SOLUTION

$2000 = A\left[(F/P, 6\%, 4) + (F/A, 6\%, 4)\right]$

8 $F = 100\left[(F/P, 6\%, 10) + (F/P, 6\%, 8) + (F/P, 6\%, 6)\right]$

$= 100\,(1.7908 + 1.5938 + 1.4185)$

$= 480.31$

9 $r = .06$

$\phi = \dfrac{.06}{12} = .005$

$N = 5(12) = 60$

$F = 500\,(1.005)^{60} = 674.43$

10 $120 = 80\,(F/P, i, 7)$

$(F/P, i, 7) = \dfrac{120}{80} = 1.5$

SEARCHING THE TABLES, $i = 6\%$

CONCENTRATES

1 $EUAC = (17000 + 5000)(A/P, 6\%, 5)$

$- (14000 + 2500)(A/F, 6\%, 5) + 200$

$= (22000)(.2374) - (16500)(.1774) + 200$

$= 2495.70$

2 ASSUME THE BRIDGE WILL BE THERE FOREVER.

KEEP OLD BRIDGE

THE GENERALLY ACCEPTED METHOD IS TO CONSIDER THE SALVAGE VALUE AS A BENEFIT LOST (COST). SEE P. 2-6.

$EUAC = (9000 + 13000)(A/P, 8\%, 20)$

$- 10{,}000\,(A/F, 8\%, 20) + 500$

$= (22000)(.1019) - (10{,}000)(.0219) + 500$

$= 2522.80$

REPLACE

$EUAC = 40{,}000\,(A/P, 8\%, 25)$

$- 15000\,(A/F, 8\%, 25) + 100$

$= (40{,}000)(.0937) - 15000(.0137) + 100$

$= 3642.50$

KEEP OLD BRIDGE

3 $D = \dfrac{150{,}000}{15} = 10{,}000$

$0 = -150{,}000 + (32000)(1-.48)(P/A, i, 15)$

$- 7530\,(1-.48)(P/A, i, 15)$

$+ 10000\,(.48)(P/A, i, 15)$

$150{,}000 = \left[16640 - 3915.60 + 4800\right](P/A, i, 15)$

$(P/A, i, 15) = \dfrac{150{,}000}{17524.40} = 8.5595$

SEARCHING THE TABLES, $i = 8\%$

4 a) $\dfrac{1{,}500{,}000 - 300{,}000}{1{,}000{,}000} = 1.2$

b) $1{,}500{,}000 - 300{,}000 - 1{,}000{,}000$

$= 200{,}000$

**5** ANNUAL RENT IS $(12)(75) = 900$

$F = (14000 + 1000)(F/P, 10\%, 10)$
$+ (150 + 250 - 900)(F/A, 10\%, 10)$
$= 15000(2.5937) - 500(15.9374)$
$= 30936.80$

**6** $2000 = 89.30(P/A, ?, 30)$

$(P/A, ?, 30) = \frac{2000}{89.30} = 22.396$

$? = 2\%$ PER MONTH

$i = (1.02)^{12} - 1 = .2682$ OR $26.82\%$

**7** SL $D = \frac{500{,}000 - 100{,}000}{25} = 16000$

SOYD $T = \frac{1}{2}(25)26 = 325$

$D_1 = \frac{25}{325}(500{,}000 - 100{,}000) = 30769$

$D_2 = \frac{24}{325}(400{,}000) = 29538$

$D_3 = \frac{23}{325}(400{,}000) = 28308$

DDB $D_1 = \frac{2}{25}(500{,}000) = 40{,}000$

$D_2 = \frac{2}{25}(500{,}000 - 40{,}000) = 36{,}800$

$D_3 = \frac{2}{25}(500{,}000 - 40{,}000 - 36{,}800)$
$= 33{,}856$

**8** $P = -12000 + 2000(P/F, 10\%, 10)$
$-1000(P/A, 10\%, 10) - 200(P/G, 10\%, 10)$
$= -12000 + 2000(.3855) - 1000(6.1446)$
$-200(22.8913)$
$= -21951.86$

$EUAC = 21951.86(A/P, 10\%, 10)$
$= 21951.86(.1627) = 3571.56$

**9** ASSUME THAT THE PROBABILITY OF FAILURE IN ANY OF THE N YEARS IS $1/N$

$EUAC(9) = 1500(A/P, 6\%, 20) + \frac{1}{9}(.35)(1500)$
$+ (.04)(1500)$
$= 1500\left[.0872 + (.35)\left(\frac{1}{9}\right) + .04\right] = 249.13$

$EUAC(14) = 1600\left[.1272 + (.35)\left(\frac{1}{14}\right)\right] = 243.52$

$EUAC(30) = 1750\left[.1272 + (.35)\left(\frac{1}{30}\right)\right] = 243.01$

$EUAC(52) = 1900\left[.1272 + (.35)\left(\frac{1}{52}\right)\right] = 254.47$

$EUAC(86) = 2100\left[.1272 + (35)\left(\frac{1}{86}\right)\right] = 275.67$

CHOOSE THE 30 YEAR PIPE

**10** $EUAC(7) = .15(25000) = 3750$

$EUAC(8) = 15000(A/P, 10\%, 20)$
$+ .10(25000)$
$= 15000(.1175) + .10(25000)$
$= 4262.50$

$EUAC(9) = 20000(.1175) + .07(25000)$
$= 4100.00$

$EUAC(10) = 30{,}000(.1175) + .03(25000)$
$= 4275$

CHEAPEST TO DO NOTHING

## TIMED

**1** $EUAC(1) = 10{,}000(A/P, 20, 1) + 2000$
$- 8000(A/F, 20\%, 1)$
$= 10{,}000(1.2000) + 2000 - 8000(1.0000) = 6000$

$EUAC(2) = 10{,}000(A/P, 20, 2) + 2000$
$+ 1000(A/G, 20\%, 2) - 7000(A/F, 20\%, 2)$
$= 10{,}000(.6545) + 2000 + 1000(.4545)$
$-7000(.4545) = 5818.00$

$EUAC(3) = 10{,}000(A/P, 20\%, 3) + 2000$
$+ 1000(A/G, 20\%, 3) - 6000(A/F, 20\%, 3)$
$= 10000(.4747) + 2000 + 1000(.8791)$
$-6000(.2747) = 5977.90$

$EUAC(4) = 10{,}000(A/P, 20, 4) + 2000$
$+ 1000(A/G, 20\%, 4) - 5000(A/F, 20\%, 4)$
$= 10{,}000(.3863) + 2000 + 1000(1.2742)$
$-5000(.1863) = 6205.7$

TIMED #1, CONTINUED

$$EUAC(5) = 10{,}000\,(A/P, 20\%, 5) + 2000 + 1000\,(A/G, 20\%, 5) - 4000\,(A/F, 20\%, 5)$$
$$= 10{,}000\,(.3344) + 2000 + 1000\,(1.6405) - 4000\,(.1344) = 6446.4$$

SELL AT END OF 2ND YEAR

b) FROM EQN 2.8, COST OF KEEPING 1 MORE YEAR IS

$$6000 + (0.20)(5000) + (5000 - 4000) = 8000$$

2 THE MAN SHOULD CHARGE HIS COMPANY ONLY FOR THE COSTS DUE TO THE BUSINESS TRAVEL:

INSURANCE $300 - 200 = 100$

MAINTENANCE $200 - 150 = 50$

SALVAGE VALUE REDUCTION $(1000 - 500) = 500$

$$500\,(A/F, 10\%, 5) = 500(.1638) = 81.90$$

GASOLINE $\dfrac{5000\,(.60)}{15} = 200$

$$\text{EUAC PER MILE} = \frac{100 + 50 + 81.9 + 200}{5000} = \$.0864$$

a) YES. $\$.10 > \$.0864$ SO IT IS ADEQUATE

b) $$(.10)X = 5000\,(A/P, 10\%, 5) + 250 + 200 - 800\,(A/F, 10\%, 5) + \frac{X}{15}(.60)$$
$$= 5000\,(.2638) + 250 + 200 - 800(.1638) + .04X$$

$$.10X = 1637.96 + .04X$$
$$.06X = 1637.96$$
$$X = 27299 \text{ miles}$$

3 USE EUAC SINCE LIVES ARE DIFFERENT

$$P\{A\} = -80{,}000 - 5000\,(P/F, 10\%, 10) + 7000\,(P/F, 10\%, 20) - 2000\,(P/A, 10\%, 20) + 500\,(P/A, 10\%, 10) + 500\,(P/A, 10\%, 5)$$
$$= -80{,}000 - 5000(.3855) + 7000(.1486) - 2000(8.5136) + 500(6.1446 + 3.7908)$$
$$= -92947$$

$$EUAC\{A\} = 92947\,(A/P, 10\%, 20) = 92947\,(.1175) = 10921$$

$$P\{B\} = -35000 - 4000\,(P/A, 10\%, 10) + 1000\,(P/A, 10\%, 5)$$
$$= -35000 - 4000(6.1446) + 1000(3.7908)$$
$$= -55788$$

$$EUAC\{B\} = 55788\,(A/P, 10\%, 10) = 55788\,(.1627) = 9077$$

**B HAS THE LOWEST COST**

4 LET X BE THE NUMBER OF MILES DRIVEN PER YEAR. THE EUAC FOR BOTH ALTERNATIVES ARE

$$EUAC(A) = .15X$$
$$EUAC(B) = .04X + 500 + 5000\,(A/P, 10\%, 3) - 1200\,(A/F, 10\%, 3)$$
$$= .04X + 500 + 5000(.4021) - 1200\,(.3021)$$
$$= .04X + 2148$$

SETTING THESE EQUAL,

$$.15X = .04X + 2148$$

**X = 19527 MILES**

5 a) THE OPERATING COSTS PER YEAR ARE

A: $(10.50)(365)(24) = 91980$

B: $(8)(365)(24) = 70080$

$$EUAC\{A\} = (13000 + 10000)(A/P, 7\%, 10) + 91980 - 5000\,(A/F, 7\%, 10)$$
$$= (23000)(.1424) + 91980 - 5000(.0724)$$
$$= 94893$$

THE COST PER TON-YEAR IS

$$\text{COST}(A) = \frac{94893}{50} = \boxed{1898}$$

NOW WORK WITH B. EXCLUSIVE OF OPERATING COSTS, THE PRESENT WORTH IS

$$P\{B\} = -8000 - (7000 + 2200 - 2000)\,(P/F, 7\%, 5) + 2000\,(P/F, 7\%, 10)$$
$$= -8000 - (7200)(.7130) + 2000\,(.5083)$$
$$= -12117$$

(MORE)

TIMED #5 CONTINUED

$$EUAC\{B\} = 70080 + (12117)(A/P, 7\%, 10)$$
$$= 70080 + (12117)(.1424)$$
$$= 71805$$

THE COST PER TON-YEAR FOR B IS

$$\text{COST B} = \frac{71805}{20} = \boxed{3590}$$

(b)

| | COST OF USING A'S | COST OF USING B'S | CHEAPEST |
|---|---|---|---|
| 0-20 | 94893 (1X) | 71805 (1X) | B |
| 20-40 | 94893 (1X) | 143610 (2X) | A |
| 40-50 | 94893 (1X) | 215415 (3X) | A |
| 50-60 | 189,786 (2X) | 215415 (3X) | A |
| 60-80 | 189,786 (2X) | 287220 (4X) | A |

6 First year cost = (1)(37,440) = 37,440

Gradient = (0.1)(37,440) = 3,744

(a) $EUAC = 60{,}000\,(A/P, 7\%, 20) + 37{,}440 + 3744\,(A/G, 7\%, 20) - 10{,}000\,(A/F, 7\%, 20)$

$= 60{,}000\,(0.0944) + 37{,}440 + 3744\,(7.3163) - 10{,}000\,(0.0244)$

$= \$70{,}252.23$

for 80,000 passengers, required fare is

$$\frac{70{,}252.23}{80{,}000} = \boxed{\$0.878} \text{ per passenger trip}$$

(b) $\$0.878 = \$0.35 + \text{gradient}\,(A/G, 7\%, 20)$

$\text{gradient} = (\$0.878 - 0.35)/7.3163$

$= \boxed{\$0.072}$ (7.2 ¢) increase per year

(c) AS IN PART (b), THE SUBSIDY SHOULD BE

SUBSIDY = COST − REVENUE

$$P = 0.878 - (0.35 + 0.05(A/G, 7\%, 20))$$
$$= 0.878 - 0.35 - (0.05)(7.3163)$$
$$= \boxed{\$0.162}$$

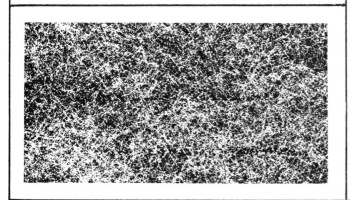

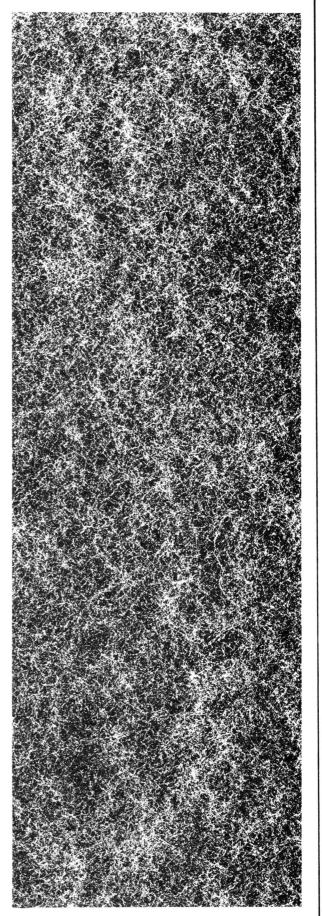

# Fluid Statics

WARM-UPS

1 FROM PAGE 3-36 FOR AIR AT 14.7 PSIA AND 80°F,

$\mu = 3.85\text{ EE-7 } \frac{\text{LB-SEC}}{\text{FT}^2}$ {INDEPENDENT OF PRESSURE}

$\rho = P/RT = \frac{(70)(144)}{(53.3)(80+460)} = .350$

FROM EQN 3.12,

$\nu = \frac{\mu g_c}{\rho} = \frac{(3.85\text{ EE-7 } \frac{\text{LBF-SEC}}{\text{FT}^2})(32.2 \frac{\text{LBM-FT}}{\text{LBF-SEC}^2})}{.35 \frac{\text{LBM}}{\text{FT}^3}}$

$= 3.54\text{ EE-5 FT}^2/\text{SEC}$

2 $P = 14.7 - 8.7 = 6.0$ PSIA

3 FROM EQN 3.25,

$P_1 - P_2 = 7(.491) - 7(.0361) = 3.184$ PSI

4 THE PERIMETER IS

$p = \pi d = \pi(10) = 31.42'$

ASSUME AN ELLIPTICAL CROSS SECTION.

FROM PAGE 1-2,

$b = \frac{1}{2}(7.2) = 3.6$

THEN

$31.42 = 2\pi\sqrt{\frac{1}{2}(a^2 + (3.6)^2)}$

SO $a = 6.09''$

THE AREA IN FLOW IS

$A = \frac{1}{2}(\pi ab) = \frac{1}{2}(\pi)(6.09)(3.6) = 34.44\text{ IN}^2$

THEN FROM EQN 3.65

$r_H = \frac{34.44}{\frac{1}{2}(31.42)} = 2.19''$

5 ASSUME SCHEDULE 40 PIPE AND 70°F WATER

$D_i = 6.065'' = .5054$ FT

$A_i = .2006\text{ FT}^2$

$V = \frac{1.5\text{ FT}^3/\text{SEC}}{.2006\text{ FT}^2} = 7.478$ FT/SEC

FROM PAGE 3-37 AT 70°F

$\nu = 1.059\text{ EE-5 FT}^2/\text{SEC}$

FROM EQN 3.62

$N_{Re} = \frac{DV}{\nu} = \frac{(.5054)\text{FT}(7.478)\text{FT/SEC}}{(1.059\text{ EE-5})\text{ FT}^2/\text{SEC}} = 3.57\text{ EE5}$

6 FROM PAGE 3-20, $\epsilon = 0.0002$

$\epsilon/D = \frac{0.0002}{.5054} = 0.0004$

FROM PAGE 3-20, $f \approx 0.0175$

FROM EQN 3.71,

$h_f = \frac{(.0175)(1200)\text{FT}(7.478)^2\text{ FT}^2/\text{SEC}^2}{(.5054)\text{ FT }(2)(32.2)\text{ FT/SEC}^2}$

$= 36.1$ FT

7 ASSUME SCHEDULE 40 STEEL PIPE.

AT A
$D_A = .5054$ FT
$A_A = .2006\text{ FT}^2$
$V_A = \frac{5\text{ FT}^3/\text{SEC}}{.2006\text{ FT}^2} = 24.925$ FT/SEC
$P_A = 10$ PSIA $= 1440$ PSF
$Z_A = 0$

AT B
$D_B = 1.4063$ FT
$A_B = 1.5533\text{ FT}^2$
$V_B = 5/1.5533 = 3.219$ FT/SEC
$P_B = 7$ PSIA $= 1008$ PSF
$Z_B = 15$

IGNORE THE MINOR LOSSES AND ASSUME $\rho = 62.3$ AT 70°F. THEN, THE TOTAL ENERGY AT A AND B FROM EQN 3.56 IS

$(TH)_A = \frac{1440}{62.3} + \frac{(24.925)^2}{(2)(32.2)} + 0 = 32.76$

$(TH)_B = \frac{1008}{62.3} + \frac{(3.219)^2}{(2)(32.2)} + 15 = 31.3$

FLOW IS FROM HIGH ENERGY TO LOW.

$A \rightarrow B$

8 FROM EQUATION 16.38,

$r = \frac{V_o^2 \sin 2\phi}{g}$

$= \frac{(50)^2 \sin(2(45))}{32.2} = 77.64$ FT

9 THE WEIGHT OF WATER AT 70°F IS

$(100)\frac{\text{FT}^3}{\text{SEC}}(62.4)\frac{\text{LBM}}{\text{FT}^3} = 6240 \frac{\text{LBM}}{\text{SEC}}$

THE ENERGY LOSS IN FT OF WATER IS

$\frac{(30+5)\frac{\text{LBF}}{\text{IN}^2}(144)\frac{\text{IN}^2}{\text{FT}^2}}{62.4\text{ LBM/FT}^3} = 80.77$ FT

THE HORSEPOWER IS

$\frac{6240 \frac{\text{LBM}}{\text{SEC}}(80.77)\text{ FT}}{550 \frac{\text{FT-LBF}}{\text{HP-SEC}}} = 916.37$ HP

10 ASSUME SCHEDULE 40 PIPE AND 70°F WATER

$D_1 = 7.981''$

$D_2 = 6.065''$

(MORE)

## WARM-UP #10 CONTINUED

THE MASS FLOW IS
$$A_2 V_2 \rho = \frac{\pi}{4}\left(\frac{6.065}{12}\right)^2 (12)(62.4) = 150.2 \frac{LBM}{SEC}$$

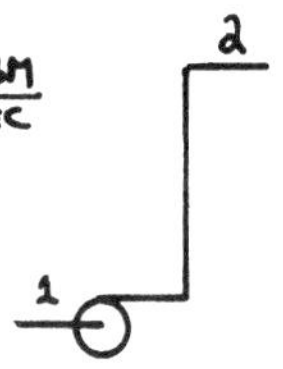

THE INLET PRESSURE IS
$$(14.7-5)(144) = 1396.8 \; LBF/FT^2$$

THE HEAD ADDED BY THE PUMP IS
$$h_A = \frac{(.70)(20)(550)}{(150.2)}$$
$$= 51.26 \; FT$$

AT 1   $P_1 = 1396.8$
$$V_1 = 12\left(\frac{6.065}{7.981}\right)^2 = 6.93$$
$$z_1 = 0$$

AT 2   $P_2 = (14.7)(144) = 2116.8$
$$V_2 = 12$$
$$z_2 = ?$$

USING BERNOULLI
$$\frac{1396.8}{62.4} + \frac{(6.93)^2}{2 \times 32.2} + 51.26 = \frac{2116.8}{62.4} + \frac{(12)^2}{2(32.2)} + z_2 + 10$$
$$z_2 = 28.2 \; FT$$

## CONCENTRATES

1  THE BUOYANT FORCE IS EQUAL TO THE WEIGHT OF THE DISPLACED AIR, WHICH HAS THE SAME VOLUME AS THE HYDROGEN.
$$V = \frac{wRT}{P} = \frac{(10,000)(766.8)(460+56)}{(30.2)(491)(144)}$$
$$= 1.853 \; EE6 \; FT^3$$

THE DISPLACED AIR WEIGHS
$$w = \frac{PV}{RT} = \frac{(30.2)(491)(144)(1.85 EE6)}{(53.3)(460+56)}$$
$$= 1.439 \; EE5 \; LBM$$

NEGLECTING THE STRUCTURAL WEIGHT, THE LIFT IS
$$F_{BUOY} - W_{He} = 1.439 \; EE5 - 10,000 = 1.339 \; EE5 \; LBF$$

2  FROM PAGE 3-40,
$$D_i = 6.065'' = .5054 \; FT$$
$$A_i = .2006 \; FT^2$$

THE VOLUME FLOW {FROM PAGE 3-37} IS
$$(500) \; gpm (.00223)\frac{FT^3}{SEC-gpm} = 1.115 \; FT^3/SEC$$

THE FLUID VELOCITY IS
$$v = \frac{Q}{A} = \frac{1.115}{.2006} = 5.558 \; FT/SEC$$

AT 100°F, $\nu = .739 \; EE-5 \; FT^2/SEC$ (PAGE 3-36)
$$N_{Re} = \frac{vD}{\nu} = \frac{(5.558)(.5054)}{.739 \; EE-5} = 3.8 \; EE5$$

---

SO FROM PAGE 3-20, $\varepsilon = .0002$
$$\frac{\varepsilon}{D} = \frac{.0002}{.5054} = .0004$$

FROM PAGE 3-20,  $f = .0175$

THE APPROXIMATE EQUIVALENT LENGTHS OF THE FITTINGS ARE

|          |              |        |
|----------|--------------|--------|
| ELBOWS   | 2 × 8.9 =    | 17.8   |
| GATE V   | 2 × 3.2 =    | 6.4    |
| 90° V    | 1 × 63.0 =   | 63.0   |
| BF LIM   | 1 × 63.0 =   | 63.0   |
|          |              | 150.2  |

THE FRICTION LOSS ACROSS THE 300 FEET IS
$$L_e = 300 + 150.2 = 450.2$$
$$h_f = \frac{(.0175)(450.2)(5.558)^2}{(2)(.5054)(32.2)} = 7.48 \; FT$$

THE PRESSURE DIFFERENCE FOR 100°F WATER IS
$$p = \rho h = 62.0 \; (7.48+20) = 1703.8 \; PSF$$

3  FOR 70°F AIR, (ASSUMED INCOMPRESSIBLE)
$$\nu \approx 16.15 \; EE-5 \; FT^2/SEC$$

THEN $N_{Re} = \frac{(60)(.5054)}{16.15 \; EE-5} = 1.88 \; EE5$

USING $\varepsilon/D = .0004$ FROM P. 3-20,
$$f = .0185$$

THE FRICTION LOSS FROM EQN 3.71 IS
$$h_f = \frac{(.0185)(450.2)(60)^2}{(2)(.5054)(32.2)} = 921.2 \; FT$$

ASSUMING $\rho_{AIR} \approx .075 \; LBM/FT^3$, THE PRESSURE DIFFERENCE IS
$$\Delta P = \rho h = (.075)(921.20 + 20) = 70.59 \; PSF$$

4  FROM PAGE 3-37,
$$(2000) \; gpm (.00223)\frac{FT^3}{SEC-gpm} = 4.46 \frac{FT^3}{SEC}$$

ASSUMING SCHEDULE 40 PIPE, FROM PAGE 3-40
$$D_1 = .9948 \; FT$$
$$A_1 = .7773 \; FT^2$$
$$D_2 = .6651$$
$$A_2 = .3474$$

$$P_1 = [14.7 - (6)(.491)]144 = 1692.6 \; PSF$$

$$P_2 = [14.7 + (20)]144 + (4)(1.2)(62.4) = 5296.3 \; PSF$$

(MORE)

CONCENTRATE #4 CONTINUED

$V_1 = \frac{4.46}{.7773} = 5.74$ FPS

$V_2 = \frac{4.46}{.3474} = 12.84$ FPS

THE TOTAL HEADS AT 1 AND 2 {EQN 3.56} ARE

$$(TH)_1 = \frac{1692.6}{(62.4)(1.2)} + \frac{(5.7)^2}{(2)(32.2)} = 23.11$$

$$(TH)_2 = \frac{5296.3}{(62.4)(1.2)} + \frac{(12.84)^2}{(2)(32.2)} = 73.29$$

THE PUMP MUST ADD $(73.29 - 23.11) = 50.18$ FT OF HEAD. THE POWER REQUIRED IS

$$P = \Delta h \dot{m}$$

$$= \frac{(50.18)\text{ FT}(1.2)(62.4)\frac{\text{LBM}}{\text{FT}^3}(4.46)\text{ FT}^3/\text{SEC}}{550\ \frac{\text{FT-LBF}}{\text{HP-SEC}}}$$

$$= 30.47 \text{ HP}$$

THE INPUT HORSEPOWER IS

$$\frac{30.47}{.85} = 35.85 \text{ HP}$$

5 THIS CANNOT BE SOLVED CORRECTLY WITH THE ENGINEERING REVIEW MANUAL. REFER TO PAGE 2-13 OF CRANE'S TECHNICAL PAPER 410.

$$K_B = K_1 + (N-1)\left(0.25 f_T \pi \left(\frac{r}{D}\right) + .5K_1\right)$$

ASSUME TYPE K COPPER TUBING. THEN

$$D_i = 2.125 - 2(.083) = 1.959'' = .1633 \text{ FT}$$

$$\frac{r}{D} = \frac{1}{.1633} = 6.126$$

FROM PAGE A-29 OF CRANE FOR A 90° BEND

$K_1 \approx 17.4 f$

$N = \#90°$ BENDS $= (4)(5) = 20$

ASSUMING SMOOTH COPPER PIPE {PAGE 3-20}

$\epsilon = .000005$

$$\frac{\epsilon}{D} = \frac{.000005}{.1633} = .000031$$

THE FULLY-TURBULENT FRICTION FACTOR IS

$$f_T = 0.0095$$

THEN, $K_1 = (17.4)(0.0095) = 0.165$

$$K_B = 0.165 + (20-1)\left[(0.25)(0.0095)(\pi)(6.126) + (.5)(0.165)\right]$$

$$= 2.6$$

$$h_f = K_B \frac{v^2}{2g} = (2.6)\frac{(10)^2}{(2)(32.2)} = \boxed{4.04 \text{ ft}}$$

6 $\dot{Q} = \frac{\pi}{4}\left(\frac{2}{12}\right)^2 \text{FT}^2 (40)\text{ FT/SEC} = .8727 \text{ FT}^3/\text{SEC}$

FROM EQN 3.168

$$\dot{Q}' = \frac{40-15}{40}(.8727) = .5454 \text{ FT}^3/\text{SEC}$$

FROM EQNS 3.171 AND 3.172

$$F_x = \frac{-(.5454)(62.4)}{32.2}(40-15)(1-\cos 60°)$$

$= -13.2$ LBF {FORCE OF FLUID IS TO THE LEFT}

$$F_y = \frac{(.5454)(62.4)}{32.2}(40-15)(\sin 60°)$$

$= 22.9$ {FORCE ON FLUID IS UPWARDS}

$$F = \sqrt{(13.2)^2 + (22.9)^2} = 26.4 \text{ LBF}$$

7 FOR DYNAMIC SIMILITUDE, USE EQN 3.190

$$N_{Re,\text{MODEL}} = N_{Re,\text{TRUE}}$$

$$\left(\frac{vL}{\nu}\right)_M = \left(\frac{vL}{\nu}\right)_T$$

$$\left(\frac{vL\rho}{\mu g}\right)_M = \left(\frac{vL\rho}{\mu g}\right)_T$$

BUT $\rho = p/RT$ FOR IDEAL GASES

$$\left(\frac{vLp}{\mu g RT}\right)_M = \left(\frac{vLp}{\mu g RT}\right)_T$$

$V_M = V_t$

$L_M = L_t/20$

$T_M = T_t$

$g = g$

$R = R$

$\mu_M = \mu_t$ {INDEPENDENT OF PRESSURE}

SO $P_M = 20 P_t$ {UNITS DEPEND ON UNITS OF $P_t$}

8 FROM EQN 3.190

$$(N_{Re})_M = (N_{Re})_t$$

$\left(\frac{VD}{\nu}\right)_M = \left(\frac{VD}{\nu}\right)_t$ WHERE D IS THE IMPELLER (NOT PIPE) DIAMETER

V IS A TANGENTIAL VELOCITY, AND

$V \propto (\text{RPM})(\text{DIAMETER})$

SO $V_M \propto 2(\text{RPM})_M$

$V_t \propto (1)(1000)$

$D_M = 2D_t$

$\nu_{AIR}, 68° = 16.0$ EE-5 FT²/SEC {FROM P 3-38} (MORE)

CONCENTRATE # 8 CONTINUED

$\nu_{\text{CASTOR OIL}} \approx 1110 \text{ EE-5 FT}^2/\text{SEC}$

THEN,

$$\frac{(2)(\text{RPM})_m (2) D_t}{16 \text{ EE-5}} = \frac{(1000) D_t}{1110 \text{ EE-5}}$$

$$(\text{RPM})_M = 3.6$$

9 THE SPECIFIC GRAVITY OF BENZENE AT 60°F IS $\approx .885$ {PAGE 3-38}

THE DENSITY IS $(.885)(62.4) = 55.2 \ \frac{\text{LBM}}{\text{FT}^3}$

$= .0319 \ \frac{\text{LBM}}{\text{IN}^3}$

THE PRESSURE DIFFERENCE IS

$$\Delta p = 4(.491 - .0319) = 1.836 \text{ PSI}$$
$$= 264.4 \text{ PSF}$$

FROM EQN 3.143, RECOGNIZING $\left(\frac{A_2}{A_1}\right)^2 = \left(\frac{D_2}{D_1}\right)^4$

$$F_{va} = \frac{1}{\sqrt{1 - \left(\frac{3.5}{8}\right)^4}} = 1.019$$

$$A_2 = \frac{\pi}{4}\left(\frac{3.5}{12}\right)^2 = .0668 \text{ FT}^2$$

FROM EQN 3.142,

$$\dot{V} = (1.019)(.99)(.0668)\sqrt{\frac{(2)(32.2)(264.4)}{55.2}}$$

$$= 1.184 \text{ FT}^3/\text{SEC}$$

10 ASSUME SCHED 40 PIPE, THEN,

$D_i = .9948$

$A_i = .7773 \text{ FT}^2$

$$V_1 = \frac{Q}{A} = \frac{10}{.7773} = 12.87 \text{ FT/SEC}$$

FOR 70°F WATER, $\nu = 1.059$ EE-5

$$N_{Re} = \frac{VD}{\nu} = \frac{(12.87)(.9948)}{(1.059 \text{ EE-5})} = 1.21 \text{ EE6}$$

USE EQN 3.139.

$$\dot{V} = C_f A_o \sqrt{\frac{2g(p_1 - p_2)}{\rho}}$$

$$10 = C_f A_o \sqrt{(2)(32.2)(25)}$$

$$C_f A_o = .249$$

BOTH $C_f$ AND $A_o$ DEPEND ON $D_o$.

ASSUME $D_o = 6"$

$$A_o = \frac{\pi}{4}\left(\frac{6}{12}\right)^2 = .1963 \text{ FT}^2$$

FROM FIGURE 3.25,

$$\frac{A_o}{A_1} = \frac{.1963}{.7773} = .25$$

AND $N_{Re} = 1.21$ EE6,

$C_f = .63$

$C_f A_o = (.63)(.1963) = .124$ TOO SMALL

ASSUME $D_o = 9"$

$$A_o = \frac{\pi}{4}\left(\frac{9}{12}\right)^2 = .442$$

$$\frac{A_o}{A_1} = \frac{.442}{.7773} = .57$$

FROM FIGURE 3.25, $C_f \approx .73$

$C_f A_o = (.73)(.442) = .322$ TOO HIGH

INTERPOLATING

$$D_o = 6" + (9-6)\frac{.249 - .124}{.322 - .124} = 7.9$$

FURTHER ITERATIONS YIELD

$D_o \approx 8.1$ INCHES

$C_f A_o \approx .243$

TIMED

1

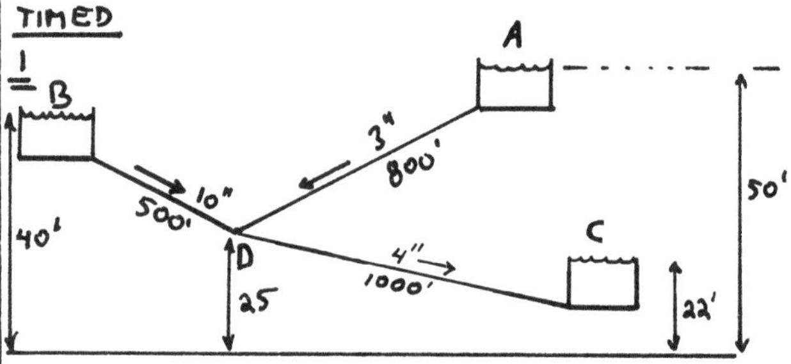


WE ASSUME THE RESERVOIRS A, B, FLOW TOWARDS D AND THE FLOW IS TOWARDS C, THEN

$$Q_{A-D} + Q_{B-D} - Q_{D-C} = 0$$

$$A_A V_{A-D} + A_B V_{B-D} - A_C V_{D-C} = 0$$

ASSUME SCHEDULE 40 PIPE SO THAT

$A_A = .05134 \text{ FT}^2$ $\quad D_A = .2557$ FT

$A_B = .5476 \text{ FT}^2$ $\quad D_B = .8350$ FT

$A_C = .08841 \text{ FT}^2$ $\quad D_C = .3355$ FT

SO

$$.05134 V_{A-D} + .5476 V_{B-D} - .08841 V_{D-C} = 0 \quad (1)$$

IGNORING THE VELOCITY HEADS, THE CONSERVATION OF ENERGY EQN 3.70 BETWEEN POINTS A AND D IS

$$z_A = \frac{p_D}{\rho} + z_D + h_{f,A-D}$$

TIMED #1 CONTINUED

$$50 = \frac{P_D}{62.4} + 25 + \frac{(.02)(800)(V_{A-D})^2}{(2)(.2557)(32.2)}$$

OR,

$$V_{A-D} = \sqrt{25.73 - .0165 P_D} \quad (2)$$

SIMILARLY FOR B-D:

$$40 = \frac{P_D}{62.4} + 25 + \frac{(.02)(500)(V_{B-D})^2}{(2)(.8350)(32.2)}$$

$$\text{OR,} \quad V_{B-D} = \sqrt{80.66 - .0862 P_D} \quad (3)$$

AND FOR D-C:

$$22 = \frac{P_D}{62.4} + 25 - \frac{(.02)(1000)(V_{D-C})^2}{(2)(.3355)(32.2)}$$

$$\text{OR,} \quad V_{D-C} = \sqrt{3.24 + .0173 P_D} \quad (4)$$

ASSUME $P_D$ = 935 PSF, WHICH IS THE LARGEST IT CAN BE TO KEEP EQUATIONS 2 AND 3 REAL. THEN,

$V_{A-D} = 3.21$
$V_{B-D} = .25$
$V_{D-C} = 4.41$

FROM EQN 1:

$$(.05134)(3.21) + (.5476)(.25) - (.08841)(4.41) = -.088$$

THIS IS REPEATED USING INTERPOLATION.

| TRIAL | $P_D$ | RIGHT-HAND SIDE OF EQN 1 | HOW WAS $P_D$ CHOSEN? |
|---|---|---|---|
| 1 | 935 | −.088 | TO KEEP 2,3 REAL |
| 2 | 900 | .745 | ARBITRARILY |
| 3 | 931.3 | .116 | INTERPOLATION WITH TRIALS 1,2 |
| 4 | 933.4 | .02 | INTERPOLATION, 1+3 |
| 5 | 933.7 | .005 | INTERPOLATION, 1+4 |

IF $P_D$ = 933.7,

$V_{AD}$ = 3.213 FT/SEC; $V_{BD}$ = 0.4184; FT/SEC
$V_{DC}$ = 4.404 FT/SEC

FLOW IS OUT OF B. IF IT HAD BEEN ASSUMED THAT RESERVOIR A FED BOTH B AND C, IT WOULD HAVE BEEN IMPOSSIBLE TO SATISFY THE CONTINUITY EQUATION

$$A_A V_{AD} + A_B V_{BD} - A_C V_{DC} = 0$$

**2** ASSUME 70°F WATER. FROM PAGE 3.36,

$E$ = 320 EE3 PSI

$\rho = 62.3$

ALTHOUGH $E$ VARIES CONSIDERABLY WITH THE CLASS OF THE CAST IRON, USE $E$ = 20 EE6 {P. 14-3}

THE COMPOSITE MODULUS OF ELASTICITY IS

$$E = \frac{E_{water}\, t_{pipe}\, E_{pipe}}{t_{pipe} E_{pipe} + d_{pipe} E_{water}}$$

FROM TABLE 14.1, $E_{CAST\ IRON} = 20 \times 10^6$ psi.

$$E = \frac{(320\times10^3 \text{ psi})(0.75 \text{ in})(20\times10^6 \text{ psi})}{(0.75 \text{ in})(20\times10^6 \text{ psi}) + (24 \text{ in})(320\times10^3 \text{ psi})}$$

$$= 2.12 \times 10^5 \text{ psi}$$

$$c = \sqrt{\frac{E g_c}{\rho}} = \sqrt{\frac{(2.12\times10^5)(32.2)(144)}{62.3}} = 3972 \text{ ft/sec}$$

{THE ANSWER IS 4880 ft/sec IF PIPE ELASTICITY IS DISREGARDED.}

FROM EQN 3.177, THE PRESSURE INCREASE IS

$$\Delta p = \frac{\rho c \Delta V}{g_c} = \frac{\left(62.3 \frac{\text{lbm}}{\text{ft}^3}\right)\left(3972 \frac{\text{ft}}{\text{sec}}\right)\left(6 \frac{\text{ft}}{\text{sec}}\right)}{32.2 \frac{\text{ft-lbm}}{\text{lbf-sec}^2}}$$

$$= 46{,}110 \text{ psf}$$

FROM EQN. 3.176, THE MAXIMUM CLOSURE TIME WITHOUT CHANGING $\Delta p$ IS

$$t = \frac{2L}{c} = \frac{(2)(500 \text{ FT})}{3972 \frac{\text{FT}}{\text{SEC}}} = 0.252 \text{ SEC}$$

**3** (a) Using Eq. 3.184,

$$D = \tfrac{1}{2} C_D A \rho v^2 / g$$

from appendix B, p. 3-36, 68°F

$\rho = 0.0752$ lb/ft³

$$v = \frac{(55)(5280)}{3600} = 80.67 \text{ ft/sec}$$

$$D = \frac{(\tfrac{1}{2})(0.42)(28)(.0752)(80.67)^2}{32.2} = 89.4 \text{ lbf}$$

Total resisting force = 89.4 + (0.01)(3300) = 122.4

Using equation 3.173,

Power consumed = $F \cdot \dot{s} = F \cdot v$ = (122.4)(80.67)

$$= \frac{9850 \text{ ft-lb/sec}}{778 \text{ ft-lb/BTU}}$$

$$= 12.66 \text{ BTU/sec}$$

Available power from fuel = (0.28)(115,000)

= 32,200 BTU/gal

$$\text{Consumption @ 55 mph} = \frac{12.66}{32{,}200}$$

= 3.94 EE-4 gal/sec

$$\text{In 1 sec., car travels } \frac{55 \text{ mi}}{\text{hr}} \times \frac{1 \text{ hr}}{3600 \text{ sec}}$$

= 0.0153 mi/sec

So fuel consumption is

$$\frac{3.94 \text{ EE-4 gal/sec}}{0.0153 \text{ mi/sec}} = \boxed{2.58 \text{ EE-2 gal/mi.}}$$

(this corresponds to 38.8 mpg)

(b) At 65 mph, v = 95.33, D = 124.8 lb

resisting force = 124.8 + 33 = 157.8

P = (157.5)(95.33) = 150,145 = 19.3 BTU/sec

Consumption at 65 mph = 6 EE-4 gal/sec;

in 1 sec., car travels $\frac{65}{3600}$ = 0.0181 mi/sec

Fuel consumption is $\frac{6EE-4}{0.0181}$ = 3.32EE-2 gal/mi

(corresponding to 30.1 mpg)

% increase is $\frac{3.32EE-2 - 2.58EE-2}{2.58EE-2}$ = 0.289 = **28.9 %** incr. in fuel consumed per mile

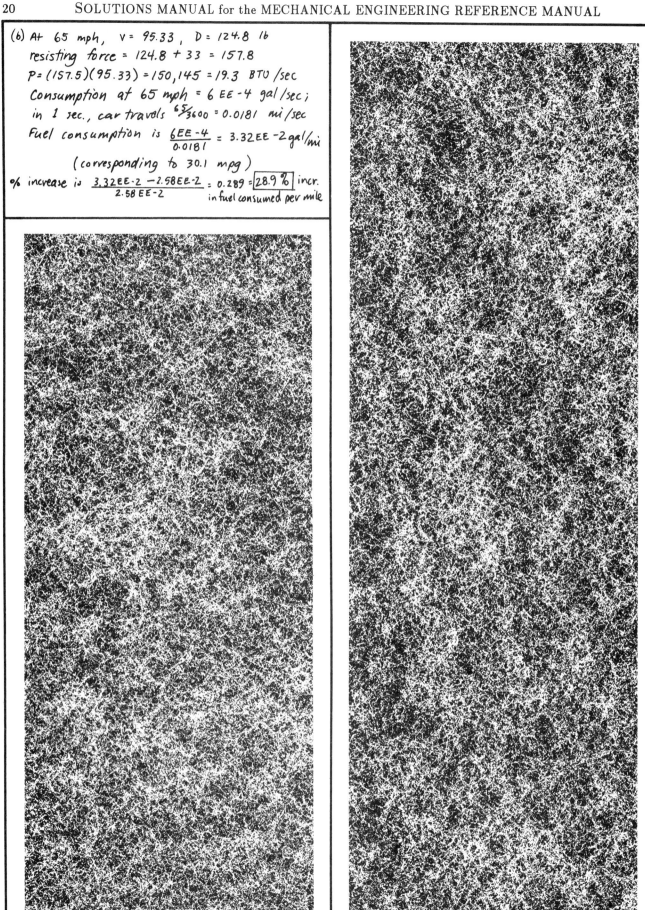

# Hydraulic Machines

WARM-UPS

1 FROM PAGE 1-40,

$(72)\ \text{gpm}\ (2.228\ EE{-3})\ \frac{\text{CFS}}{\text{gpm}} = .1604\ \text{CFS}$

2 FROM TABLE 4.2,

$$HP = \frac{Q\,H\,(SG)_{oil}}{3960}$$

$Q = 37\ \text{gpm}$

$$H = \frac{\Delta P}{\rho} = \frac{(40)\ \frac{\text{LBF}}{\text{IN}^2}\ (144)\ \frac{\text{IN}^2}{\text{FT}^2}}{(SG)_{oil}\ (62.4)\ \frac{\text{LBM}}{\text{FT}^3}} = \frac{92.308}{(SG)_{oil}}$$

THEN,

$$HP = \frac{\frac{(37)(92.308)\ (SG)}{SG}}{3960} = .862\ \text{HP}$$

3 FROM EQN 4.33,

$(HP)_2 = (.5)\left(\frac{2000}{1750}\right)^3 = .746$ (SAY 3/4 hp motor.)

4 USE EQN 4.27.

$N = 900$ RPM

$Q = (300)\ \frac{\text{GAL}}{\text{SEC}}\ (60)\ \frac{\text{SEC}}{\text{MIN}} = 18{,}000\ \text{gpm}$

$H = 20$ FT

$$N_s = \frac{(900)\sqrt{18{,}000}}{(20)^{.75}} = 12{,}768$$

5 ASSUME EACH STAGE ADDS 150 FEET OF HEAD. THE SUCTION LIFT IS 10 FEET. FROM PAGE 4-18 FOR SINGLE SUCTION,

$N_s \approx 2050$

CONCENTRATES

1 FROM PAGE 3-39,

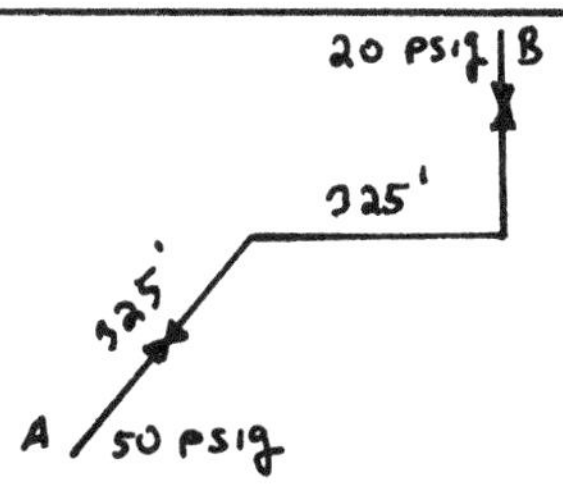


$D_i = .3355$ ft

$A_i = .08841\ \text{ft}^2$

SO $V = \frac{Q}{A} = \frac{1.25}{.08841} = 14.139$

ASSUME REGULAR, SCREWED, STEEL FITTINGS. THE APPROXIMATE EQUIVALENT LENGTHS ARE.

| FITTING | LENGTH |
|---|---|
| ELBOWS | 2 × 13 = 26 ft |
| GATE V. | 1 × 2.5 = 2.5 |
| CHECK V. | 1 × 38. = 38.0 |
| | 66.5 ft |

ASSUME 70°F, SO $\nu = 1.059\ EE{-5}$.

$$N_{Re} = \frac{D\,v}{\nu} = \frac{(.3355)(14.139)}{1.059\ EE{-5}} = 4.479\ EE5$$

FROM PAGE 3-20, $\epsilon = .0002$ FT, SO

$$\frac{\epsilon}{D} = \frac{.0002}{.3355} \approx .0006$$

FROM PAGE 3-20, $f = .0185$. FROM EQN 3.71,

$$h_f = \frac{(.0185)(700 + 66.5)(14.139)^2}{(2)(.3355)(32.2)} = 131.20\ \text{FT}$$

FROM EQN 3.70, THE HEAD ADDED IS

$$h_A = \frac{P_2}{\rho} + \frac{V_2^2}{2g} + z_2 + h_f - \frac{P_1}{\rho} - \frac{V_1^2}{2g} - z_1$$

BUT $V_1 = 0$ AND $z_1 = 0$.

$$h_A = \frac{(20+14.7)144}{62.3} + \frac{(14.139)^2}{(2)(32.2)} + 50 + 131.20 - \frac{(50+14.7)(144)}{62.3} = 114.96\ \text{FT}$$

ASSUME 70°F WATER. $\rho = 62.3\ \text{LBM/FT}^3$

$$\dot{m} = (62.3)\ \frac{\text{LBM}}{\text{FT}^3}\ (1.25)\ \frac{\text{FT}^3}{\text{SEC}} = 77.875\ \text{LBM/sec}$$

$$HP = \frac{(77.875)\ \frac{\text{LBM}}{\text{SEC}}\ (114.96)\ \text{FT}}{550\ \frac{\text{FT-LBF}}{\text{HP-SEC}}} = 16.28\ \text{hp}$$

2

THE PROPELLER IS LIMITED BY CAVITATION. DISREGARDING A SAFETY FACTOR, THIS WILL OCCUR WHEN

$$h_{ATMOS} - h_{VELOCITY} < h_{VAPOR\ PRESSURE}$$

ASSUME THE DENSITY OF SEA WATER IS 64.0 $\text{LBM/FT}^3$

$$h_{ATMOS} = \frac{(14.7)(144)}{64.0} = 33.075'$$

$h_{DEPTH} = 8$

$$h_{VELOCITY} = \frac{V_{PROP}^2}{2g} = \frac{(4.2\,V_{boat})^2}{(2)(32.2)} = .2739\,V_{boat}^2$$

THE VAPOR PRESSURE OF 70°F FRESH WATER IS FOUND FROM PAGE 3-36 OR THE STEAM TABLES.

$$h_v = \frac{(.3631)\ \frac{\text{LBF}}{\text{IN}^2}\ (144)}{62.3} = 0.839\ \text{FT}$$

RAOULT'S LAW PREDICTS THE ACTUAL VAPOR PRESSURE OF THE SOLUTION:

$$P_{VAPOR,\ SOLUTION} = P_{VAPOR,\ SOLVENT}\ (\text{MOLE FRACTION OF SOLVENT})$$

ASSUME 2½% SALT BY WEIGHT. OUT OF 100 POUNDS OF SEA WATER, WE'LL HAVE 2.5 LBM SALT AND 97.5 LBM WATER. THE MOLECULAR WEIGHT OF SALT IS $(23.0 + 35.5) = 58.5$. SO, THE # OF MOLES OF SALT IN 100 POUNDS OF SEA WATER IS

$$N_s = \frac{2.5}{58.5} = .043$$

SIMILARLY, THE MOLECULAR WEIGHT OF WATER IS 2(1) + 16 = 18. THE NUMBER OF MOLES OF WATER IS

$$N_w = \frac{97.5}{18} = 5.417$$

THE MOLE FRACTION OF WATER IS

$$\frac{5.417}{5.417 + .043} = .992$$

SO, $P_{\text{VAPOR SEA WATER}} = (.992)(.839) = 0.832 \text{ FT}$

then

$$8 + 33.075 - .2739\, v^2 = 0.832$$

or $v = 12.12$ ft/sec

3

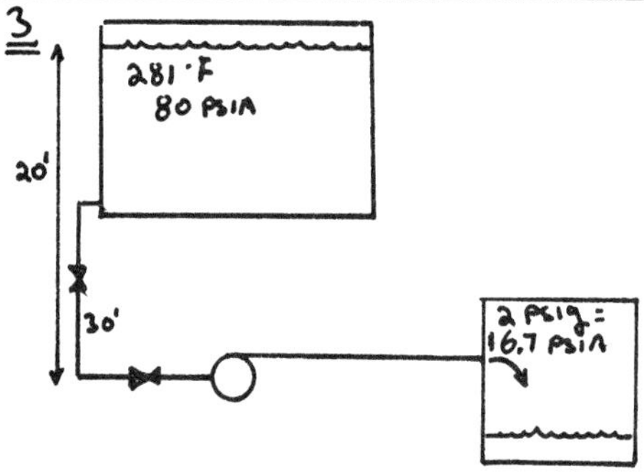


ASSUME SCHEDULE 40 PIPE. FROM PAGE 3-39,

$D_i = .1342$

$A_i = .01414$

FROM PAGE 3-37

$(100)\text{gpm} = (100)(2.228\ \text{EE-3}) = .2228\ \text{FT}^3/\text{SEC}$

$$V = \frac{Q}{A} = \frac{.2228}{.01414} = 15.76\ \text{FT/SEC}$$

ASSUME REGULAR, SCREWED, STEEL FITTINGS. THE APPROXIMATE FITTING EQUIVALENT LENGTHS ARE

| | | |
|---|---|---|
| INLET | | 3.1 (SQUARE) |
| ELBOWS | 2 × 3.4 | = 6.8 |
| GATE V | 2 × 1.2 | = 2.4 |
| | | 12.3 |

SO, THE TOTAL EQUIVALENT LENGTH IS

30 + 12.3 = 42.3 FT

FROM PAGE 3-20,

$\epsilon = .0002$ FT

$$\frac{\epsilon}{D} = \frac{.0002}{.1342} = .0015$$

AT 281°F, $\nu = .239\ \text{EE-5}$ {INTERPOLATION, P10-23}

$$N_{Re} = \frac{VD}{\nu} = \frac{(15.76)(.1342)}{.239\ \text{EE-5}} = 8.85\ \text{EE}5$$

FROM PAGE 3-20, $f = .022$

$$h_f = \frac{(.022)(42.3)(15.76)^2}{(2)(.1342)(32.2)} = 26.74\ \text{FT}$$

AT 281°, $\rho = 1/.01727 = 57.9\ \frac{\text{LBM}}{\text{FT}^3}$ {PAGE 6-29}

ALSO FROM PAGE 6-29, $P_{\text{VAPOR}} = 50.04$ psi

FROM EQN. 4.12,

$$\text{NPSHA} = \frac{(80)\text{PSIA}\,(144)\,\frac{\text{IN}^2}{\text{FT}^2}}{57.9\ \frac{\text{LBM}}{\text{FT}^3}} + 20 - 26.74 - \frac{(50.04)(144)}{57.9}$$

$= 67.77$ FT

SINCE NPSHR = 10, THE PUMP WILL NOT CAVITATE AT ANY TIME AS LONG AS THE PRESSURE REMAINS AT 80 PSI. NOTICE THAT THE DISCHARGE LINE DOES NOT AFFECT NPSHA.

4

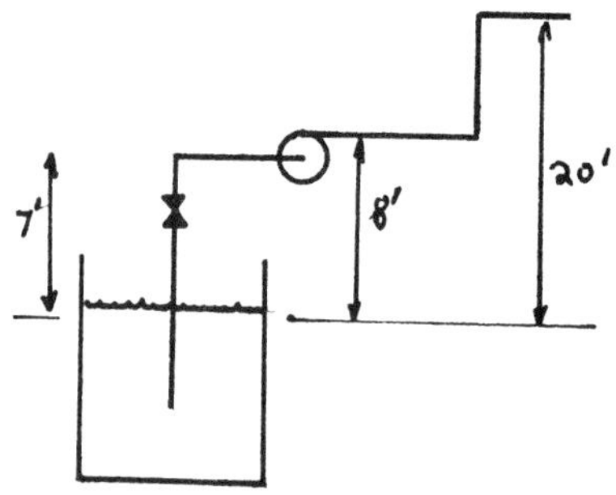


ASSUME SCREWED STEEL FITTINGS. THE APPROXIMATE FITTING EQUIVALENT LENGTHS ARE

| | |
|---|---|
| INLET | 8.5 |
| CHECK VALVE | 19.0 |
| ELBOWS (3 × 3.6) | 10.8 |

THE TOTAL EQUIVALENT LENGTH OF THE 2" LINE IS

$$L_e = 12 + 8.5 + 10.8 + 19 + 80 = 130.3$$

ASSUME SCHEDULE 40 PIPE. FROM PAGE 3-39,

$D_i = .1723$

$A_i = .0233$

SINCE THE FLOW RATE IS UNKNOWN, IT MUST BE ASSUMED TO FIND V.

ASSUME 90 gpm.

$$(90)\text{gpm}(.00228)\frac{\text{FT}^3}{\text{SEC-gpm}} = .20052\ \text{FT}^3/\text{SEC}$$

$$V = \frac{Q}{A} = \frac{.20052}{.0233} = 8.606\ \text{FT/SEC}$$

$\nu = 1.059\ \text{EE-5}$ (PAGE 3-36)

$$N_{Re} = \frac{DV}{\nu} = \frac{(.1723)(8.606)}{1.059\ \text{EE-5}} = 1.4\ \text{EE}5$$

FROM PAGE 3-20, $\epsilon = .0002$ FT

$$\frac{\epsilon}{D} = \frac{.0002}{.1723} = .0012$$

FROM PAGE 3-20, $f = .022$

THE FRICTION LOSS IN THE LINE AT 90 gpm IS

$$h_f = \frac{(.022)(130.3)(8.606)^2}{(2)(.1723)(32.2)} = 19.1$$

AT 90 gpm, THE VELOCITY HEAD IS

$$h_v = \frac{v^2}{2g} = \frac{(8.606)^2}{(2)(32.2)} = 1.2$$

IN GENERAL, THE VELOCITY HEAD IS $1.2\left(\frac{Q_2}{90}\right)^2$

THE TOTAL SYSTEM HEAD IS

$$H = (\Delta z) + h_v + h_f$$

$$= 20 + (1.2 + 19.1)\left(\frac{Q_2}{90}\right)^2$$

| $Q_2$ | H | $Q_2$ | H |
|---|---|---|---|
| 0 | 20.0 | 60 | 29.0 |
| 10 | 20.3 | 70 | 32.3 |
| 20 | 21.0 | 80 | 36.0 |
| 30 | 22.3 | 90 | 40.3 |
| 40 | 24.0 | 100 | 45.0 |
| 50 | 26.3 | 110 | 50.3 |

THE INTERSECTION POINT IS PROBABLY NOT IN AN EFFICIENT RANGE FOR THE PUMP. A DIFFERENT PUMP SHOULD BE USED.

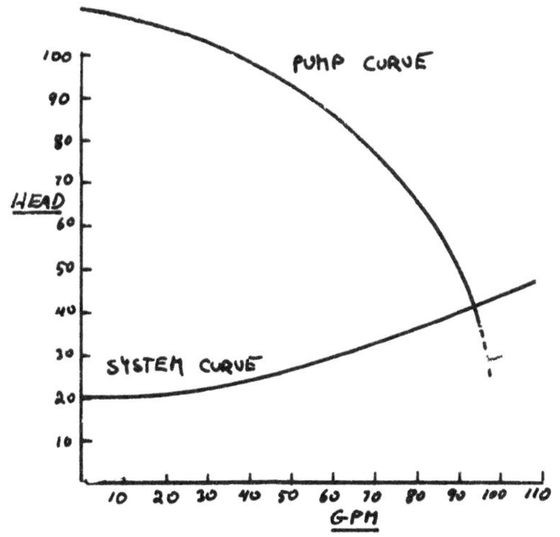


**5** 8 MGD $= \dfrac{(8\text{ EE}6)\frac{\text{GAL}}{\text{DAY}}(.002228)\frac{\text{FT}^3}{\text{SEC-GPM}}}{(24)\frac{\text{HR}}{\text{DAY}}(60)\frac{\text{MIN}}{\text{HR}}}$

$= 12.377 \ \frac{\text{FT}^3}{\text{SEC}}$

ASSUME SCHEDULE 40 PIPE. FROM PAGES 3-39 AND 3-40,

| PIPE | $D_i$ | $A_i$ |
|---|---|---|
| 8 | .6651 | .3474 |
| 12 | .9948 | .7773 |
| 16 | 1.25 | 1.2272 |

THE VELOCITY IN THE INLET PIPE IS

$$v = \frac{Q}{A} = \frac{12.377}{.3474} = 35.63 \text{ FT/SEC}$$

ASSUME 70°F, SO $\nu = 1.59$ EE-5 {PAGE 3-36}

THEN

$$N_{Re} = \frac{Dv}{\nu} = \frac{(.6651)(35.63)}{1.059\text{ EE-5}} = 2.24\text{ EE}6$$

FROM PAGE 3-20 FOR STEEL PIPE,

$\epsilon = .0002$ THEN $\frac{\epsilon}{D} = \frac{.0002}{.6651} = 0.003$

FROM PAGE 3-20 $f = 0.015$ SO

$$h_{f,1} = \frac{(0.015)(1000)(35.63)^2}{(2)(.6651)(32.2)} = 444.6 \text{ FT}$$

FOR THE OUTLET PIPE,

$$v = \frac{Q}{A} = \frac{12.377}{.7773} = 15.92 \text{ FPS}$$

$$N_{Re} = \frac{(.9948)(15.92)}{1.059\text{ EE-5}} = 1.5\text{ EE}6$$

USING $\epsilon = .0002$, $\frac{\epsilon}{D} = \frac{.0002}{.9948} = 0.0002$

SO $f = 0.014$

$$h_{f,2} = \frac{(0.014)(1500)(15.92)^2}{(2)(.9948)(32.2)} = 83.1 \text{ FT}$$

NOW, ASSUME A 50% SPLIT THROUGH THE 2 BRANCHES. IN THE UPPER BRANCH,

$$v = \frac{Q}{A} = \frac{\frac{1}{2}(12.377)}{.3474} = 17.81 \text{ FPS}$$

$$N_{Re} = \frac{(.6651)(17.81)}{(1.059\text{ EE-5})} = 1.1\text{ EE}6$$

USING $\frac{\epsilon}{D} = 0.003$, $f = 0.015$

FOR THE 16" PIPE,

$$v = \frac{Q}{A} = \frac{\frac{1}{2}(12.377)}{1.2272} = \frac{6.189}{1.2272} = 5.04$$

$\epsilon = 0.0002$ SO $\frac{\epsilon}{D} = \frac{0.0002}{1.25} = 0.00016$

$$N_{Re} = \frac{(1.25)(5.04)}{1.059\text{ EE-5}} = 5.9\text{ EE}5$$

FROM PAGE 3-20, $f \approx 0.015$

THESE VALUES OF $f$ FOR THE 2 BRANCHES ARE FAIRLY INSENSITIVE TO CHANGES IN Q, SO THEY WILL BE USED FOR THE REST OF THE PROBLEM. IN THE UPPER BRANCH, ASSUMING 4 MGD,

$$h_{f,\text{upper}} = \frac{(0.015)(500)(17.81)^2}{(2)(.6651)(32.2)} = 55.5$$

EQN 4.32 PREDICTS THE LOSS FOR ANY OTHER FLOW.

$$H_2 = H_1\left(\frac{Q_2}{Q_1}\right)^2 = 55.5\left(\frac{Q}{6.189}\right)^2 = 1.45Q^2$$

SIMILARLY FOR THE LOWER BRANCH,

IN THE 8" SECTION

$$h_{f,\text{lower},8"} = \frac{(0.015)(250)(17.81)^2}{(2)(.6651)(32.2)} = 27.8 \text{ FT}$$

IN THE 16" SECTION

$$h_{f,\text{lower},16"} = \frac{(0.015)(1000)(5.04)^2}{(2)(1.25)(32.2)} = 4.7 \text{ FT}$$

THE TOTAL LOSS IN THE LOWER BRANCH IS

$$h_{f,\text{lower}} = 27.8 + 4.7 = 32.5$$

FOR ANY OTHER FLOW, THE LOSS will be

$$H_2 = 32.5\left(\frac{Q}{6.189}\right)^2 = 0.85(Q)^2$$

LET X BE THE FRACTION FLOWING IN THE UPPER BRANCH. THEN, BECAUSE THE FRICTION LOSSES ARE EQUAL,

$$1.45\left[(x)12.377\right]^2 = 0.85\left[(1-x)12.377\right]^2$$

OR $x = .432$

THEN, $Q_{\text{upper}} = .432(12.377) = 5.347$ CFS

$Q_{\text{lower}} = (1-.432)(12.377) = 7.03$ CFS

$$H_2 = 1.45(5.347)^2 = 41.5 \text{ FT}$$

$$h_{f,\text{total}} = 444.6 + 41.5 + 83.1 = 569.2$$

6 a) AT 70°F, THE HEAD DROPPED IS

$$H = \frac{(500-30)\frac{\text{LBF}}{\text{IN}^2}(144)\frac{\text{IN}^2}{\text{FT}^2}}{62.4\frac{\text{LBM}}{\text{FT}^3}} = 1084.6 \text{ FT}$$

FROM EQN 4.52, THE SPECIFIC SPEED OF A TURBINE IS

$$N_s = \frac{N\sqrt{bhp}}{(H)^{1.25}} = \frac{1750\sqrt{250}}{(1084.6)^{1.25}} = 4.445$$

SINCE THE LOWEST SUGGESTED VALUE OF $N_s$ FOR A REACTION TURBINE IS 10 {PAGE 4-16} WE RECOMMEND AN IMPULSE {PELTON} WHEEL.

b) FROM PAGE 3-30 {APPLICATION E}

$$Q = Av = \frac{\pi}{4}\left(\frac{4}{12}\right)^2(35) = 3.054 \text{ CFS}$$

$$Q' = \frac{(35-10)(3.054)}{35} = 2.181 \text{ CFS} \quad \{\text{EQN } 3.168\}$$

ASSUMING 62.4 LBM/FT³,

$$F_x = \frac{(2.181)(62.4)}{32.2}(35-10)(\cos 80° - 1) = -87.32 \text{ LBF}$$

$$F_y = \frac{(2.181)(62.4)}{32.2}(35-10)(\sin 80°) = 104.06 \text{ LBF}$$

$$R = \sqrt{(87.32)^2 + (104.06)^2} = 135.84 \text{ LBF}$$

7

80 gpm = (80)(.002228) = .1782 CFS

$$A = \frac{\pi}{4}\left(\frac{2}{12}\right)^2 = .0218$$

$$V = \frac{Q}{A} = \frac{.1782}{.0218} = 8.174 \text{ FPS}$$

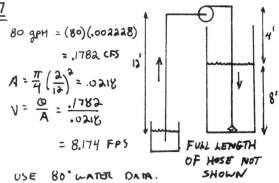

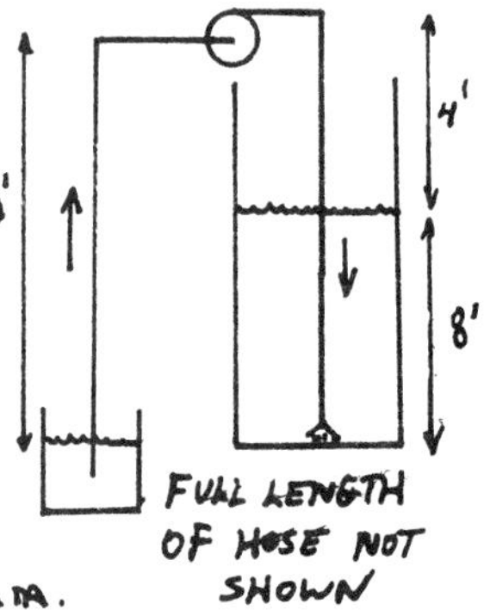


USE 80° WATER DATA.

$\nu = .93$ EE-5 {PAGE 3-36}

$$N_{Re} = \frac{VD}{\nu} = \frac{(8.174)(2/12)}{(.93 \text{ EE-5})} = 1.46 \text{ EE5}$$

ASSUME THE RUBBER HOSE IS SMOOTH. THEN, FROM PAGE 3-20, $f = .016$.

$$h_f = \frac{(.016)(50)(8.174)^2}{(2)\left(\frac{2}{12}\right)(32.2)} = 5'$$

$$h_v = \frac{(8.174)^2}{2(32.2)} = 1.0$$

$$h_A = 5 + 1 + 12 - 4 = 14 \text{ FT}$$

THIS NEGLECTS ENTRANCE LOSSES

NOTE THAT THE HOSE IS 50' LONG EVEN THOUGH THE LIFT IS ONLY 12'.

TIMED

1 THIS IS SIMILAR TO EXAMPLE 4.10.

a) FROM EQN 3.56, THE TOTAL HEAD ENTERING IS

$$H = 92.5 + \frac{(12)^2}{(2)(32.2)} + 5.26 = 100'$$

b) THE WATER HORSEPOWER IS

$$\frac{\dot{m}H}{550} = \frac{(25)\ \text{FT}^3/\text{sec}\ (62.4)\ \text{LBM}/\text{FT}^3\ (100)\ \text{FT}}{550} = 283.6 \text{ HP}$$

$$\eta = \frac{250}{283.6} = .881$$

c) FROM EQN 4.32,

$$N_2 = N_1\sqrt{H_2/H_1} = 610\sqrt{\frac{225}{100}} = 915 \text{ RPM}$$

d) FROM EQN 4.40,

$$HP_2 = HP_1\left(\frac{H_2}{H_1}\right)^{1.5} = 250\left(\frac{225}{100}\right)^{1.5} = 843.75\ HP$$

e) FROM EQN 4.39,

$$Q_2 = Q_1\sqrt{H_2/H_1} = 25\sqrt{\frac{225}{100}} = 37.5\ CFS$$

2 (a) 1,000,000 BTUH SYSTEM

THE ENTHALPY DROP ACROSS THE RADIATOR IS

$$(167.99 - 147.92) = 20.07\ BTU/LBM$$

THE MASS FLOW RATE THROUGH THE SYSTEM IS

$$\frac{(1{,}000{,}000)\frac{BTU}{HR}}{(60)\frac{MIN}{HR}(20.07)\frac{BTU}{LBM}} = 830.4\ \frac{LBM}{MIN}$$

AT THE BULK TEMPERATURE OF 190°F,

$$\rho = \frac{1}{.01657} = 60.35\ \frac{LBM}{FT^3}$$

THE VOLUME FLOW RATE IS

$$Q = \frac{(830.4)\frac{LBM}{MIN}(7.48)\frac{GAL}{FT^3}}{(60.35)\frac{LBM}{FT^3}} = 102.9\ gpm$$

NEXT, PLOT THE PUMP CURVES AND GET THE HEAD LOSSES AT 102.9 gpm

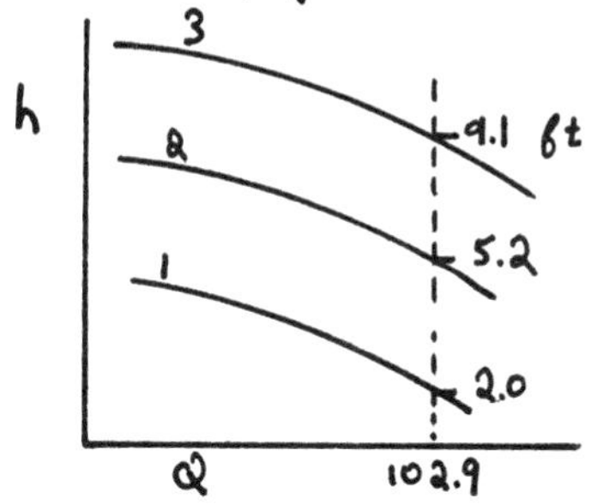


THE LOSS PER FOOT FOR THESE 3 PUMPS IS

PUMP 1 $(2.0)(12)/420 = .057$ IN/FT - TOO LOW

PUMP 2 $(5.2)(12)/420 = .12$ IN/FT - TOO LOW

PUMP 3 $(9.1)(12)/420 = .26$ IN/FT - OK

SO, CHOOSE **PUMP #3**

SINCE THE PIPE MATERIAL WAS NOT SPECIFIED, ASSUME $f = .020$ (GOOD FOR TURBULENT FLOW IN STEEL PIPE 1" TO 3" DIAMETER).

KEEP VELOCITY BETWEEN 2 AND 4 FPS FOR NOISE

$$h_f = \frac{fLv^2}{2Dg} \quad \text{SO}$$

$$D = \frac{fLv^2}{2h_f g} = \frac{(.02)(420)(4)^2}{(2)(9.1)(32.2)} = .229\ FT$$

**SAY D = 3"**

TRY A DIFFERENT APPROACH

$$A = \frac{Q}{V} = \frac{(102.9)\frac{GAL}{MIN}}{(60)\frac{SEC}{MIN}(7.48)\frac{GAL}{FT^3}(4)\frac{FT}{SEC}}$$

$$= .057\ FT^2$$

$$D = \sqrt{\frac{4A}{\pi}} = \sqrt{\frac{(4)(.057)}{\pi}} = .27\ FT$$

SAY **D = $3\frac{1}{2}$"** {SIZE MAY NOT BE AVAILABLE}

THE ACTUAL PIPE INSIDE DIAMETER WILL DEPEND ON WHETHER COPPER, STEEL, OR CAST IRON PIPE IS USED.

(b) 300,000 BTUH

PROCEED SIMILARLY.

$$Q = \frac{(300{,}000)(7.48)}{(60)(20.07)(60.35)} = 30.88\ gpm$$

THE HEAD LOSS AT 30.88 gpm FOR EACH PUMP IS

PUMP 1 $\frac{(5.5)(12)}{420} = .157"$ TOO LOW

PUMP 2 $\frac{(9.0)(12)}{420} = .257$ OKAY

PUMP 3 $\frac{(13.4)(12)}{420} = .38$ OKAY

**PUMPS 2 AND 3** MEET THE SPECIFICATIONS

$$A = \frac{30.88}{(60)(7.48)(4)} = .0172\ FT^2$$

$$D = \sqrt{\frac{(4)(.0172)}{\pi}} = .148\ FT$$

SAY, **$1\frac{3}{4}$" PIPE** {SIZE MAY NOT BE AVAILABLE}

3. THE EQUIVALENT LENGTHS ARE ASSUMED TO BE IN FEET OF THE SECTIONS' DIAMETERS. SINCE THE DIAMETERS ARE DIFFERENT, THE EQUIVALENT LENGTHS CANNOT BE ADDED TOGETHER. IT IS NECESSARY TO CONVERT THE LENGTHS TO PRESSURE DROPS. ALTHOUGH THE DARCY EQUATION COULD BE USED, IT IS MORE CONVENIENT TO USE STANDARD TABLES FOR TYPE L TUBING, FROM THE ASHRAE GUIDE AND DATA BOOK (APPLICATIONS) 1966-1967, PAGE 157,

| CIRCUIT | LOSS PER FT | EQUIV. LENGTH | TOTAL LOSS |
|---|---|---|---|
| 5-1-P-2 | 0.3" WG | 40 | 12" |
| 2-4 | 0.47 | 70 | 32.9 |
| 2-3 | 0.40 | 55 | 22 |
| 3-4 | 0.47 | 65 | 30.6 |
| 3-5 | 0.47 | 60 | 28.2 |
| 4-5 | 0.40 | 50 | 20 |

THESE HEAD LOSSES ARE FOR COLD WATER, AND ARE GREATER THAN HOT WATER LOSSES. THE REQUIRED HORSEPOWER WILL BE LESS FOR HOT WATER. HOWEVER, THE SYSTEM MAY BE TESTED WITH COLD WATER, AND THE HORSEPOWER SHOULD BE ADEQUATE FOR THIS.

NOW, REDRAW THE PIPE NETWORK.

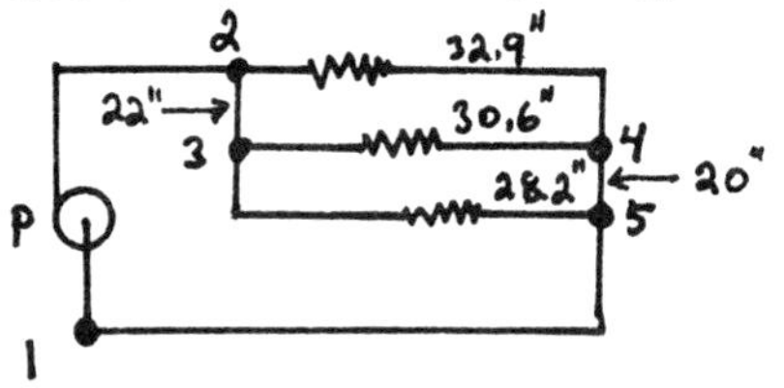


THE PUMP MUST BE ABLE TO HANDLE THE LONGEST RUN.

RUN 2-4-5 : 32.9 + 20 = 52.9
2-3-4-5: 22 + 30.6 + 20 = 72.6
2-3-5: 22 + 28.2 = 50.2

THE LONGEST RUN IS 2-3-4-5 WITH A LOSS OF 72.6". THE TOTAL CIRCUIT LOSS IS

$$h_t = \frac{72.6 + 12}{12} = 7.05 \text{ FT}$$

NOTE THAT THE LOSS BETWEEN POINTS 2 AND 5 MUST BE THE SAME REGARDLESS OF THE PATH. THEREFORE, FLOW RESTRICTION VALVES MUST EXIST IN THE OTHER TWO RUNS IN ORDER TO MATCH THE LONGEST RUN'S LOSS.

THE WATER HORSEPOWER IS

$$\text{whp} = \frac{(7.05)\text{ft}\,(60)\text{gpm}}{3960} = 0.107$$

$$\text{bhp} = \frac{\text{whp}}{\eta} = \frac{0.107}{0.45} = 0.24$$

**USE A 1/4 hp MOTOR OR LARGER**

$$\dot{m} = (60)\text{ gpm}\,(0.1337)\frac{\text{ft}^3}{\text{gal}}(60.1)\frac{\text{lbm}}{\text{ft}^3}$$

$$= 482 \text{ LBM/MIN}$$

$$q = \dot{m} C_p \Delta T$$

$$= (482)\frac{\text{LBM}}{\text{MIN}}(60)\frac{\text{MIN}}{\text{HR}}(1)\frac{\text{BTU}}{\text{LBM-}^\circ\text{F}}(20)\,^\circ\text{F}$$

$$= \boxed{578{,}400 \text{ BTUH}}$$

4 ASSUME REGULAR 90° ELBOWS.

THE EQUIVALENT LENGTH OF PIPE AND FITTINGS IS

$$500 + 6(4.4) + 2(2.8) = 532 \text{ FT}$$

FOR 3" STEEL SCHEDULE 40 PIPE,

$$A_i = .05134 \text{ FT}^2$$

$$D_i = .2557 \text{ FT}$$

ASSUME 100 GPM. (THE ACTUAL VALUE YOU CHOOSE FOR THIS FIRST ESTIMATE IS NOT IMPORTANT).

$$v = \frac{Q}{A} = \frac{(100)\text{gpm}\,(.1337)\frac{\text{FT}^3}{\text{GAL}}}{(.05134)\text{ FT}^2\,(60)\frac{\text{SEC}}{\text{MIN}}} = 4.34 \text{ FT/SEC}$$

$$N_{Re} = \frac{Dv}{\nu} = \frac{(.2557)(4.34)}{6\text{EE-}6} = 1.85 \text{ EE}5$$

USE $\epsilon = .0002$ FT

SO $\frac{\epsilon}{D} = \frac{.0002}{.2557} = .0008$

FROM THE MOODY CHART,

$f \approx .02$ (GOES TO .018 AT FULL TURBULENCE) WHICH THIS WILL BE AT HIGHER FLOWS. SO, USE $f = .018$.

$$h_f = \frac{(.018)(532)(4.34)^2}{(2)(.2557)(32.2)} = 10.95 \text{ FT OF GASOLINE}$$

ALTHOUGH IT CAN BE INCLUDED, THIS NEGLECTS THE SMALL VELOCITY HEAD.

USING $\frac{h_1}{h_2} = \left(\frac{Q_1}{Q_2}\right)^2$ THE OTHER SYSTEM CURVE POINTS CAN BE FOUND

$$h_2 = \frac{10.95}{(100)^2} Q_2^2 = .001095\, Q_2^2$$

| Q | $h_f$ | h + 60 |
|---|---|---|
| 100 | 10.95 | 71.0 |
| 200 | 43.8 | 103.8 |
| 300 | 98.6 | 158.6 |
| 400 | 175.2 | 235.2 |
| 500 | 273.8 | 333.8 |
| 600 | 394.2 | 454.2 |

THE HEAD DOES NOT DEPEND ON THE FLUID BEING PUMPED. THEREFORE, THE WATER CURVE CAN BE USED.

PLOTTING THE SYSTEM AND PUMP CURVES GIVES

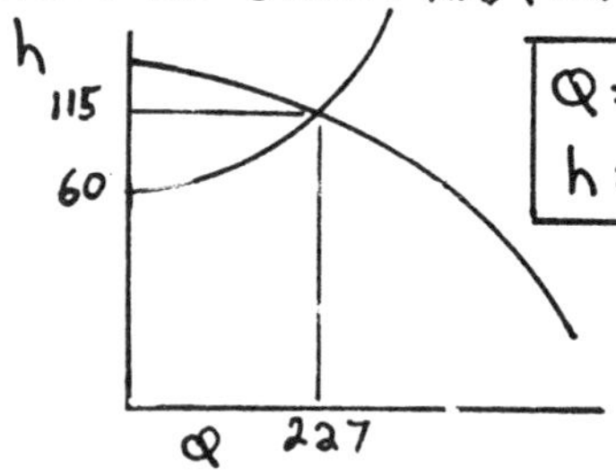


**Q = 227 gpm**
**h = 115 ft**

THE COST PER HOUR IS

$$(.045)\frac{\$}{\text{KW-HR}}\left[\frac{(227)\text{gpm}\,(.1337)\frac{\text{FT}^3}{\text{GAL}}\,(62.4)\frac{\text{LBM}}{\text{FT}^3}\,(.7)}{(33{,}000)(.88)(.88)\frac{\text{FT-LB}}{\text{HP-MIN}}} \times (115)\text{ FT}\,(.7457)\frac{\text{KW}}{\text{HP}}(1)\text{ HR}\right]$$

$$= \boxed{\$.20}$$

**5**

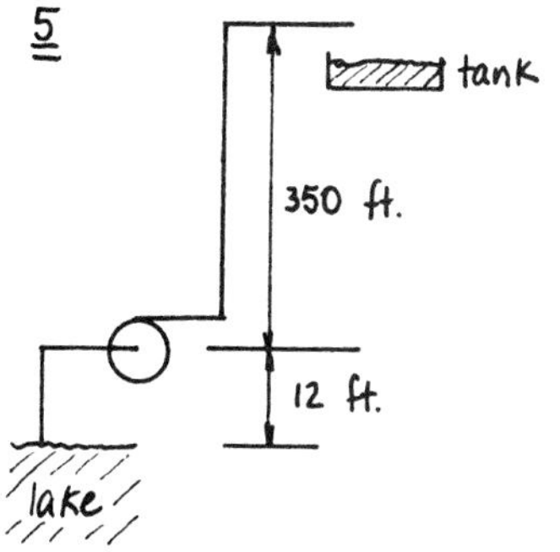


$Q = 10{,}000$ gal/hr
$= 1337\ \text{ft}^3/\text{hr}$
(appendix D, p. 3-37)
$\eta = 0.7$

Assume 60°F and 14.7 psia.

For 4" schedule 40 steel pipe, from appendix F, p. 3-39, I.D. = 0.3355 ft.

$A = 0.08841\ \text{ft}^2$

$$V = \frac{Q}{A} = \frac{(1337\ \text{ft}^3/\text{hr})\left(\frac{1\ \text{hr}}{3600\ \text{sec}}\right)}{0.08841\ \text{ft}^2}$$

$= 4.2$ ft/sec $= V_d$

using equation 4.2,

$$h_f = \frac{f L_e v^2}{2 D g_c}$$

from appendix A, p. 3-36, at 60°F, $\nu = 1.217\ \text{EE}-5\ \text{ft}^2/\text{sec}$

$$N_{Re} = \frac{D_e V}{\nu} = \frac{(0.3355)(4.2)}{1.217\ \text{EE}-5} = 1.16\ \text{EE}5$$

from table 3.8, for welded and seamless steel, $\epsilon \approx 0.0002$ ft

$$\frac{\epsilon}{D} = \frac{0.0002}{0.3355} = 5.96\ \text{EE}-4 \approx 0.0006$$

using Moody friction chart (figure 3.13), $f = 0.02$, then

$$h_f = \frac{(0.02)(7000\ \text{ft})(4.2\ \text{ft/sec})^2}{2(0.3355\ \text{ft})(32.2\ \text{lbm-ft/lbf-sec}^2)} = 114.3\ \text{ft}$$

$$h_v = \frac{v^2}{2g_c} = \frac{(4.2)^2}{2\ (32.2)} = 0.27\ \text{ft}$$

USE EQUATION 4.18, THE HEAD ADDED BY THE PUMP IS

$$h_A = 12 + 350 + 0.27 + 114.3 = 476.6\ \text{FT}$$

Then, from table 4.2,

$$\text{whp} = \frac{h_A Q\ (\text{S.G.})}{3960} = \frac{(476.6\ \text{ft})(10{,}000\ \text{gal/hr})\left(\frac{1\ \text{hr}}{60\ \text{min}}\right)(1)}{3960}$$

$= 20.06$ hp

Using equation 4.22,

$$\text{ehp} = \frac{\text{whp}}{\eta} = \frac{20.06}{0.7} = 28.7\ \text{hp}$$

equation 4.26,

Power $= (0.7457)\ 28.7 = 21.4$ kW

At \$0.04/kW-hr, power costs

(21.4)(\$0.04) = **\$0.856 per hour**

(b) motor hp ≈ ehp = 28.7 ~ **30 hp motor required**

(c) using equation 4-12,

$$\text{NPSHA} = h_a + h_s - h_{f(s)} - h_{vp}$$

using equation 4.4, and $\rho = 62.37$ from appendix A, p. 3-36

$$h_a = \frac{p_a}{\rho} = \frac{(14.7)(144)}{62.37} = 33.94\ \text{ft}$$

$$h_f = \left(\frac{300}{7000}\right) h_{f,\text{total}} = \left(\frac{300}{7000}\right) 114.3 = 4.9\ \text{FT}$$

FROM APPENDIX A AT 60°F, OR FROM STEAM TABLES, $h_{vp} = 0.59$ FT

$$\text{NPSHA} = 33.94 - 12 - 4.9 - 0.59 = 16.45\ \text{FT}$$

**6**

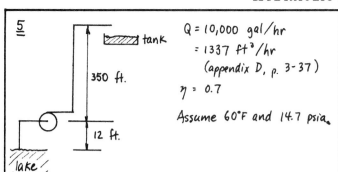

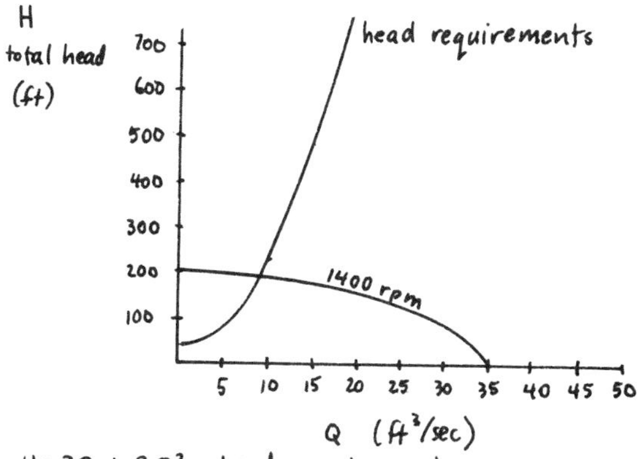


$H = 30 + 2Q^2$ head requirements

(a) from the intersection of the graphs, at 1400 rpm, THE FLOW RATE IS APPROXIMATELY 9 CFS,

$$Q = (9\ \text{CFS})\left(448.8\ \frac{\text{GPM}}{\text{CFS}}\right) = 4039\ \text{GPM}$$

$$h_A = 30 + 2(9)^2 = 192\ \text{FT}$$

(b) FROM TABLE 4.2,

$$\text{WHP} = \frac{(192)(9)(1)}{8.814} = 196\ \text{HP}$$

FROM EQN 4.27,

$$n_s = \frac{n\sqrt{Q}}{(h_A)^{0.75}} = \frac{1400\sqrt{4039}}{(192)^{0.75}} = 1725$$

FROM FIG. 4.13 $\eta \approx 83\%$. THE MINIMUM PUMP MOTOR POWER SHOULD BE

$$\frac{196\ \text{HP}}{0.83} = 236\ \text{HP}\ \ (\text{use } 250\ \text{HP})$$

(c) IF PUMP IS TURNED AT 1200 RPM, FROM EQN 4.31,

$$Q_2 = \frac{N_2 Q_1}{N_1} = \frac{(1200)(4039)}{1400} = 3462\ \text{GPM}$$

# Fans and Ductwork

**1** FROM EQN 5.30 WITH $R=4$, THE SHORT SIDE IS

$$a = \frac{18''(4+1)^{.25}}{1.3\,(4)^{.625}} = 8.705''$$

THE LONG SIDE IS

$$b = (R)(a) = (4)(8.705) = 34.82$$

**2** DISREGARDING TERMINAL PRESSURE, OTHER POINTS ON THE SYSTEM CURVE ARE FOUND FROM EQN. 5.11,

$$\frac{P_2}{P_1} = \left(\frac{Q_2}{Q_1}\right)^2 \quad \text{OR} \quad Q_2 = Q_1\sqrt{P_2/P_1}$$

$$Q_2 = \frac{(10{,}000)\sqrt{P_2}}{\sqrt{4}} = 5000\sqrt{P_2}$$

| $P_2$ | Q | $P_2$ | Q |
|---|---|---|---|
| 1" | 5000 CFM | 4" | 10,000 |
| 2" | 7070 | 5" | 11,180 |
| 3" | 8660 | 6" | 12,250 |

**3** AT STANDARD CONDITIONS, $\rho = 0.075\ \text{LBM/FT}^3$, USING DATA FROM P. 8-19,

$$\rho_{5000} = \frac{P}{RT} = \frac{(12.225)(144)}{(53.3)(500.9)} = 0.06594\ \text{LBM/FT}^3$$

FROM EQN 5.7,

$$(AHP)_1 = \frac{(40)\ \text{CFM}\ (.5)''\ \text{WG}}{6356\ \frac{\text{CM-WG}}{\text{HP}}} = 3.147\ \text{EE-3}$$

FROM EQN 5.17,

$$(AHP)_2 = (3.147\ \text{EE-3})\left(\frac{1}{\frac{1}{8}}\right)^5\left(\frac{\frac{1}{2}(300)}{300}\right)^3\left(\frac{.06594}{.075}\right) = 11.3\ \text{HP}$$

NOTE THAT SUCH A LARGE EXTRAPOLATION IS NOT RECOMMENDED BECAUSE ACTUAL EFFICIENCY WILL PROBABLY BE MUCH LOWER.

**4** FROM PAGE 5-7, THE FRICTION LOSS IS 0.47" WATER PER 100 FT AND V=2700 FPM. THE FRICTION LOSS DUE TO THE DUCT IS

$$h_{f,1} = (.47)\ \text{IN WG}\left(\frac{750}{100}\right) = 3.525''\ \text{WG}$$

FROM TABLE 5.3, THE EQUIVALENT LENGTH OF EACH ELBOW IS 12 D. FOR THE TWO,

$$L_e = 4\left[(12)\left(\frac{20\ \text{IN}}{12\ \text{IN/FT}}\right)\right] = 80'$$

THIS CREATES A FRICTION LOSS OF

$$h_{f,2} = (.47)\left(\frac{80}{100}\right) = .376''\ \text{WG}$$

ALSO, FROM TABLE 5.3, FOR $\frac{E}{D} = \frac{2}{20} = .10$, $C_1 = .2$. USING EQN 5.33,

$$h_{f,3} = 2\left[(.2)\left(\frac{2700}{4005}\right)^2\right] = .182''\ \text{WG}$$

THE TOTAL LOSS IS

$$\Sigma h_f = 3.525 + 0.376 + 0.182 = 4.083''\ \text{WG}$$

**5** IN THE 18" DUCT,

$$A = \frac{\frac{\pi}{4}(18)^2}{144} = 1.767\ \text{FT}^2$$

$$\text{SO } V_1 = \frac{Q}{A} = \frac{1500}{1.767} = 848.9\ \text{FPM}$$

IN THE 14" DUCT, $A = \frac{\frac{\pi}{4}(14)^2}{144} = 1.069\ \text{FT}^2$

$$\text{SO } V_2 = \frac{(1500-400)}{1.069} = 1029$$

SINCE $V_2 > V_1$, THERE IS NO REGAIN. USE $R = 1.1$ TO ACCOUNT FOR A 10% FRICTION LOSS.

$$\Delta P_s = 1.1\left(\frac{(848.9)^2 - (1029)^2}{(4005)^2}\right) = -0.023''\ \text{WG LOSS}$$

## CONCENTRATES

**1**

FROM TABLE 5.5, SELECT 1600 FPM AS THE MAIN DUCT VELOCITY. FROM PAGE 5-7 WITH 1600 FPM AND 1500 CFM

$$h_f = .27''\ \text{WG PER } 100'$$

$$D_{\text{FAN-A}} = 13''$$

FROM EQN 5.30,

$$a = \frac{D\,(1.5+1)^{.25}}{1.3\,(1.5)^{.625}} = .75\,D$$

$$b = Ra = (1.5)(.75\,D) = 1.125\,D$$

SO, $a_{\text{FAN-A}} = (.75)(13) = 9.75$

AND

$$b_{\text{FAN-A}} = (1.125)(13) = 14.63$$

PROCEEDING SIMILARLY FOR ALL SECTIONS, THE FOLLOWING TABLE IS OBTAINED:

| SECTION | Q | D | a × b * |
|---|---|---|---|
| FAN-A | 1500 | 13 | 9.8 × 14.6 |
| A-B | 1200 | 11.8 | 8.9 × 13.3 |
| B-C | 900 | 11.0 | 8.3 × 12.4 |
| C-D | 600 | 9.1 | 6.8 × 10.2 |
| D-E | 400 | 8.0 | 6 × 9 |
| E-F | 200 | 6.0 | 4.5 × 6.8 |

ASSUME $r/D = 1.5$ FOR THE BENDS. FROM TABLE 5.3, $L_e = 12D$. THEN, FOR THE 2 ELBOWS,

$$L_e = 12\left[\frac{13}{12} + \frac{9.1}{12}\right] = 22.1\ \text{FT}$$

* ACTUAL VALUES WOULD BE ROUNDED TO WHOLE INCHES.

THE DISTANCE FROM FAN TO F IS

15 + 45 + 30 + 30 + 20 + 10 + 20 + 20 = 190

THEN

$$h_f = \left(\frac{190+22.1}{100}\right)\left(.27'' \text{ wg}/100 \text{ FT}\right) = .57'' \text{ wg}$$

THE FAN MUST SUPPLY THE TERMINAL PRESSURE ALSO

.57 + .25 = .82" wg

THE EQUAL FRICTION METHOD GENERALLY IGNORES VELOCITY HEAD CONTRIBUTION TO STATIC PRESSURE. IF IT IS INCLUDED, ADD

$$\left(\frac{1600}{4005}\right)^2 = 0.98'' \text{ WG}$$

2 CHOOSE 1600 FPM AS THE MAIN DUCT VELOCITY.

$Q_{FAN} = (12)(300) = 3600$ CFM

FROM PAGE 5-7 {WITH Q = 3600 AND V = 1600 FPM},

h = .16" wg PER 100 FT

$D_{FAN-1ST}$ = 20"

PROCEEDING ACCORDING TO THE EQUAL FRICTION METHOD RESULTS IN THE FOLLOWING TABLE:

| SECTION | Q | D | SECTION | Q | D |
|---|---|---|---|---|---|
| FAN-1ST | 3600 | 20 | 2ND-E | 1200 | 13.2 |
| 1ST-2ND | 2400 | 17.2 | E-F | 900 | 12 |
| 2ND-3RD | 1200 | 13.2 | F-G | 600 | 10.1 |
| 1ST-A | 1200 | 13.2 | G-H | 300 | 7.9 |
| A-B | 900 | 12 | I-J | 900 | 12 |
| B-C | 600 | 10.1 | J-K | 600 | 10.1 |
| C-D | 300 | 7.9 | K-L | 300 | 7.9 |

THE LONGEST RUN IS (FAN-L). ITS LENGTH IS

$L_{LONGEST}$ = 25 + 35 + 20 + 20 + 10 + 20 + 20 + 20 = 170

FROM TABLE 5.3, THE ELBOWS HAVE $L_e = 14.5$ D {INTERPOLATED}. SO

$$L_e = 170 + 14.5\left(\frac{20'' + 13.2''}{12 \frac{\text{IN}}{\text{FT}}}\right) = 210.1 \text{ FT}$$

THE FAN SUPPLIES

$$\left(\frac{210.1}{100}\right)\left(.16'' \text{ wg PER } 100'\right) + .15 = .486''$$

3 CHOOSE 1600 FPM AS THE MAIN DUCT VELOCITY. THEN Q = 3600 CFM, AND FROM PAGE 5-7

$h_f$ = .16" wg PER 100 FEET

D = 20"

AS IN PROBLEM 2, THE BENDS HAVE $L_e$ = 14.5 D. THEN, THE LOSS UP TO THE FIRST TAKE-OFF IS

$$h_{f,\text{FAN-1ST}} = \frac{.16\left[25 + 35 + 14.5\left(\frac{20}{12}\right)\right]}{100} = .135'' \text{ wg}$$

AFTER THE FIRST TAKE OFF,

Q = 3600 − 1200 = 2400

$L_e$ = 20

$$L_e/Q^{.61} = \frac{20}{(2400)^{.61}} = .173$$

FROM FIGURE 5.6 OR EQN 5.40

$V_2$ = 1390. PROCEEDING SIMILARLY, THE FOLLOWING TABLE IS DEVELOPED:

| SECTION | Q | $L_e$ | $L_e/Q^{.61}$ | V ** | D |
|---|---|---|---|---|---|
| FAN-1ST BR | 3600 | 80.8 | .547 | 1600 | 20 |
| 1ST BR-2ND BR | 2400 | 20 | .173 | 1390 | 18 (17.8) |
| 2ND - I * | 1200 | 49.3 | .65 | 960 | 15 (15.1) |
| I-J | 900 | 20 | .31 | 800 | 15 (14.4) |
| J-K | 600 | 20 | .40 | 650 | 13 |
| K-L | 300 | 20 | .62 | 500 | 11 (10.5) |

* $L_e$ FOUND ASSUMING D = 16

$$L_e = 20 + 14.5\left(\frac{16}{12}\right) + 10 = 49.3 \text{ FT}$$

** SOLVED GRAPHICALLY.

THE FAN MUST SUPPLY

.135" + .15" = .285" wg

TIMED

1 NO DAMPER IS NEEDED IN DUCT A.

THE AREA OF SECTION A IS

$$\frac{\pi}{4}\left(\frac{12}{12}\right)^2 = .7854 \text{ FT}^2$$

THE VELOCITY IN SECTION A IS

$$V_A = \frac{Q}{A} = \frac{3000 \text{ FT}^3/\text{MIN}}{(60)\frac{\text{SEC}}{\text{MIN}}(.7854)\text{ FT}^2} = 63.66 \text{ FT/SEC} = 3819.7 \text{ FT/MIN}$$

FROM PAGE 32, FIGURE 8, 1969 ASHRAE EQUIPMENT GUIDE + DATA BOOK, FOR 4-PIECE ELLS, WITH R/d = 1.5

$L_e \approx 14$ D

ASSUME D = 10".

THE TOTAL EQUIVALENT LENGTH OF RUN C IS

$$L_e = 50 + 10 + 10 + 2\left[14\left(\frac{10''}{12\ \frac{\text{in}}{\text{ft}}}\right)\right] = 93.3 \text{ FT}$$

FOR ANY DIAMETER, D, IN INCHES OF SECTION C, THE VELOCITY WILL BE

$$V_C = \frac{Q}{A} = \frac{2000\ \frac{\text{FT}^3}{\text{MIN}}}{\left(\frac{\pi}{4}\right)\left(\frac{D}{12}\right)^2 \text{FT}^2\ 60\ \frac{\text{SEC}}{\text{MIN}}} = \frac{6111.5}{D^2}$$

THE FRICTION LOSS IN SECTION C WILL BE (EQN 3.71)

$$h_{f,C} = \frac{(.02)(93.3)(6111.5/D^2)}{(2)\left(\frac{D}{12}\right)(32.2)} = \frac{1.3\text{ EE7}}{D^5} \text{ FT OF AIR}$$

THE REGAIN BETWEEN A AND C WILL BE

$$h_{\text{REGAIN}} = .65\left[\frac{V_A^2 - V_C^2}{2g}\right]$$

$$h_{\text{REGAIN}} = .65\left[\frac{(63.66)^2 - \left(\frac{6111.5}{D^2}\right)^2}{(2)(32.2)}\right]$$

$$= 40.9 - \frac{3.77\text{ EE5}}{D^4} \text{ FT OF AIR}$$

THE PRINCIPLE OF STATIC REGAIN IS THAT

$h_f = h_{\text{REGAIN}}$, SO

$$\frac{1.3\text{ EE7}}{D^5} = 40.9 - \frac{3.77\text{ EE5}}{D^4}$$

BY TRIAL AND ERROR, D = 13.5".

SINCE D = 10" WAS ASSUMED TO FIND THE EQUIVALENT LENGTH OF THE ELLS, THIS PROCESS SHOULD BE REPEATED.

$$L_e = 70 + 2\left[(14)\left(\frac{13.5}{12}\right)\right] = 101.5$$

$$h_f = \frac{(.02)(101.5)(6111.5/D^2)^2}{(2)\left(\frac{D}{12}\right)(32.2)} = \frac{1.41\text{ EE7}}{D^5}$$

THEN

$$\frac{1.41\text{ EE7}}{D^5} = 40.9 - \frac{3.77\text{ EE5}}{D^4}$$

BY TRIAL + ERROR, D = 13.63" (SAY 14")

THIS RESULTS IN A FRICTION LOSS OF

$$h_f = \frac{1.41\text{ EE7}}{(14)^5} = 26 \text{ FT OF AIR.}$$

HOWEVER, THE REGAIN CANCELS THIS, SO THE PRESSURE LOSS FROM A TO C IS ZERO. NO DAMPERS ARE NEEDED IN DUCT C.

FOR ANY DIAMETER, D, IN INCHES IN SECTION B, THE VELOCITY WILL BE

$$V_B = \frac{Q}{A} = \frac{1000}{\left(\frac{\pi}{4}\right)\left(\frac{D}{12}\right)^2(60)} = \frac{3055.8}{D^2}$$

THE FRICTION LOSS IN SECTION B WILL BE

$$h_f = \frac{(.02)(10)\left(\frac{3055.8}{D^2}\right)^2}{(2)\left(\frac{D}{12}\right)(32.2)} = \frac{3.48\text{ EE5}}{D^5} \text{ FT OF AIR}$$

THE 45° TAKE-OFF WILL ALSO CREATE A LOSS. FROM EQN 5.33,

$$h_f = C_b\left(\frac{V_A}{4005}\right)^2 \text{ INCHES OF WATER}$$

AT THIS POINT, ASSUME $\frac{V_B}{V_A} = 1.00$. THEN,

FROM TABLE 5.4 FOR A 45° FITTING, $C_b = .5$

SO

$$h_f = .5\left(\frac{3819.7}{4005}\right)^2 = .455 \text{ INCHES WATER} = 31.5 \text{ FT OF AIR}$$

THE REGAIN BETWEEN A AND B WILL BE

$$h_{\text{REGAIN}} = .65\left[\frac{V_A^2 - V_B^2}{2g}\right]$$

$$h_{\text{REGAIN}} = .65\left[\frac{(63.66)^2 - (3055.8/D^2)^2}{(2)(32.2)}\right]$$

$$= 40.9 - \frac{9.42\text{ EE4}}{D^4}$$

THEN SETTING REGAIN EQUAL TO LOSS

$$40.9 - \frac{9.42\text{ EE4}}{D^4} = \frac{3.48\text{ EE5}}{D^5} + 31.5$$

BY TRIAL AND ERROR,

$D_B = 10.77$ (SAY, 11")

THIS MAKES $V_B = \frac{3055.8}{(11)^2} = 25.25$ FPS

AND SINCE $\frac{V_B}{V_A} = \frac{25.25}{63.6} \approx .4$, THE VALUE OF $C_b$ IS STILL ABOUT .5

FOR SECTION B, THE FRICTION THAT IS CANCELLED BY THE REGAIN IS

$$h_f = 31.5 + \frac{3.48\text{ EE5}}{(11)^5} = 33.7 \text{ FT}$$

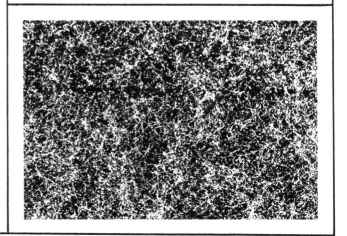

# Thermodynamics

WARM-UPS

1 FROM PAGE 6-29,

$h = 218.48 + .92(945.5) = 1088.34$ BTU/LBM

THE MOLECULAR WEIGHT OF STEAM {WATER} IS 18, SO

$H = 18(1088.34) = 19590$ BTU/pmole

2 FROM PAGE 6-30,

$h_1 = 393.84 + .95(809) = 1162.4$

FROM THE MOLLIER DIAGRAM FOR AN ISENTROPIC PROCESS,

$h_2 = 1031$ BTU/LBM

THE MAXIMUM WORK OUTPUT IS

$h_1 - h_2 = 1162.4 - 1031 = 131.4$ BTU/LBM

3 THE AVE. VALUE OF $C_p$ FOR IRON IS

$\bar{C}_p = .10$ (TABLE 6.1)

$$Q = (.10)\frac{\text{BTU}}{\text{LBM-°F}}(780-80)\text{°F} = 70.0 \text{ BTU/LBM}$$

4 FROM PAGE 10-24,

$C_p = .250$ BTU/LBM-°F

FROM TABLE 6.4, ⟶ $R = 53.3 \frac{\text{FT-LBF}}{\text{LBM-°R}}$

$$C_v = C_p - \frac{R}{J} = .250\frac{\text{BTU}}{\text{LBM-°F}} - \frac{53.3 \frac{\text{FT-LBF}}{\text{LBM-°R}}}{778 \frac{\text{FT-LBF}}{\text{BTU}}}$$

$= .1815$

$$\text{SO } K = \frac{C_p}{C_v} = \frac{.250}{.1815} = 1.377$$

5 a) FROM EQN 3.21,

$$C_{AIR} = \sqrt{KgRT}$$

$$= \sqrt{(1.4)(32.2)(53.3)(460+70)} = 1128.5 \text{ FT/SEC}$$

b) FROM TABLE 14.1

$E_{STEEL} = 29$ EE6 PSI

$\rho_{STEEL} = .283$ LBM/IN$^3$

FROM EQN 3.20,

$$C_{STEEL} = \sqrt{\frac{(29 \text{ EE6})\frac{\text{LBF}}{\text{IN}^2}(32.2)\frac{\text{FT}}{\text{SEC}^2}}{(.283)\frac{\text{LBM}}{\text{IN}^3}(12)\frac{\text{IN}}{\text{FT}}}} = 16{,}582 \text{ FT/SEC}$$

c) FROM PAGE 3-36

$E = 320$ EE3 LBF/IN$^2$

$\rho = 62.3$ LBM/FT$^3$

FROM EQN 3.20,

$$C_{WATER} = \sqrt{\frac{(320 \text{ EE3})(144)\,32.2}{62.3}} = 4880 \text{ FT/SEC}$$

6 FROM TABLE 6.4,

$R_{He} = 386.3$

FROM EQN 6.45,

$$\rho = \frac{P}{RT} = \frac{(14.7)\frac{\text{LBF}}{\text{IN}^2}(144)\frac{\text{IN}^2}{\text{FT}^2}}{(386.3)\frac{\text{FT-LBF}}{\text{LBM-°R}}(460+600)\text{°R}}$$

$= .00517$ LBM/FT$^3$

7 FROM PAGE 6-30,

$h_1 = 1187.2$ BTU/LBM

FROM THE MOLLIER DIAGRAM {PAGE 6-34},

$h_2 = 953$ BTU/LBM {AT 3 PSIA, $S_2 = S_1$}

$h_2' = 1022$ BTU/LBM

FROM EQN 7.11,

$$\eta_s = \frac{1187.2 - 1022}{1187.2 - 953} = .705$$

8 FROM PAGE 7-33,

$$(470)\text{ BTU}\left(2.93 \text{ EE-4}\right)\frac{\text{KW-HRS}}{\text{BTU}} = 1.377 \text{ EE-1 KW-HRS}$$

9 THE AIR MASS {FROM EQN 6.42}

$$M = \frac{P_1V_1}{RT_1} = \frac{(14.7)\frac{\text{LBF}}{\text{IN}^2}(144)\frac{\text{IN}^2}{\text{FT}^2}(8)\text{ FT}^3}{(53.3)\frac{\text{FT-LBF}}{\text{LBM-°R}}(460+180)\text{°R}}$$

$= .4964$ LBM

FROM EQN 6.130 FOR A CONSTANT-PRESSURE PROCESS,

$W = mR(T_2 - T_1)$

$$= (0.4964 \text{ LBM})(53.3 \tfrac{\text{FT-LBF}}{\text{LBM-°R}})(100\text{°F} - 180\text{°F})\left(1 \tfrac{\text{°R}}{\text{°F}}\right)$$

$= -2116.6$ FT-LBF

THIS IS NEGATIVE BECAUSE WORK IS DONE ON THE SYSTEM.

10 ASSUME THE BUILDING NEEDS 3 EE5 CFH OF 75°F AIR. THEN,

$$\dot{M}_{AIR} = \frac{PV}{RT} = \frac{(14.7)(144)(3 \text{ EE5})}{(53.3)(460+75)} = 2.227 \text{ EE4 LBM/HR}$$

THIS IS A CONSTANT PRESSURE PROCESS, SO

$$Q = \dot{m}C_p\Delta T = \frac{(2.227 \text{ EE4})\frac{\text{LBM}}{\text{HR}}(.241)\frac{\text{BTU}}{\text{LBM-°R}}(75-35)\text{°F}}{3600 \frac{\text{SEC}}{\text{HR}}}$$

$= 59.63$ BTU/SEC

FOR THE WATER,

$Q = \dot{m}_w C_p \Delta T = V\rho C_p \Delta T$

(MORE)

WARM-UP #10 CONTINUED

DENSITY OF WATER AT 165°F ≈ 61 LBM/FT³

SO $59.63 = (\text{gpm})(.002228)\frac{\text{FT}^3}{\text{SEC-gpm}}(61)\frac{\text{LBM}}{\text{FT}^3} \times (1)\frac{\text{BTU}}{\text{LBM-°F}}(180-150)\text{°F}$

SO gpm = 14.63

CONCENTRATES

1 a) FROM PAGE 7-33

$(5000)\text{ KW }(3412.9)\frac{\text{BTU}}{\text{HR-KW}} = 1.706\text{ EE}7\text{ BTU/HR}$

$T_{SAT}$ FOR 200 PSIA STEAM IS 381.8°F

SO THIS STEAM IS AT 381.8 + 100 = 481.8 °F

$h_1 \approx 1258$ BTU/LBM

$h_2 \approx 868$ {FROM MOLLIER, ASSUMING ISENTROPIC EXPANSION}

SO $\Delta h = 1258 - 868 = 390$ BTU/LBM

THE STEAM FLOW RATE IS

$$\dot{m} = \frac{(1.706\text{ EE}7)\text{ BTU/HR}}{(390)\text{ BTU/LBM}} = 4.374\text{ EE}4\text{ LBM/HR}$$

FROM PAGE 7-2, THE WATER RATE IS

$$WR = \frac{\dot{m}}{KW} = \frac{4.374\text{ EE}4\ \frac{\text{LBM}}{\text{HR}}}{5000\text{ KW}} = 8.749\ \frac{\text{LBM}}{\text{KW-HR}}$$

b) IF, WHEN THE LOAD DECREASES, THE STEAM FLOW IS REDUCED ACCORDINGLY, THERE WILL BE NO LOSS OF ENERGY. IF, HOWEVER, THE STEAM IS THROTTLED TO REDUCE THE AVAILABILITY,

$$\text{LOSS} = \tfrac{1}{2}(h_1 - h_2) = \tfrac{1}{2}(390) = 195\text{ BTU/LBM}$$

2 $h_1 \approx 1390$ BTU/LBM {FROM MOLLIER}

$h_2 \approx 935$ {IF EXPANSION IS ISENTROPIC}

THE 'ADIABATIC HEAT DROP' IS

$h_1 - h_2 = 1390 - 935 = 455$ BTU/LBM

3 THE WATER RATE IS

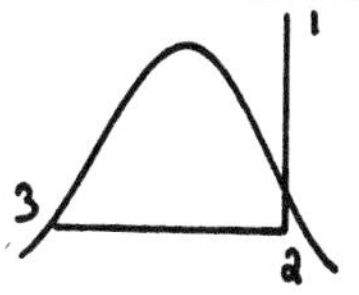


$$WR = \frac{(\dot{m})\text{ LBM/HR}}{(P)\text{ KW}}$$

OR $P = \frac{\dot{m}}{WR}$

ALSO, $\dot{m} = \frac{(P)\text{ KW }(3412.9)\frac{\text{BTU}}{\text{HR-KW}}}{(\Delta h)\text{ BTU/LBM}}$

SO, $\Delta h = \frac{3412.9}{WR} = \frac{3412.9}{20} = 170.65$

THE ACTUAL STEAM FLOW IS

$\dot{m} = (P)\text{ KW }(WR)\frac{\text{LBM}}{\text{HR-KW}} = (750)(20) = 15{,}000\text{ LBM/HR}$

$P_1 = 150$ PSIG $= 164.7$ PSIA {SAY 165}

$T_{SAT} = 366$°F FOR 165 PSIA STEAM

$T_1 = T_{SAT} + T_{SUPERHEAT} = 366 + 50 = 416$°F

$h_1 \approx 1226$ FROM MOLLIER DIAGRAM

$P_2 = (26)\text{ IN}_{Hg}(.491)\frac{\text{LBF}}{\text{IN}^3} = 12.77$ PSIA

$h_2 = 1226 - 170.65 = 1055.4$ BTU/LBM

$h_{f,3} \approx 171.6$ BTU/LBM {INTERPOLATED FROM PAGE 6-30,

SO, THE HEAT REMOVAL IS

$(1055.4 - 171.6)\text{ BTU/LBM }(15000)\text{ LBM/HR} = 1.326\text{ EE}7\text{ BTU/HR}$

THE SATURATION TEMPERATURE CORRESPONDING TO $P_2 = 12.77$ PSI IS $T_2 = T_3 \approx 204$°F

ASSUMING THE WATER AND STEAM LEAVE IN THERMAL EQUILIBRIUM, THE COOLING WATER BALANCE IS

$C_p = .999$ BTU/LBM-°F {PAGE 10-23}

$q_{IN} = \dot{m} C_p \Delta T$

$(1.326\text{EE}7) = (\dot{m}_w)(.999)(204-65)$

$\dot{m}_w = 9.55\text{ EE}4$ LBM/HR

4 $T_{SAT}$ FOR 4.45 PSIA STEAM ≈ 157°F

ASSUME COUNTER FLOW OPERATION TO CALCULATE $\Delta T_m$

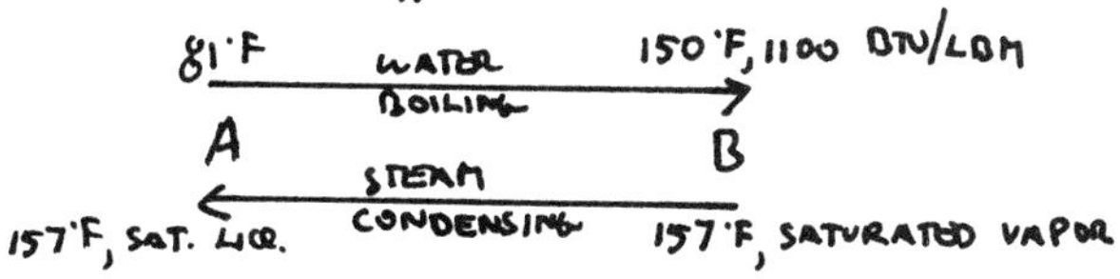


THE AVERAGE WATER TEMPERATURE IS

$\tfrac{1}{2}(81 + 150) = 115.5$, SO

$C_p \approx .999$ BTU/LBM-°F {PAGE 3-28}

ASSUMING 81°F SATURATED WATER,

$h_{w,1} \approx 49$ BTU/LBM

THE HEAT TRANSFERRED TO THE WATER IS

$Q = \dot{m}\Delta h = (332{,}000)\frac{\text{LBM}}{\text{HR}}(1100-49)\frac{\text{BTU}}{\text{LBM}}$

$= 3.489\text{ EE}8$ BTU/HR

FROM EQN 10.66

$$\Delta T_m = \frac{(157-81)-(157-150)}{\ln\left(\frac{157-81}{157-150}\right)} = 28.93$$

SINCE $T_{STEAM,A} = T_{STEAM,B}$ THE CORRECTION FACTOR FOR $\Delta T_m$ {$F_c$ IN EQN 10.68} IS ONE.

{MORE}

CONCENTRATE #4 CONTINUED

FROM EQN 10.67,

$$U = \frac{Q}{A \Delta T_M} = \frac{(3.489 \text{ EE}8) \text{ BTU/HR}}{(1850) \text{ FT}^2 (28.93) \text{°F}}$$

$$= 6519 \frac{\text{BTU}}{\text{HR-FT}^2\text{-°F}}$$

b) THE ENTHALPY OF SATURATED 4.45 PSIA STEAM IS

$h_1 \approx 1128$ BTU/LBM

THE ENTHALPY OF SATURATED 4.45 PSIA WATER

$h_2 \approx 125$ BTU/LBM

THE ENTHALPY CHANGE IS

$h_1 - h_2 = 1128 - 125 = 1003$ BTU/LBM

FROM AN ENERGY BALANCE WITH THE WATER,

$(3.489 \text{ EE}8) = \dot{m}(1003)$

$\dot{m} = 3.479 \text{ EE}5$ LBM/HR STEAM

THEN, FROM PAGE 7-2, THE EXTRACTION RATE IS

$$ER = \dot{m} h_1 = (3.479 \text{ EE}5) \frac{\text{LBM}}{\text{HR}} (1128) \frac{\text{BTU}}{\text{LBM}}$$

$$= 3.92 \text{ EE}8 \text{ BTU/HR}$$

5

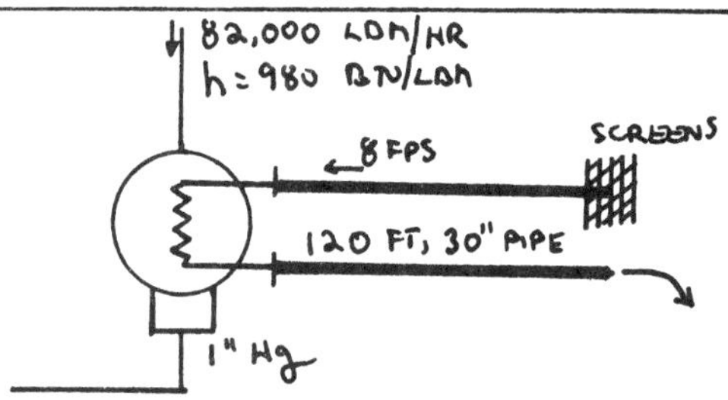


FROM PAGE 3-42 FOR X PIPE,

$D_i = 2.4167$ FT

$A_i = 4.5869$ FT$^2$

FROM PAGE 3-20,

$\epsilon = .0002$ FOR STEEL

$$\frac{\epsilon}{D} = \frac{.0002}{2.416} = 0.000083$$

FOR FULLY TURBULENT FLOW, $f \approx 0.012$.

$$\text{SO } h_f = \frac{(0.012)(120)(8)^2}{(2)(2.416)(32.2)} = 0.59 \text{ FT}$$

(THIS WILL VARY DEPENDING ON THE VALUE OF $f$ CHOSEN.)

THE SCREEN LOSS IS 6" OR 0.5 FT.

THE TOTAL HEAD ADDED BY A COOLANT PUMP (NOT SHOWN) NOT INCLUDING LOSSES INSIDE THE CONDENSER IS (INCLUDING VELOCITY HEAD)

$$0.59 + .5 + \frac{(8)^2}{2(32.2)} = 2.08 \text{ FT}$$

b) THE WATER FLOW CAN BE FOUND FROM THE PIPE FLOW VELOCITY:

$$\text{gpm} = VA = \frac{(8) \text{ FT/SEC } (4.5869) \text{ FT}^2}{(.002228) \frac{\text{FT}^3}{\text{SEC-gpm}}} = 16{,}470 \text{ gpm}$$

SINCE THIS FLOW RATE DOESN'T USE THE 10°F INFORMATION, CHECK WITH A HEAT BALANCE.

$(1'')\text{Hg} (.491) \text{ LBM/IN}^3 \approx .5$ PSI

$h_{\text{SAT. LIQ. AT .5 PSI}} = 47.6$ BTU/LBM

FOR THE STEAM,

$$Q_{out} = \dot{m}\Delta h = \frac{(82{,}000) \frac{\text{LBM}}{\text{HR}} (980 - 47.6) \frac{\text{BTU}}{\text{LBM}}}{(60) \text{ MIN/HR}}$$

$$= 1.274 \text{ EE}6 \text{ BTU/MIN}$$

FOR THE WATER,

$$(1.274 \text{ EE}6) \frac{\text{BTU}}{\text{MIN}} = \dot{m} c_p \Delta T = \dot{m} (1) \frac{\text{BTU}}{\text{LBM-°F}} (10) \text{°F}$$

$\dot{m} = 1.274 \text{ EE}5$ LBM/MIN

$$\text{gpm} = \frac{\dot{m}}{\rho} = \frac{(1.274 \text{ EE}5) \text{ LBM/MIN } (7.48) \text{ gal/FT}^3}{(62.4) \text{ LBM/FT}^3}$$

$$= 1.527 \text{ EE}4 \text{ gpm } \checkmark$$

6 $h_1 = 28.06$ BTU/LBM

$h_2 = 1150.4$

$\Delta h = 1150.4 - 28.06 = 1122.3$ BTU/LBM

$Q = \dot{m}\Delta h = (100)(1122.3) = 1.122 \text{ EE}5$ BTU/HR

FROM PAGE 7-33

$$\frac{(1.122 \text{ EE}5) \frac{\text{BTU}}{\text{HR}} (.2930) \frac{\text{WATT-HR}}{\text{BTU}}}{(1000) \frac{\text{WATTS}}{\text{KW}}} = 32.87 \text{ KW}$$

$$\text{COST} = \frac{(32.87) \text{ KW } (.04) \text{ \$/KW-HR}}{1 - 0.35} = \$2.02/\text{HR}$$

7 $P = 25 + 14.7 = 39.7$ PSIA (SAY 40 PSIA)

AT 60°F,

$h_1 = 28.06$ BTU/LBM

$h_2 = 236.03 + .98(933.7) = 1151.06$

$\Delta h = 1151.06 - 28.06 = 1123$ BTU/LBM

$$Q = \dot{m}\Delta h = \frac{(250) \text{ LBM/HR } (1123) \text{ BTU/LBM}}{(60) \text{ MIN/HR}}$$

$$= 4679.2 \text{ BTU/MIN}$$

NEXT, FIND THE VOLUME OF GAS USED AT STANDARD CONDITIONS FOR A HEATING GAS (60°F):

$$\dot{V}^{\circ} = (13.5) \text{ CFM } \frac{(460 + 60)}{(460 + 80)} \left[ \frac{(30.2)(.491) + (4)(.0361)}{14.7} \right] = 13.24 \text{ SCFM}$$

(MORE)

CONCENTRATE #7 CONTINUED

$$\eta = \frac{4679.2\ \text{BTU/MIN}}{(1324)\ \text{FT}^3/\text{MIN}\ (550)\ \text{BTU/FT}^3} = .643$$

**8**

THE TANGENTIAL BLADE SPEED IS

$$V_b = \frac{(\pi)\left(\frac{18}{12}\right)\text{FT}\,(12000)\ \text{RPM}}{(60)\ \text{SEC/MIN}} = 942.5\ \text{FT/SEC}$$

THE JET SPEED IS

$$V = \frac{942.5}{.4} = 2356.3\ \text{FT/SEC}$$

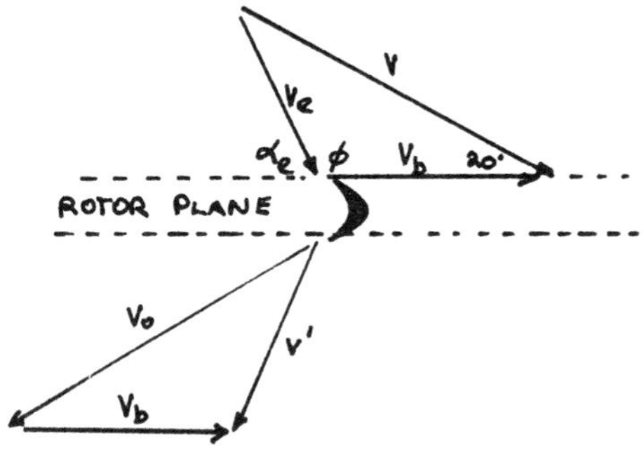


THE RELATIVE JET SPEED IS

$$V_e = \sqrt{(2356.3)^2 + (942.5)^2 - 2(2356.3)(942.5)\cos 20°}$$
$$= 1505.6\ \text{FT/SEC}$$

THE ANGLE $\phi$ CAN BE FOUND FROM THE LAW OF SINES

$$\frac{1505.5}{\sin 20°} = \frac{2356.2}{\sin \phi} \quad \text{OR} \quad \phi = 147.6°$$

SO $\alpha_e = 180 - \phi = 32.4$

THE ACTUAL STEAM EXIT VELOCITY IS

$$V' = \sqrt{(942.5)^2 + (1505.6)^2 - 2(942.5)(1505.6)\cos 32.4°}$$
$$= 871.1\ \text{FT/SEC}$$

THE REDUCTION IN KINETIC ENERGY IS

$$\Delta E_K = \frac{(2356.3)^2 - (871.1)^2}{(2)(32.2)(778)\ \frac{\text{FT-LBF}}{\text{BTU}}} = 95.67\ \text{BTU/LBM}$$

**9** STEP 1 DETERMINE THE ACTUAL GRAVIMETRIC ANALYSIS OF THE COAL AS FIRED, USE THE 'SUCCESSIVE DELETION' METHOD ON A PER-POUND BASIS. ONE POUND OF COAL CONTAINS .02 LBM MOISTURE, LEAVING .98 LBM COAL. OF THIS, 5% IS ASH, SO THE WEIGHT OF ASH IS (.05)(.98) = .049 LBM. THE REMAINDER (.98 − .049) = .931 IS ASSUMED TO BE CARBON.

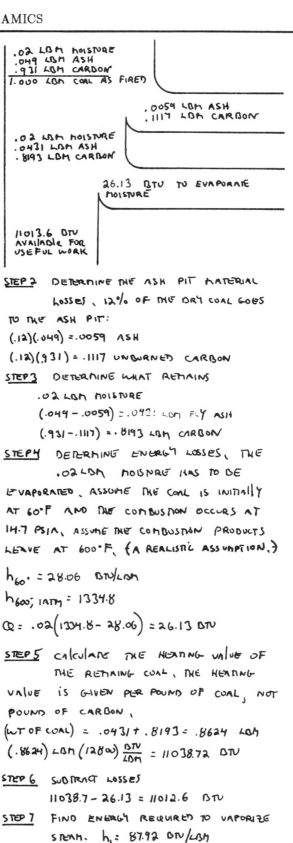

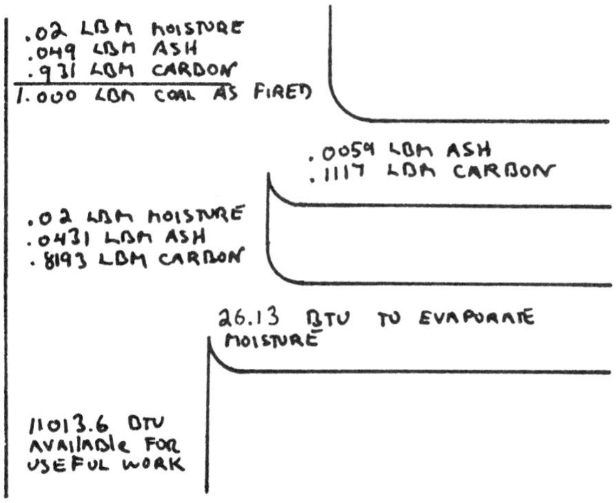


STEP 2 DETERMINE THE ASH PIT MATERIAL LOSSES, 12% OF THE DRY COAL GOES TO THE ASH PIT:

(.12)(.049) = .0059 ASH

(.12)(.931) = .1117 UNBURNED CARBON

STEP 3 DETERMINE WHAT REMAINS

.02 LBM MOISTURE

(.049 − .0059) = .0431 LBM FLY ASH

(.931 − .1117) = .8193 LBM CARBON

STEP 4 DETERMINE ENERGY LOSSES, THE .02 LBM MOISTURE HAS TO BE EVAPORATED. ASSUME THE COAL IS INITIALLY AT 60°F AND THE COMBUSTION OCCURS AT 14.7 PSIA, ASSUME THE COMBUSTION PRODUCTS LEAVE AT 600°F. (A REALISTIC ASSUMPTION.)

$h_{60°} = 28.06$ BTU/LBM

$h_{600;\ 1\text{ATM}} = 1334.8$

$Q = .02(1334.8 - 28.06) = 26.13$ BTU

STEP 5 CALCULATE THE HEATING VALUE OF THE REMAINING COAL. THE HEATING VALUE IS GIVEN PER POUND OF COAL, NOT POUND OF CARBON.

(WT OF COAL) = .0431 + .8193 = .8624 LBM

$(.8624)\ \text{LBM}\ (12800)\ \frac{\text{BTU}}{\text{LBM}} = 11038.72$ BTU

STEP 6 SUBTRACT LOSSES

11038.7 − 26.13 = 11012.6 BTU

STEP 7 FIND ENERGY REQUIRED TO VAPORIZE STEAM. $h_1 = 87.92$ BTU/LBM

100 PSIG ≈ 115 PSIA

$h_2 \approx 1190$ BTU/LBM

$Q = (8.23)\ \text{LBM}\ (1190 - 87.92)\ \text{BTU/LBM} = 9070.1$ BTU

STEP 8 THE COMBUSTION EFFICIENCY IS

$$\eta = \frac{9070.1}{11012.6} = .824$$

10 STEP 1 THE INCOMING REACTANTS ON A PER-POUND BASIS ARE

.07 LBM ASH
.05 LBM HYDROGEN
.05 LBM OXYGEN
.83 LBM CARBON

THIS IS AN ULTIMATE ANALYSIS (SEE P. 9-2). IT IS ASSUMED THAT ALL OF THE OXYGEN AND 1/8 OF THE HYDROGEN IS IN THE FORM OF WATER (SEE P. 9-7). THE REACTANTS AS COMPOUNDS ARE

.07 LBM ASH
.05625 LBM MOISTURE
.04375 LBM HYDROGEN
.83 LBM CARBON

THE AIR IS 23.15% OXYGEN BY WEIGHT (SEE TABLE 9.9) SO OTHER REACTANTS ARE

$(.2315)(26) = 6.019$ LBM OXYGEN

$(.7685)(26) = 19.981$ LBM NITROGEN

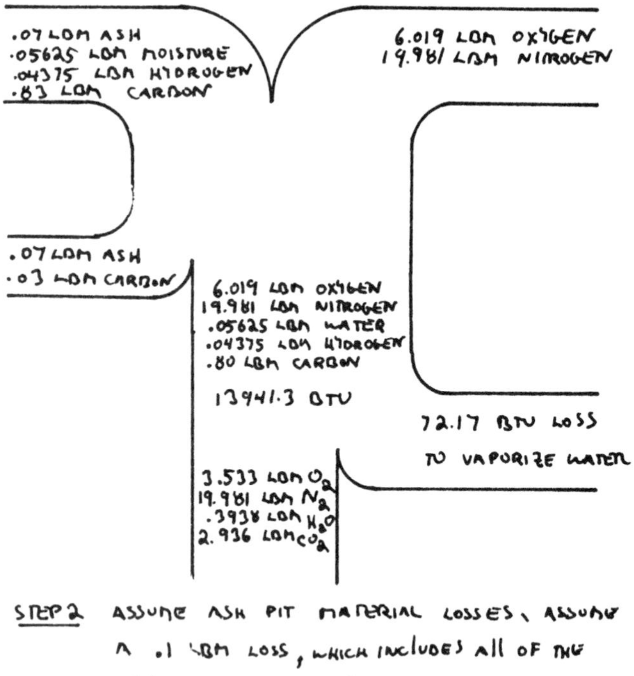


STEP 2 ASSUME ASH PIT MATERIAL LOSSES. ASSUME A .1 LBM LOSS, WHICH INCLUDES ALL OF THE ASH.
.07 LBM ASH
.03 LBM UNBURNED COAL

STEP 3 DETERMINE WHAT REMAINS

STEP 4 DETERMINE THE ENERGY LOSS IN VAPORIZING THE MOISTURE. ASSUME THE COAL IS INITIALLY AT 60°F AND THAT COMBUSTION OCCURS AT 14.7 PSIA.

$h_{60°} = 28.06$ BTU/LBM

$h_{550} = 1311$ BTU/LBM

$Q = .05625(1311 - 28.06) = 72.17$ BTU

STEP 5 CALCULATE THE HEATING VALUE OF THE REMAINING FUEL COMPONENTS. FROM PAGE 9-22

$Q_{CARBON} = (.80)\text{LBM}(14093)\frac{\text{BTU}}{\text{LBM}} = 11274.4$ BTU

$Q_{HYDROGEN} = (.04375)(60958) = 2666.9$

$\phantom{Q} = 13941.3$ BTU

THE HEATING VALUE AFTER THE COAL MOISTURE IS EVAPORATED IS

$13941.3 - 72.17 = 13869.13$ BTU

STEP 6 DETERMINE THE COMBUSTION PRODUCTS. FROM TABLE 9.11, THE CARBON NEEDS

$.8(2.67) = 2.136$ LBM OXYGEN

AND PRODUCES

$.8(3.67) = 2.936$ LBM $CO_2$

THE HYDROGEN NEEDS

$(.04375)(8) = .35$ LBM OXYGEN

AND PRODUCES

$(.04375)(9) = .3938$ LBM WATER

THE REMAINING OXYGEN IS

$6.019 - 2.136 - .35 = 3.533$ LBM

STEP 7 THE GASEOUS PRODUCTS MUST BE HEATED FROM 70° TO 550°. THE AVERAGE TEMPERATURE IS $\frac{1}{2}(70 + 550) = 310°F$ OR 770°R. $C_p$ FOR THESE GASES CAN BE FOUND FROM TABLE 9.14.

| | $C_p$ |
|---|---|
| OXYGEN | .228 |
| NITROGEN | .252 |
| WATER | .460 |
| $CO_2$ | .225 |

$Q_{HEATING} = [(3.533)(.228) + (19.981)(.252) + (.3938)(.460) + (2.936)(.225)](550 - 70)$

$= 3207.6$ BTU

STEP 8 THE % LOST IS

$$\frac{3207.6 + 72.17}{13941.3} = .235$$

TIMED

1 FROM PAGE 7-33, THE DRILL POWER IS

$(.25)\text{HP}(2544.9)\frac{\text{BTU}}{\text{HP-HR}}$

$= 636.23$ BTU/HR

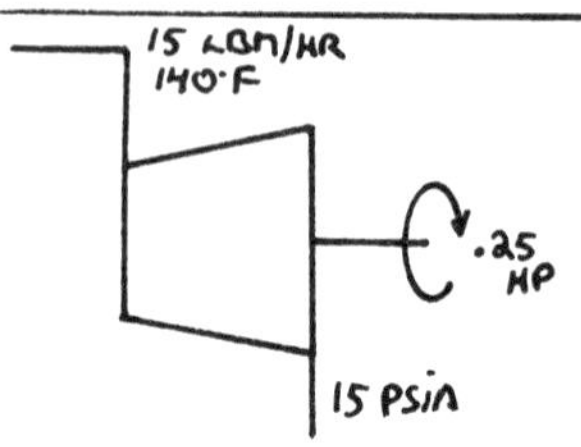


(MORE)

TIMED #1 CONTINUED

FROM PAGE 6-35 FOR $T = (460 + 140) = 600°R$,

$h_1 = 143.47$ BTU/LBM

$p_{r,1} = 2.005$

$\phi_1 = .62607$

THEN, SINCE $W_{actual} = \dot{m}(h_1 - h_{2,actual}) = \dot{m}\eta(h_1 - h_{2,ideal})$

$$h_{2,ideal} = h_1 - \frac{W}{\dot{m}\eta}$$

$$= \left(143.47 \frac{BTU}{LBM}\right) - \frac{\left(636.23 \frac{BTU}{HR}\right)}{\left(15 \frac{LBM}{HR}\right)(.60)}$$

$$= 72.778 \text{ BTU/LBM}$$

SEARCHING KEENAN AND KAYES <u>GAS TABLES</u> FOR $h = 72.778$ GIVES

$T = 305°R$

$p_{r,2} = .18851$

a) HOWEVER, DUE TO THE IRREVERSIBILITY OF THE EXPANSION,

$$h_2' = 143.47 - .60(143.47 - 72.778) = 101.05 \text{ BTU/LBM}$$

SEARCHING THE <u>GAS TABLES</u> AGAIN GIVES

$T_2' = 423°R$

$\phi_2 = .54228$

b) SINCE $\frac{P_1}{P_2} = \frac{P_{r,1}}{P_{r,2}}$

$$P_1 = (15 \text{ PSIA})\left(\frac{2.005}{.18851}\right) = 159.5 \text{ PSIA}$$

THE RATIOS OF PRESSURES IS VALID ONLY FOR ISENTROPIC PROCESSES. $P_{r,2}$ MUST BE USED, NOT $P_{r,2}'$.

c) FROM EQN 6.29,

$$S_2 - S_1 = .54228 - .62607 - \left(\frac{53.3}{778}\right) \ln\left(\frac{15}{159.5}\right)$$

$$= .07816 \text{ BTU/LBM-°R}$$

2 FOR THE STEAM,

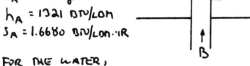

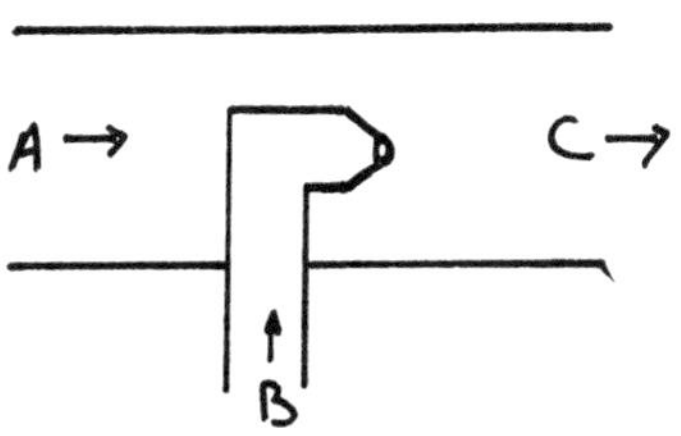


$T_A = 600°F$

$P_A = 200$ PSIG $\approx 215$ PSIA

$h_A = 1321$ BTU/LBM

$S_A = 1.6680$ BTU/LBM-°R

FOR THE WATER,

$T_B = 82°F$

$h = 50.03$ BTU/LBM

CORRECTION FOR COMPRESSION (PAGE 6-33) IS $\approx .55$ BTU/LBM

$h_B = 50.03 + .55 = 50.58$ BTU/LBM

$S_B = .0969$ BTU

FROM AN ENERGY BALANCE,

$$h_A M_A + h_B M_B = h_C (M_A + M_B)$$

OR

$$h_C = \frac{(1000)(1321) + (50)(50.58)}{1000 + 50} = 1260.5$$

SOME OF THIS ENERGY GOES INTO KINETIC ENERGY,

$$E_K = \frac{\left(2000 \frac{FT}{SEC}\right)^2}{(2)\left(32.2 \frac{FT}{SEC^2}\right)\left(778 \frac{FT\text{-}LBF}{BTU}\right)}$$

$$= 79.8 \text{ BTU}$$

$$h_C' = 1260.5 - 79.8 = 1180.7 \text{ BTU/LBM}$$

FOR 100 PSIA STEAM,

$h_f = 298.4$

$h_{fg} = 888.8$

SO $X = \frac{1180.7 - 298.4}{888.8} = .993$

$T_C$ IS THE SATURATION TEMPERATURE FOR 100 PSIA STEAM $= 327.8°F$

FROM EQN 6.32,

$$S_C = .4740 + .993(1.1286) = 1.5947 \frac{BTU}{LBM\text{-}°R}$$

THE ENTROPY PRODUCTION IS

$$\left(1050 \frac{LBM}{HR}\right)\left(1.5947 \frac{BTU}{LBM\text{-}°R}\right) - (1000)(1.6680) - (50)(.0969) = 1.59 \frac{BTU}{HR\text{-}°R}$$

3 (a) ASSUME THE TANK IS ORIGINALLY AT 70°F. FROM TABLE 6.4, $R = 11.9$

$$m = \frac{pV}{RT} = \frac{(20)(144)(100)}{(11.9)(460+70)}$$

$$= \boxed{45.66 \text{ LBM}}$$

(b) FOR XENON, THE CRITICAL PROPERTIES ARE

$P_c = 855.3$ PSIA

$T_c = 521.9$ °R

AT THE TANK CONDITIONS,

$$P_r = \frac{3800}{855.3} = 4.44$$

$$T_r = \frac{530}{521.9} = 1.02$$

FROM FIGURE 6.11, $Z = .6$

$$M = \frac{pV}{ZRT} = \frac{(3800)(144)(100)}{(.6)(11.9)(530)}$$

$$= 14{,}460.1 \text{ LBM}$$

$$\dot{M} = 14{,}460 - 45.66 = \boxed{14414 \frac{LBM}{HR}}$$

TIMED PROBLEM #3 CONTINUED

(c)

SINCE A HEAT EXCHANGER WAS MENTIONED, ASSUME ISOTHERMAL COMPRESSION

$$W = p_1V_1 \ln\left(\frac{V_2}{V_1}\right) = mRT_1 \ln\left(\frac{p_1}{p_2}\right)$$

$$= \frac{(14414)(11.9)(460+70)\ \ln\left(\frac{20}{3800}\right)}{(778)(3413)}$$

$$= \boxed{180 \text{ KW-HR}}$$

$$\text{COST} = (.045)(180) = \boxed{\$8.10}$$

THIS CALCULATION ASSUMES THE COMPRESSOR EFFICIENCY IS CONSTANT OVER THE ENTIRE RANGE OF RECEIVER PRESSURES.

NOTICE THAT Z DOES NOT AFFECT THE WORK EQUATION. Z ONLY AFFECTS THE MASS FLOW RATE.

4 CHOOSE THE CONTROL VOLUME TO INCLUDE THE AIR EXTERNAL TO THE TANK THAT IS PUSHED INTO THE TANK, AS WELL AS THE TANK VOLUME.

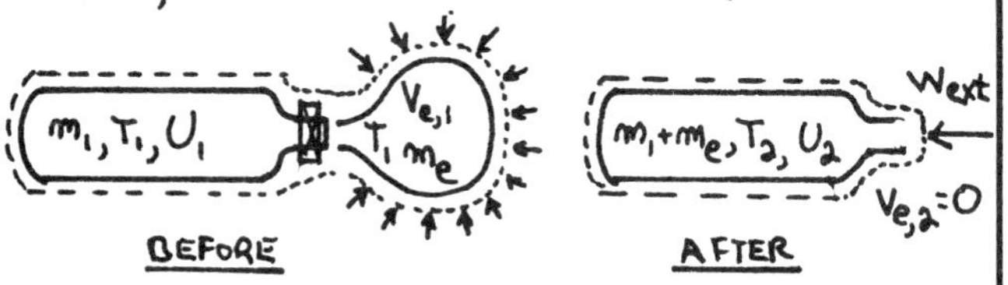


$m_1$ = AIR IN TANK WHEN EVACUATED

$$= \frac{P_1V_1}{RT_1} = \frac{(1 \text{ PSIA})\left(144 \frac{\text{IN}^2}{\text{FT}^2}\right)(20 \text{ FT}^3)}{\left(53.3 \frac{\text{FT-LBF}}{\text{LBM-}°\text{R}}\right)(70+460)°\text{R}}$$

$$= 0.102 \text{ LBM}$$

ASSUME $T_2 = 80°$F TO GET STARTED.

$$m_1 + m_e = \frac{(14.7)(144)(20)}{(53.3)(80+460)} = 1.471 \text{ lbm}$$

$$m_e \approx 1.471 - 0.102 = 1.369 \text{ lbm}$$

THE INITIAL VOLUME OF THE EXTERNAL AIR IS

$$V_{e,1} \approx \frac{mRT}{p} = \frac{(1.369)(53.3)(70+460)}{(14.7)(144)}$$

$$= 18.27 \text{ FT}^3$$

THIS IS A CLOSED SYSTEM. THE FIRST LAW OF THERMODYNAMICS FOR CLOSED SYSTEMS IS

$$Q = \Delta U + W = U_2 - U_1 + W$$

SINCE THE SYSTEM IS ADIABATIC (FOR THE FIRST FEW INSTANTS), Q = 0. SO,

$$W_{ext} = \Delta U$$

THE EXTERNAL WORK IS A CONSTANT PRESSURE PROCESS. $W_{ext}$ IS NEGATIVE BECAUSE THE SURROUNDINGS DO WORK ON THE SYSTEM.

$$W_{ext} = p(V_{e,2} - V_{e,1}) = \frac{(14.7)(144)(0 - 18.27)}{778 \frac{\text{ft-lbf}}{\text{BTU}}}$$

$$= -49.71$$

SINCE $\Delta U = mC_v\Delta T$ FOR THE AIR AND $C_p = C_v$ FOR THE TANK MATERIAL,

$$W_{ext} = \left[(m_1+m_e)C_v + m_{tank}C_p\right](T_1 - T_2)$$

$$-49.71 = \left[(1.471)(0.171) + (40)(0.11)\right](70 - T_2)$$

$$T_2 = \boxed{80.69°\text{F}}$$

THIS IS CLOSE ENOUGH TO THE ASSUMED VALUE OF $T_2$ THAT A SECOND ITERATION IS UNNECESSARY.

5 ASSUME PRESSURES ARE LOW ENOUGH TO IGNORE COMPRESSIBILITY.

CALCULATE THE FLOW RATES:

$$\dot{m}_C = \frac{pV}{RT} = \frac{(80)(144)(100)}{(53.3)(460+85)} = 39.66 \frac{\text{LBM}}{\text{MIN}}$$

$$\dot{m}_D = \frac{(85)(144)(120)}{(53.3)(460+80)} = 51.03 \frac{\text{LBM}}{\text{MIN}}$$

$$\dot{m}_E = 8 \frac{\text{LBM}}{\text{MIN}} \quad \text{(GIVEN)}$$

$$\dot{m}_{total} = 39.66 + 51.03 + 8 = 98.69 \frac{\text{LBM}}{\text{MIN}}$$

THE INPUT FROM COMPRESSOR A IS

$$\dot{m}_A = \frac{(14.7)(144)(600)}{(53.3)(460+80)} = 44.13 \frac{\text{LBM}}{\text{MIN}}$$

SO, $\dot{m}_B = 98.69 - 44.13 = 54.56$ LBM/MIN

$$\dot{V}_B = \frac{54.56}{44.13}(600) = \boxed{742 \text{ CFM}}$$

6

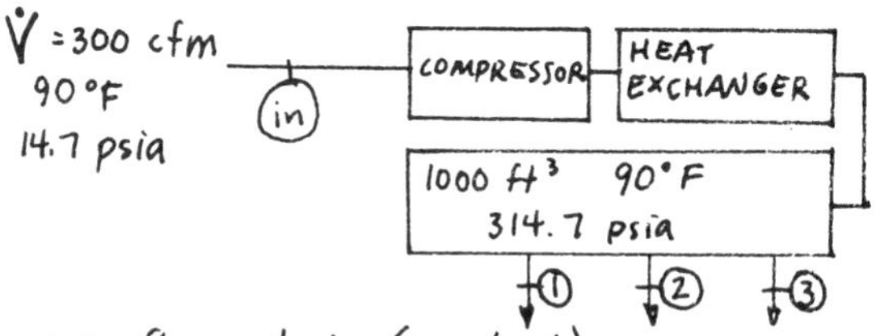


mass flow rate in (constant):

using equation 6.42, assume steady flow, constant properties

$$pV = mRT \Rightarrow m = \frac{pV}{RT}$$

$$\frac{dm}{dt} = \frac{d}{dt}\left(\frac{pV}{RT}\right) = \frac{p}{RT}\left(\frac{dV}{dt}\right)$$

$$= \frac{(14.7)(144)(300)}{(53.3)(550)} = 21.7 \text{ lbm/min}$$

total mass flow rate in

TIMED #6 CONTINUED

mass initially in storage:

$$m = \frac{pV}{RT} = \frac{(314.7)(144)(1000)}{(53.3)(550)} = 1545.9 \text{ lbm}$$

mass leaving system (assume each tool operates at its minimum pressure)

Tool #1 $\dot{m}_{out\,1} = \frac{(104.7)(144)(40)}{(53.3)(550)}$

$= 20.57$ lbm/min

Tool #2 $\dot{m}_{out\,2} = \frac{(64.7)(144)(15)}{(53.3)(545)} = 4.81$ lbm/min

Tool #3 $\dot{m}_{out\,3} = 6$ lbm/min

Total mass flow rate leaving system

$\dot{m}_{out} = 20.57 + 4.81 + 6 = 31.38$ lbm/min

"Critical" pressure is 90 psig = 104.7 psia. Mass in tank when "critical" pressure achieved:

$$m_{tank,critical} = \frac{(104.7)(144)(1000)}{(53.3)(550)} = 514.3 \text{ lbm}$$

Net flow rate of air to tank $= \dot{m}_{in} - \dot{m}_{out}$

$= 21.7 - 31.38$

$= -9.68$ lbm/min

(rate of depletion of tank)

Amount to be depleted $= m_{initial} - m_{final}$

$= 1545.9 - 514.3$

$= 1031.6$ lbm

Time system can run $= \frac{1031.6 \text{ lbm}}{9.68 \text{ lbm/min}}$

$= 106.57$ min

$= 1.78$ hour

7

$\textcircled{1}$ → air heater → $\textcircled{2}$

$T_1 = 540°F$

$= 1000°R$

$p_1 = 100$ psia

$T_2 = 2000°R$

$p_2 = 80$ psia

$T_0 = 560°R$

using equation 6.111,

$$W_{max} = \Phi_1 - \Phi_2$$

and equation 6.112

$$\Phi = h - T_0 s$$

Since pressures are low (< 300 psia) and temperatures are much higher than 235.8°R (see table 6.7 for air), appendix F can be used. For $T_1 = 1000°R$, $h_1 = 240.98$ BTU/lbm, $\phi_1 = 0.75042$ BTU/lbm°R

$T_2 = 2000°R$, $h_2 = 504.71$ BTU/lbm, $\phi_2 = 0.93205$ BTU/lbm°R

$$W_{max} = h_1 - h_2 + (s_2 - s_1)T_0$$

For no pressure drop, using equation 6.29,

$s_2 - s_1 = \phi_2 - \phi_1$, so

$$W_{max} = 240.98 - 504.71 + (0.93205 - 0.75042)560$$

$= -162.0172$ BTU/lbm

With a pressure drop from 100 to 80 psia (again, equation 6.29)

$$s_2 - s_1 = \phi_2 - \phi_1 - \left(\frac{R}{J}\right)\ln\left(\frac{p_2}{p_1}\right)$$

Then

$$W_{max,\,p\,loss} = 240.98 - 504.71 + \left[0.93205 - 0.75042 - \frac{53.3}{778}\ln\left(\frac{80}{100}\right)\right]560$$

$= -153.4563$ BTU/lbm

$$\frac{-153.4563 - (-162.0172)}{-162.0172} = -0.0528$$

$= 5.28\%$ loss

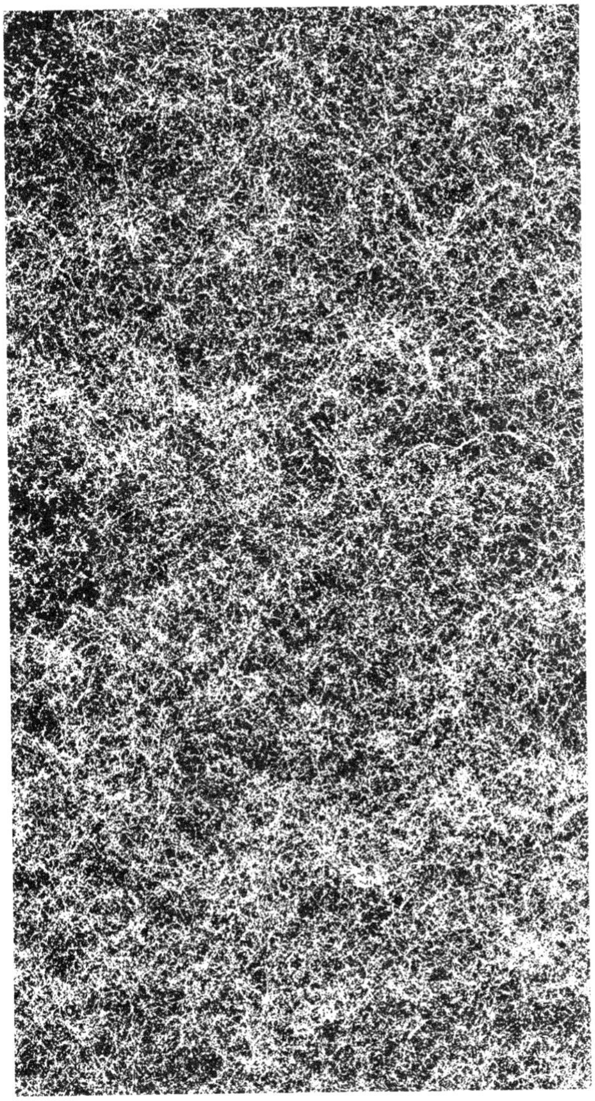

# Power Cycles

WARM-UPS

**1** FROM EQN 7.30,

$$\eta_{th} = \frac{(650+460)-(100+460)}{650+460} = .495$$

**2** FROM EQN 7.151,

$$COP = \frac{700+460}{(700+460)-(40+460)} = 1.76$$

**3** USE THE PROCEDURE ON PAGE 7-9, REFER TO FIGURE 7.12.

AT a: $T_a = 650°F$
$h_a = 696.4$ (INTERPOLATED)
$s_a = .8833$

AT b: $T_b = 650$
$h_b = 1117.5$ (INTERPOLATED)
$s_b = 1.2631$

AT c: $T_c = 100°F$
$s_c = s_b = 1.2631$

$$x_c = \frac{1.2631-.1295}{1.8531} = .612$$

$$h_c = 67.97 + .612(1037.2) = 702.7$$

AT d: $T_d = 100°F$
$s_d = s_a = .8833$

$$x_d = \frac{.8833-.1295}{1.8531} = .407$$

$$h_d = 67.97 + .407(1037.2) = 490.1$$

DUE TO THE INEFFICIENCIES (FROM EQUATIONS 7.31 AND 7.32.)

$$h_c' = 1117.5 - .9(1117.5-702.7) = 744.2$$

$$h_a' = 490.1 + \frac{696.4-490.1}{.8} = 748.0$$

FROM EQN 7.30,

$$\eta_{th} = \frac{(1117.5-744.2)-(748.0-490.1)}{1117.5-748.0} = .312$$

**4** FROM EQN 7.150,

$$COP = \frac{50+460}{(90+460)-(5+460)} = 5.47$$

$$HP = \frac{(1)\ \text{TON}\ (200)\frac{\text{BTU}}{\text{MIN-TON}}\ (778)\frac{\text{FT-LBF}}{\text{BTU}}}{(5.47)\frac{\text{BTU OUT}}{\text{BTU IN}}\ (33000)\frac{\text{FT-LBF}}{\text{MIN-HP}}} = .862\ \text{HP}$$

FROM EQN 7.148,

$$EER = \frac{(200)\ \text{BTU/MIN}\ (60)\ \text{MIN/HR}}{(.862)\ \text{HP}\ (745.7)\ \text{WATTS/hp}} = 18.7$$

**5** FROM EQN 3.7, THE SPECIFIC GRAVITY IS

$$SG = \frac{141.5}{131.5+40} = .825$$

FROM EQN. 7.76,

$$HHV = 22320 - 3780(SG)^2$$
$$= 22320 - 3780(.825)^2$$
$$= 19749\ \text{BTU/LBM}$$

**6** THE ACTUAL HORSEPOWER IS

$$hp = \frac{(\text{RPM})(\text{TORQUE IN FT-LBF})}{5252}$$

$$= \frac{(200)(600)}{5252} = 22.85\ \text{HP}$$

THE NUMBER OF POWER STROKES PER MINUTE IS

$$N = \frac{(2)(200)(2)}{(4)} = 200$$

THE STROKE IS $\frac{18}{12} = 1.5$ FT

THE BORE AREA IS $\frac{\pi}{4}(10)^2 = 78.54\ \text{IN}^2$

THE IDEAL HORSEPOWER (FROM EQN 7.89) IS

$$hp = \frac{(95)(1.5)(78.54)(200)}{33000} = 67.83$$

THE FRICTION HORSEPOWER IS

$$67.83 - 22.85 = 44.98$$

**7** FROM EQUATIONS 7.156 AND 7.157,

$$R = \frac{65}{14.7} = 4.42$$

$$\eta_v = 1 - \left((4.42)^{\frac{1}{1.33}} - 1\right)(.07) = .856$$

THE MASS OF AIR DISPLACED PER MINUTE IS $\frac{48}{.856} = 56.07$ LBM/MIN

**8** THE VALVE IS OPEN

$180 + 40 = 220°$

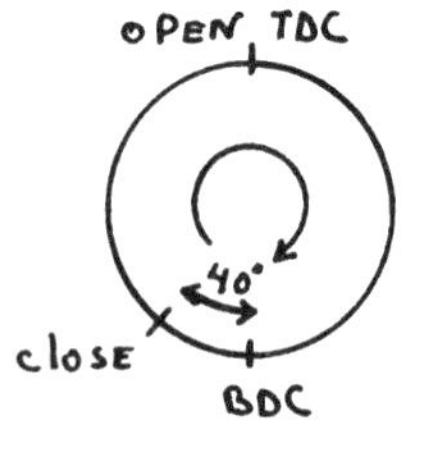


THE TIME THE VALVE IS OPEN IS

$$\left(\frac{220}{360}\right)\left(\frac{\text{TIME PER}}{\text{REVOLUTION}}\right) =$$

$$\left(\frac{220}{360}\right)\left(\frac{60\ \text{SEC/MIN}}{4000\ \text{RPM}}\right) = (9.167\ \text{EE-3})\ \text{SEC}$$

THE DISPLACEMENT IS

$$\left(\frac{\pi}{4}\right)\left(\frac{3.1}{12}\right)^2\left(\frac{3.8}{12}\right) = .0166\ \text{FT}^3$$

THE ACTUAL VOLUME INCOMING IS

$$V = (.65)(.0166) = .01079\ \text{FT}^3 \text{ PER INTAKE STROKE}$$

THE AREA IS

$$A = \frac{V}{vt} = \frac{.01079\ \text{FT}^3}{(100)\ \text{FT/SEC}\ (9.167\ \text{EE-3})\ \text{SEC}} = .0118\ \text{FT}^2$$
$$= 1.69\ \text{IN}^2$$

**9** METHOD 1: IDEAL GAS RELATIONSHIPS

FROM PAGES 6-24 AND 6-25,

$$V_2 = (10)\frac{FT^3}{SEC}\left(\frac{200\ PSIA}{50\ PSIA}\right)^{\frac{1}{1.4}} = 26.918\ CFS$$

$$T_2 = (1500+460)°R\left(\frac{50}{200}\right)^{\frac{1.4-1}{1.4}} = 1319.0\ °R$$

FROM PAGES 6-24 AND 6-25,

$$\Delta h = (.241)\frac{BTU}{LBM\text{-}°R}(1319.0-1960) = -154.5\ \frac{BTU}{LBM}$$

METHOD 2: USING PAGE 6-35,

AT 1960°R,

$h_1 = 493.64$

$P_{r,1} = 160.37$

$v_{r,1} = 4.527$

AFTER EXPANSION,

$$P_{r,2} = 160.37\left(\frac{50}{200}\right) = 40.09$$

SEARCHING THE TABLE FOR THIS VALUE OF $P_r$,

$T_2 = 1375$

$h_2 = 336.39$

$v_{r,1} = 12.721$

SO $V_2 = (10)\left(\frac{12.721}{4.527}\right) = 28.1$

$\Delta h = 493.64 - 336.39 = 157.25$

**10** ALTHOUGH THE IDEAL GAS LAWS COULD BE USED, IT IS EXPEDIENT TO USE AIR TABLES.

FROM PAGE 6-35 AT (460 + 500) = 960

$h_1 = 231.06$

$P_{r,1} = 10.610$

$\phi_1 = .7403$

FOR ISENTROPIC COMPRESSION,

$$P_{r,2} = 6P_{r,1} = 6(10.610) = 63.66$$

SEARCHING THE AIR TABLE YIELDS

$T_2 = 1552$

$h_2 = 382.95$

THE ACTUAL ENTHALPY IS

$$h_2' = 231.06 + \frac{382.95 - 231.06}{.65} = 464.74$$

WHICH CORRESPONDS TO 1855°R, AND

$\phi_2' = .91129$

$W = \Delta h = 464.74 - 231.06 = 233.68$

FROM EQUATION 6.29

$$\Delta S = .91129 - .7403 - \left(\frac{53.3}{778}\right)\ln(6) = .04824$$

CONCENTRATES

**1** FIRST, ASSUME ISENTROPIC COMPRESSION AND EXPANSION, REFER TO FIGURE 7.14.

AT a: $P_a = 100$ PSIA

$h_a = 298.40$

AT b: $P_b = 100$ PSIA

$h_b = 1187.2$

$s_b = 1.6026$

AT c: $P_c = 1$ ATM

$s_c = s_b = 1.6026$

$$x_c = \frac{1.6026 - .3120}{1.4446} = .893$$

$h_c = 180.07 + .893(970.3) = 1046.5$

AT d: $T_d = 80°F$ {SUBCOOLED}

$h_d = 48.02$

$P_d = 14.7$ PSIA

$v_d = .01608$

h AND v ARE ESSENTIALLY INDEPENDENT OF PRESSURE

AT e: $P_e = P_a = 100$ PSIA

$$h_e = 48.02 + \frac{.01608(100-14.7)(144)}{778} = 48.2$$

NOW, DUE TO THE INEFFICIENCIES,

$h_c' = 1187.2 - .80(1187.2 - 1046.5) = 1074.6$

{FROM EQN 7.40}

$$h_e' = 48.02 + \frac{48.27 - 48.02}{.6} = 48.44$$

{FROM EQN 7.41}

FROM EQN 7.39

$$\eta_{th} = \frac{(1187.2 - 1074.6) - (48.44 - 48.02)}{(1187.2 - 48.44)} = .0985$$

**2** REFER TO FIGURE 7.16.

AT d: $P_d = 500$

$T_d = 1000°F$

$h_d = 1519.6$

$s_d = 1.7363$

AT e: $P_e = 5$ PSIA

$s_e = s_d = 1.7363$

$$x_e = \frac{1.7363 - .2347}{1.6094} = .933$$

$h_e = 130.13 + .933(1001) = 1064.1$

AT f: $h_f = 130.13$

BUT, BECAUSE THE TURBINE IS 75% EFFICIENT,

$h_e' = 1519.6 - .75(1519.6 - 1064.1) = 1178$

THE MASS FLOW RATE IS

$$\dot{m} = \frac{(200{,}000)\,KW\,(1000)\,\frac{W}{KW}\,(.05692)\,\frac{BTU}{MIN\text{-}W}}{(1519.6 - 1178)\,\frac{BTU}{LBM}\,(60)\,\frac{SEC}{MIN}} = 555.7\ \frac{LBM}{SEC}$$

$$Q_{out} = (555.7)(1178 - 130.13) = 5.82\ EE5\ BTU/sec$$

**3** USE THE PROCEDURE ON PAGE 7-12 AND REFER TO FIGURE 7.18.

AT b: $P_b = 600$ PSIA

$T_b = 486.21\,°F$

$h_b = 471.6$

AT c: $h_c = 1203.2$

AT d: $P_d = 600$ PSIA

$T_d = 600\,°F$

$h_d = 1289.9$

AT e: $P_e = 200$ PSIA

$h_e = 1187$ {FROM MOLLIER ASSUMING ISENTROPIC EXPANSION}

$h'_e = 1289.9 - .88(1289.9 - 1187)$

$= 1199.3$

AT f: $P_f = 200$ PSIA

$T_f = 600\,°F$

$h_f = 1322.1$

$s_f = 1.6767$

AT g: $T_g = 60\,°F$

$s_g = s_f = 1.6767$

$$x_g = \frac{1.6767 - .0555}{2.0393} = .795$$

$h_g = 28.06 + .795(1059.9) = 870.7$

$h'_g = 1322.1 - .88(1322.1 - 870.7) = 924.9$

AT h: $h_h = 28.06$

$P_h = .2563$

$v_h = .01604$

AT a: $P_a = 600$

$$h'_a = 28.06 + \frac{(.01604)(600 - .2563)(144)}{.96\,(778)}$$

$= 29.9$

FROM EQN 7.59,

$$\eta_{th} = \frac{(1289.9 - 29.9) + (1322.1 - 1199.3) - (924.9 - 28.06)}{(1289.9 - 29.9) + (1322.1 - 1199.3)}$$

$= 0.351$

**4** REFER TO PAGE 7-12 AND THE FOLLOWING DIAGRAM:

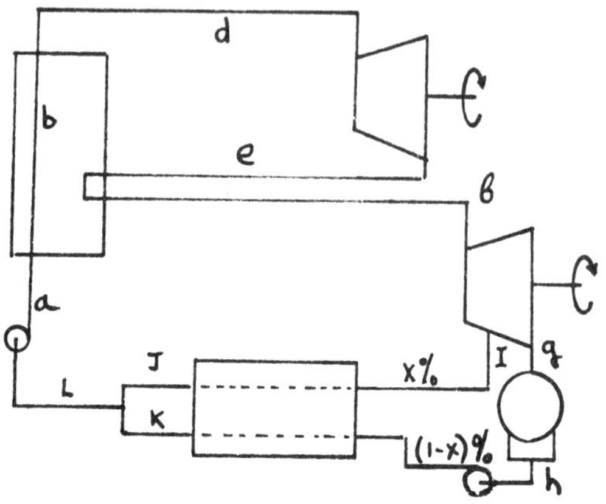


FROM PROBLEM 3

$h_b = 471.6$

$h_d = 1289.9$

$h'_e = 1199.3$

$h_f = 1322.1$

$h'_g = 924.9$

$h_h = 28.04$

AT I: THE TEMPERATURE IS 270 °F. USING THE MOLLIER DIAGRAM AND ASSUMING ISENTROPIC EXPANSION TO 270 °F,

$h_I \approx 1170$ {SATURATED}

$h'_I = 1322.1 - .88(1322.1 - 1170) = 1188.3$

AT J: THE WATER IS ASSUMED TO BE SATURATED FLUID AT 270 °F.

$h_J = 238.84$

AT K: THE TEMPERATURE IS $(270 - 6) = 264\,°F$. SINCE THE WATER IS SUBCOOLED ENTHALPY IS A FUNCTION OF TEMPERATURE ONLY.

$h_K = 232.83$ {INCLUDES CONDENSATE PUMP WORK}

FROM AN ENERGY BALANCE IN THE HEATER,

$(1-x)(h_K - h_h) = x(h'_I - h_J)$

$(1-x)(232.83 - 28.04) = x(1188.3 - 238.84)$

$204.79 = x(1154.25)$

$x = .177$

AT L: $h_L = x(h_J) + (1-x)h_K$

$= .177(238.84) + (1 - .177)232.83$

$= 233.89$ BTU/LBM

SINCE THIS IS A SUBCOOLED LIQUID,

$P_L = 38.5$ PSIA

$v_L = .017132$

$T_L = 265\,°F$

AT a: $P_a = 600$ PSIA

$$h_a = 233.89 + \frac{.017132(600 - 38.5)\,144}{(778)(.96)}$$

$= 235.7$

$$\eta_{th} = \frac{W_{out} - W_{in}}{Q_{in}}$$

$$= \frac{(h_d - h'_e) + (h_f - h'_I) + (1-x)(h'_I - h'_g) - (h_a - h_L)}{(h_d - h_a) + (h_f - h'_e)}$$

$$= \frac{(1289.9 - 1199.3) + (1322.1 - 1188.3) + (1 - .177)(1188.3 - 924.9)}{(1289.9 - 235.7) + (1322.1 - 1199.3)}$$

$$- \frac{(235.7 - 233.89)}{(1289.9 - 235.7) + (1322.1 - 1199.3)} = 0.373$$

**5** REFER TO FIGURE 7.30:

AT a: $V = 11\ FT^3$

$T = 460 + 80 = 540\,°R$

$P = 14.2$ PSIA $= 2044.8$ PSFA

$$M = \frac{PV}{RT} = \frac{(2044.8)(11)}{(53.3)(540)} = .781\ LBM$$

{MORE}

AT b: $V_b = \frac{1}{10}V_a = 1.1\ \text{FT}^3$

$T_b = 540\left(\frac{11}{1.1}\right)^{1.4-1} = 1356.4$

AT c: $T_c = T_b + \frac{q_{in}}{c_v m}$

$= 1356.4 + \frac{160}{(.1724)(.781)}$

$= 2544.7\ °\text{R} = 2084.7\ °\text{F}$

$\eta_{th} = 1 - \frac{1}{(10)^{1.4-1}} = .602$ {EQN 7.87}

6 STEP 1: FIND THE IDEAL MASS OF AIR INGESTED. THE IDEAL MASS OF AIR TAKEN IN PER SECOND IS

$\dot{V}_i = \left(\begin{matrix}\text{SWEPT}\\\text{VOLUME}\end{matrix}\right)\left(\begin{matrix}\text{\# INTAKE STROKES}\\\text{PER SECOND}\end{matrix}\right)$

FROM EQN 7.80, THE NUMBER OF POWER STROKES PER SECOND IS

$\frac{(2)(1200)\ \text{RPM}\ (6)\ \text{CYLINDERS}}{(60)\ \text{SEC/HR}\ (4)\ \text{STROKES}} = 60\ 1/\text{SEC}$

THE SWEPT VOLUME IS

$V_s = \left(\frac{\pi}{4}\right)\left(\frac{4.25}{12}\right)^2\left(\frac{6}{12}\right) = .04926\ \text{FT}^3$

SO $\dot{V}_1 = (60)(.04926) = 2.956\ \text{FT}^3/\text{SEC}$

FROM $PV = mRT$, THE MASS OF THE AIR IS

$m = \frac{(14.7)(144)(2.956)}{(53.3)(530)} = .2215\ \text{LBM AIR/SEC}$

STEP 2: FIND THE $CO_2$ VOLUME IN THE EXHAUST ASSUMING COMPLETE COMBUSTION WHEN THE AIR/FUEL RATIO IS 15.

FROM TABLE 9.9, AIR IS 76.85% NITROGEN, SO THE NITROGEN/FUEL RATIO IS

$N/F = (.7685)(15) = 11.528$

FROM $PV = wRT$, THE NITROGEN VOLUME PER POUND OF FUEL BURNED IS

$V_N = \frac{mRT}{P} = \frac{(11.528)(55.2)(530)}{(14.7)(144)} = 159.3\ \text{FT}^3$

THE OXYGEN FUEL RATIO IS

$O/F = (.2315)(15) = 3.472$

THE OXYGEN VOLUME PER POUND OF FUEL BURNED IS

$V_O = \frac{(3.472)(48.3)(530)}{(14.7)(144)} = 41.99\ \text{FT}^3$

WHEN OXYGEN FORMS CARBON DIOXIDE, THE CHEMICAL EQUATION IS

$C + O_2 \longrightarrow CO_2$

SO, IT TAKES 1 VOLUME OF OXYGEN TO FORM 1 VOLUME OF $CO_2$. THE % $CO_2$ IN THE EXHAUST IS FOUND FROM

$\%CO_2 = \frac{(\text{vol } CO_2)}{(\text{vol } CO_2) + (\text{vol } O_2) + (\text{vol } N_2)}$

NOW, $(\text{vol } N_2) = 159.3$

$(\text{vol } CO_2) = X$ {UNKNOWN}

$(\text{vol } O_2) = 41.99 -$ OXYGEN USED TO MAKE $CO_2$

$= 41.99 - X$

OR

$.137 = \frac{X}{X + 41.99 - X + 159.3}$

$X = 27.58\ \text{FT}^3$

ASSUMING COMPLETE COMBUSTION, THIS VOLUME OF $CO_2$ WILL BE CONSTANT REGARDLESS OF THE AMOUNT OF AIR USED.

STEP 3 CALCULATE THE EXCESS AIR IF $\%CO_2 = 9$

$\%CO_2 = \frac{(\text{vol } CO_2)}{(\text{vol } CO_2) + (\text{vol } O_2 - \text{vol } CO_2) + (\text{vol } N_2) + (\text{vol EXCESS AIR})}$

$.09 = \frac{27.58}{27.58 + 41.99 - 27.58 + 159.3 + (\text{vol EXCESS})}$

$(\text{vol EXCESS AIR}) = 105.2\ \text{FT}^3$

FROM $PV = mRT$,

$w_{EXCESS} = \frac{(14.7)(144)(105.2)}{(53.3)(530)} = 7.883\ \text{LBM}$

STEP 4: THE ACTUAL AIR FUEL RATIO IS

$15 + 7.883 = 22.883$ LBM AIR/LBM FUEL

THE ACTUAL AIR MASS PER SECOND IS

$\frac{(22.883)\frac{\text{LBM AIR}}{\text{LBM FUEL}}(28)\ \text{LBM FUEL/HR}}{(3600)\ \text{SEC/HR}} = .178\ \text{LBM/SEC}$

STEP 5: $\eta_v = \frac{.178}{.2215} = .804$

7 USE THE PROCEDURE ON PAGE 7-20.*

STEP 1: 1 - 60°F, 14.7 PSIA

2 - 5000 FT ALTITUDE

STEP 2: $IHP_1 = \frac{1000}{.80} = 1250$

STEP 3: NOT NEEDED*

STEP 4: $\rho_1 = P/RT = \frac{(14.7)(144)}{(53.3)(520)} = .0764\ \text{LBM/FT}^3$

$\rho_2 = .06592$ AT 5000′ {P. 8-20}

STEP 5: $IHP_2 = 1250\left(\frac{.06592}{.0764}\right) = 1078.5$

STEP 6: $(0.80)(1078.5) = 862.8$*

STEP 7: THE ORIGINAL FLOW RATE OF FUEL IS

$\dot{w}_{F,1} = (BHP_1)(BSFC_1) = (1000)(.45)$

$= 450\ \text{LBM/HR}$

STEP 8: $\dot{w}_{F,2} = \dot{w}_{F,1} = 450\ \text{LBM/HR}$

STEP 9: $BSFC_2 = \frac{450}{862.8} = 0.522$

* THE PROBLEM SAYS $\eta$ IS CONSTANT, NOT THE FRICTION HORSEPOWER.

**8** REFER TO FIGURE 7.36. AS WITH OTHER IC ENGINES, THE 'COMPRESSION RATIO' IS A RATIO OF VOLUMES. FIRST, ASSUME ISENTROPIC OPERATION.

AT a: $P_a = 14.7 \text{ PSIA} = 2116.8 \text{ PSFA}$

$T_a = 60°F = 520°R$

FOR 1 POUND,

$$V_a = \frac{(1)(53.3)(520)}{2116.8} = 13.09 \text{ FT}^3$$

AT b: $V_b = \frac{13.09}{5} = 2.618$

$$T_b = 520\left(\frac{13.09}{2.618}\right)^{1.4-1} = 989.9°R$$

$$P_b = \frac{(1)(53.3)(989.9)}{2.618} = 20153 \text{ PSF}$$

AT C: $T_c = 1500°F = 1960°R$

$P_c = 20153 \text{ PSF}$

AT d: $P_d = 14.7 \text{ PSIA} = 2116.8 \text{ PSF}$

$$T_d = 1960\left(\frac{2116.8}{20153}\right)^{\frac{1.4-1}{1.4}} = 1029.5$$

NOW, INCLUDE THE INEFFICIENCIES

$T_a = 520°R$

$$T_b' = 520 + \frac{989.9-520}{.83} = 1086°R$$

$T_c = 1960°R$

$T_d' = 1960 - .92(1960 - 1029.5) = 1103.9$

FROM EQN 7.124, ASSUMING AN IDEAL GAS,

$$\eta_{th} = \frac{(1960-1086)-(1102.9-520)}{1960-1086} = .332$$

ALTERNATIVELY THE PROBLEM CAN BE SOLVED USING AIR TABLES. IN THAT CASE, THE CORRESPONDING VALUES ARE:

$h_a = 124.27$ AT $520°R$

$h_b = 236.02$ AT $980°R$; $h_b' = 258.9$

$h_c = 493.64$

$h_d = 264.49$ AT $1094°R$; $h_d' = 282.8$

$\eta_{th} = 0.325$

**9** REFER TO FIGURE 7.38.

FROM PROBLEM 8,

$T_a = 520°R$

$T_b' = 1086$

$T_d = 1960$

$T_e' = 1103.9$

FROM EQN 7.125, ASSUMING AN IDEAL GAS,

$$.65 = \frac{T_c - T_b'}{T_e' - T_b'} = \frac{T_c - 1086}{1103.9-1086}$$

OR $T_c = 1097.6$

THEN FROM EQN 7.126 ASSUMING AN IDEAL GAS,

$$\eta_{th} = \frac{(1960-1103.9)-(1086-520)}{1960-1097.6}$$

$= .336$

**10** THIS IS A HARD ONE TO VISUALIZE. SHOWN BELOW IS ONE OF N LAYERS. EACH LAYER CONSISTS OF 24 TUBES, ONLY 3 OF WHICH ARE SHOWN.

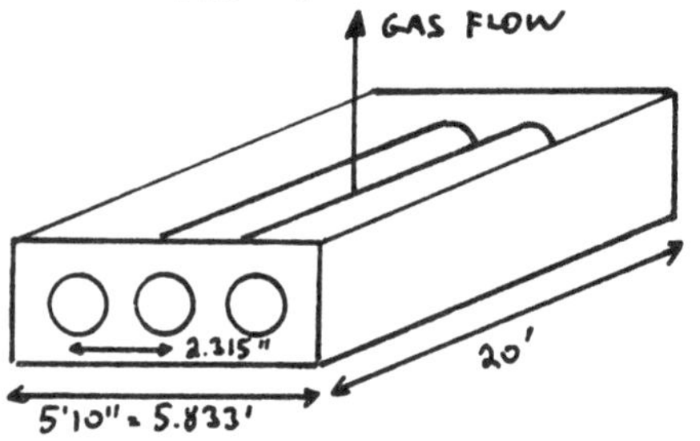


APPROACH: $q = UA\Delta T_M$

THE LOG MEAN TEMPERATURE DIFFERENCE (ASSUMING COUNTER FLOW OPERATION) IS

| | | |
|---|---|---|
| 635 | GASES → | 470 |
| $\Delta T = 350$ | | $\Delta T = 258$ |
| 285 | ← WATER | 212 |

$$\Delta T_M = \frac{350-258}{\ln\left(\frac{350}{258}\right)} = 301.7°F$$

SINCE NO INFORMATION IS GIVEN ABOUT THE STACK GASES, ASSUME THAT THEY CONSIST OF PRIMARILY NITROGEN. THE AVERAGE GAS TEMPERATURE IS

$$\tfrac{1}{2}(635+470) = 552.5°F = 1012.5°R$$

FROM TABLE 9.14

$\bar{C}_{p,\text{NITROGEN}} \approx .255$ BTU/LBM-°F

SO

$$q = \dot{m} C_p \Delta T = (191{,}000)\frac{\text{LBM}}{\text{HR}}(.255)\frac{\text{BTU}}{\text{LBM-°F}} \times (635-470) = 8.036\text{EE}6 \text{ BTU/HR}$$

SINCE NOT ENOUGH INFORMATION IS GIVEN, $h_i$ AND $h_o$ CANNOT BE EVALUATED, SO U MUST BE ASSUMED. ASSUME $U_o = 10$ BTU/HR-FT²-°F

THEN,

$$A_o = \frac{q}{U_o \Delta T_M} = \frac{8.036\text{EE}6}{(10)(301.7)} = 2663.6 \text{ FT}^2$$

THE TUBE AREA PER BANK IS

$$(24)\text{TUBES}\,(\pi)\left(\frac{1.315}{12}\right)\text{FT}\,(20)\,\text{FT} = 165.2 \text{ FT}^2, \text{ SO}$$

$$\#\text{ LAYERS} = \frac{2663.6}{165.2} = 16.1 \quad \text{(SAY 17)}$$

*TIMED*

**1** a) FROM EQN 7.80, THE NUMBER OF POWER STROKES PER MINUTE IS

$$\frac{(2)(4600)\text{ RPM }(8)\text{ CYLINDERS}}{(4)\text{ STROKES/CYCLE}} = 18400 \text{ 1/MIN}$$

THE NET WORK PER CYCLE IS

$1500 - 1200 = 300$ FT-LBF/CYCLE

(MORE)

TIMED #1, CONTINUED

THE HORSEPOWER IS

$$IHP = \frac{(18400)\ 1/\text{MIN}\ (300)\ \text{FT-LBF}}{33000\ \frac{\text{FT-LBF}}{\text{HP-MIN}}} = 167.27\ \text{HP}$$

b) THE THERMAL EFFICIENCY IS

$$\frac{(300)\ \text{FT-LBF}}{(1.27)\ \text{BTU}\ (778)\ \frac{\text{FT-LBF}}{\text{BTU}}} = .304\ (30.4\%)$$

c) METHOD 1 ASSUME AN AIR FUEL RATIO OF 15. ASSUME 14.7 PSIA AND 70°F AIR.

THE SWEPT VOLUME PER CYLINDER IS

$$\frac{(265)\ \text{IN}^3}{(8)\ (12)^3} = 0.01917\ \text{FT}^3$$

THE AIR DENSITY IS

$$\rho = P/RT = \frac{(14.7)(144)}{(53.3)(460+70)} = .07493\ \text{LBM/FT}^3$$

THE AIR WEIGHT PER HOUR IS

$$\dot{W}_A = (.01917)\ \text{FT}^3\ (18400)\ 1/\text{MIN}\ (60)\frac{\text{MIN}}{\text{HR}}\ (.07493)\frac{\text{LBM}}{\text{FT}^3}$$

$= 1585.8$ LBM/HR

THE FUEL WEIGHT PER HOUR IS

$$\dot{W}_F = \frac{\dot{W}_A}{R_{A/F}} = \frac{1585.8}{15} = 105.7\ \text{LBM/HR}$$

d) THE SPECIFIC FUEL CONSUMPTION IS

$$ISFC = \frac{105.7\ \text{LBM/HR}}{167.27\ \text{HP}} = .632\ \text{LBM/HP-HR}$$

METHOD 2

c) ASSUME THE HEATING VALUE OF GASOLINE IS LHV = 18,900 BTU/LBM {TABLE 7.3}

THEN, THE FUEL CONSUMPTION IS

$$\dot{W}_F = \frac{(1.27)\ \text{BTU/cycle}\ (18400)\ 1/\text{MIN}\ (60)\ \text{MIN/HR}}{(18,900)\ \text{BTU/LBM}}$$

$= 74.18$ LBM/HR

d) $ISFC = \frac{74.18}{167.27} = 0.443$ LBM/HP-HR

2 collect all ENTHALPIES.

AT 1: 1393.9

AT 2: 1270

AT 3: 1425.2

AT 4: 1280

AT 5: 1075

AT 6: 69.73

AT 7: 69.73 + .15 = 69.88

AT 8: 250.2

AT 9: 253.1

a) IF THE EXPANSION HAD BEEN ISENTROPIC TO 200 PSIA,

$h_2 = 1230$ BTU/LBM (FROM MOLLIER)

SO

$$\eta_{ISEN} = \frac{1393.9-1270}{1393.9-1230} = .756$$

b) LET X BE THE BLEED FRACTION. FROM AN ENERGY BALANCE IN THE HEATER,

$$h_8 = x h_4 + (1-x) h_7$$

$$250.2 = x(1280) + (1-x)69.88$$

$x = .149$

THE THERMAL EFFICIENCY IS

$$\eta_{th} = \frac{Q_{IN}-Q_{OUT}}{Q_{IN}} = \frac{(h_1-h_9)+(h_3-h_2)-(1-x)(h_5-h_6)}{(h_1-h_9)+(h_3-h_2)}$$

$$= \frac{(1393.9-253.1)+(1425.2-1270)-(1-.149)(1075-69.73)}{(1393.9-253.1)+(1425.2-1270)}$$

$= .34$

3 a) ASSUME THE ISENTROPIC EFFICIENCY IS WANTED. FROM AIR TABLES,

AT 1 $T_1 = -10°F = 450°R$

$h_1 = 107.5$

$P_{r,1} = .7329$

$P_1 = 8$ PSIA

AT 2 $P_{r,2} = \frac{P_2}{P_1}(P_{r,1}) = \left(\frac{40}{8}\right)(.7329) = 3.6645$

IF THE PROCESS WAS ISENTROPIC,

$T_2 = 712°R$

$h_2 = 170.47$

HOWEVER,

$T_2' = 315°F = 775°R$

SO $h_2' = 185.75$

$$\eta_{ISENTROPIC} = \frac{170.47-107.50}{185.73-107.50} = .805$$

b) FOR 35.7 PSIA, $T_{SAT} \approx 20°F = 480°R$

FOR 172.4 PSIA, $T_{SAT} \approx 120°F = 580°R$

FROM EQN 7.151 FOR AN IDEAL HEAT PUMP,

$$COP = \frac{580}{580-480} = 5.8$$

c) $$COP = \frac{(450)\ \text{BTUH}\ (1000)\ \text{W/KW}}{(585)\ \text{W}\ (3413)\ \text{BTUH/KW}} = .225$$

d) $$W_{OUT} = \frac{(600\ \text{EE}6)\ \text{W}\ (3413)\ \text{BTU/KW}}{(1000)\ \text{W/KW}} = 2.048\ \text{EE}9\ \text{BTUH}$$

NOW, $Q_{IN} - Q_{OUT} = W_{OUT} - W_{IN}$

BUT $W_{IN} \approx 0$, SO $Q_{IN} - Q_{OUT} = W_{OUT}$

OR $Q_{IN} = W_{OUT} + Q_{OUT}$

$= 2.048$ EE9 $+ 3.07$ EE9

$= 5.118$ EE9 {MORE}

TIMED #3 CONTINUED

THEN,

$$\eta_{th} = \frac{Q_{in} - Q_{out}}{Q_{in}} = \frac{(5.118\text{ EE9}) - (3.07\text{ EE9})}{5.118\text{ EE9}} = .400$$

e) ASSUME CARNOT CYCLE.

$$\eta_{th} = \frac{(82° + 460) - (40 + 460)}{82 + 460} = .0775$$

4) IF THE HORSEPOWERS ARE THE SAME,

a)

$$HP_1 = HP_2$$

$$(\dot{W}_{F,1})(HV_1) = (\dot{W}_{F,2})(HV_2)$$

BUT $\dot{W}_F = (SFC)(HP)$. SO

$$(SFC_1)(HV_1) = (SFC_2)(HV_2)$$

$$\frac{(SFC_2)}{(SFC_1)} = \frac{(HV_1)}{(HV_2)} = \frac{23{,}200}{11{,}930} = 1.945$$

THEN

$$\frac{(SFC_2) - (SFC_1)}{(SFC_1)} = \frac{(1.945)(SFC_1) - (SFC_1)}{(SFC_1)} = .945 \text{ (94.5\% INCREASE)}$$

b) $\dot{W} = VA\rho$, SO $A = \frac{\dot{W}}{V\rho}$

AND $\dot{W}_2 = 1.945\,\dot{W}_1$

$$\frac{A_2 - A_1}{A_1} = \frac{\frac{\dot{W}_2}{V\rho_2} - \frac{\dot{W}_1}{V\rho_1}}{\frac{\dot{W}_1}{V\rho_1}} = \frac{\frac{\dot{W}_2}{\rho_2} - \frac{\dot{W}_1}{\rho_1}}{\frac{\dot{W}_1}{\rho_1}}$$

$$= \frac{\left(\frac{1.945}{\rho_2} - \frac{1}{\rho_1}\right)}{\frac{1}{\rho_1}} = 1.945\left(\rho_1/\rho_2\right) - 1$$

USING INTERPOLATED SPECIFIC GRAVITIES AT 68°F FROM PAGE 3-45,

$$\frac{A_2 - A_1}{A_1} = 1.945\left(\frac{.724}{.789}\right) - 1 = .785$$

c) POWER IS PROPORTIONAL TO THE WEIGHT FLOW AND HEATING VALUE,

$$\frac{P_2 - P_1}{P_1} = \frac{V_2A_2\rho_2(HV)_2 - V_1A_1\rho_1(HV)_1}{V_1A_1\rho_1(HV)_1}$$

$$= \frac{\rho_2(HV)_2 - \rho_1(HV)_1}{\rho_1(HV)_1}$$

$$= \frac{(.789)(11930) - (.724)(23200)}{(.724)(23200)} = -.44$$

5(a) WORK WITH 1 LBM. ASSUME IDEAL GASES.

FOR THE 3→1 PROCESS

FIND SOME COMPOSITE PROPERTY OF THE GAS MIXTURE

$$P_1 = P_3\left(\frac{T_1}{T_3}\right)^{\frac{k}{k-1}}$$

$$14.7 = 568.6\left(\frac{520}{1600}\right)^{\frac{k}{k-1}}$$

$$.02585 = (.325)^{\frac{k}{k-1}}$$

$$\ln(.02585) = \frac{k}{k-1}\ln(.325)$$

$$\frac{k}{k-1} = 3.252 \longrightarrow k = 1.444$$

(b) FROM EQN 6.48, THE MOLAR SPECIFIC HEAT IS

$$C_{p,\text{mixture}} = \frac{R^*k}{J(k-1)} = \frac{\left(1545 \frac{\text{FT-LBF}}{\text{LBMOLE-°F}}\right)(1.444)}{\left(778 \frac{\text{FT-LBF}}{\text{BTU}}\right)(1.444 - 1)}$$

$$= 6.458 \text{ BTU/LBMOLE-°F}$$

FOR THE CONSTITUENT GASES,

$$C_{p,He} = (MW)(c_p) = \left(4 \frac{\text{LBM}}{\text{MOLE}}\right)\left(1.25 \frac{\text{BTU}}{\text{LBM-°F}}\right)$$

$$= 5.0 \text{ BTU/LBMOLE-°F}$$

$$C_{p,CO_2} = (44)(0.205) = 9.02 \text{ BTU/LBMOLE-°F}$$

FROM TABLE 6.5, MOLAR SPECIFIC HEAT IS VOLUMETRICALLY (MOLE FRACTION) WEIGHTED.

LET

$$x = \frac{n_{He}}{n_{He} + n_{CO_2}} = \text{MOLE FRACTION He}$$

$$x C_{p,He} + (1-x) C_{p,CO_2} = C_{p,\text{mixture}}$$

$$(x)\left(5.0 \frac{\text{BTU}}{\text{LBMOLE-°F}}\right) + (1-x)\left(9.02 \frac{\text{BTU}}{\text{LBMOLE-°F}}\right) = 6.458 \text{ LBMOLE-°F}$$

SOLVING,

$$x = 0.637$$

CONSIDER A 1-MOLE QUANTITY OF THE GAS MIXTURE. THE MASS OF HELIUM WOULD BE

$$m_{He} = (x)(MW) = (0.637 \text{ moles})\left(4 \frac{\text{LBM}}{\text{MOLE}}\right)$$

$$= 2.548 \text{ LBM}$$

SIMILARLY,

$$m_{CO_2} = (1 - 0.637 \text{ moles})\left(44 \frac{\text{LBM}}{\text{MOLE}}\right)$$

$$= 15.972 \text{ LBM}$$

THE MOLECULAR WEIGHT OF THE 1-MOLE OF GAS IS

$$MW = m_{He} + m_{CO_2} = 2.548 \text{ LBM} + 15.972 \text{ LBM}$$

$$= 18.52 \text{ LBM}$$

THE GRAVIMETRIC (MASS) FRACTION OF THE GASES ARE

$$G_{He} = \frac{m_{He}}{m_{He} + m_{CO_2}} = \frac{2.548}{2.548 + 15.972} = \boxed{0.138}$$

$$G_{CO_2} = 1 - G_{He} = 1 - 0.138 = \boxed{0.862}$$

FROM EQN 6.47,

$$C_v = C_p - \frac{R^*}{J}$$

$$= 6.458 \frac{\text{BTU}}{\text{LBMOLE-°F}} - \frac{1545 \frac{\text{FT-LBF}}{\text{LBMOLE-°F}}}{778 \frac{\text{FT-LBF}}{\text{BTU}}}$$

$$= 4.472 \text{ BTU/LBM-°F}$$

($C_v$ COULD ALSO BE FOUND FROM $x$, $C_{v,He}$, AND $C_{v,CO_2}$.)

FROM EQN 6.178,

$$W = c_v \Delta T = \left(\frac{C_v}{MW}\right)\Delta T$$

$$= \left(\frac{4.472 \frac{\text{BTU}}{\text{LBMOLE-°F}}}{18.52 \frac{\text{LBM}}{\text{LBMOLE}}}\right)(1600°\text{R} - 520°\text{R})$$

$$= \boxed{260.8 \text{ BTU/LBM}}$$

TIMED #5 CONTINUED

(c)

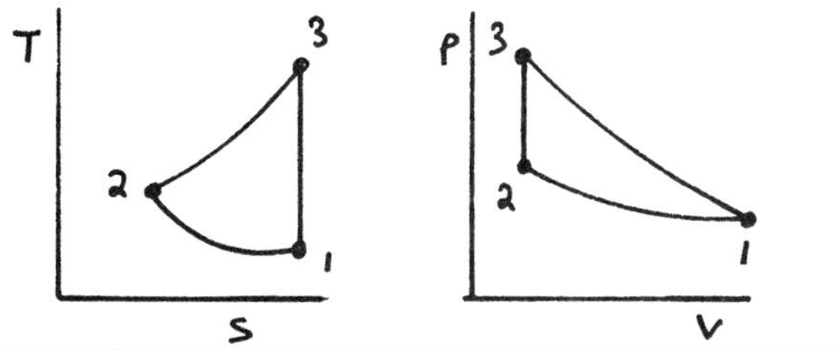


6 Follow the general steps on page 7-20

STEP 2: $IHP_1 = \frac{200}{.86} = 232.6$

STEP 3: $FHP = 232.6 - 200 = 32.6$

STEP 4: $\rho_1 = \frac{P}{RT} = \frac{(14.7)(144)}{(53.3)(460+80)} = .0735$

$\rho_2 = \frac{(12.2)(144)}{(53.3)(460+60)} = .0634$

STEP 5: $IHP_2 = 232.6\left(\frac{.0634}{.0735}\right) = 200.6$

STEP 6: $BHP_2 = 200.6 - 32.6 =$ **168**

$\eta = \frac{168}{200.6} =$ **.837**

STEP 7: $\dot{W}_{f1} = (.48)(200) = 96 \frac{LBM}{HR}$

$\dot{W}_{a1} = (22)(96) = 2112 \frac{LBM}{HR}$

$\dot{V}_{a1} = \frac{2112}{.0735} = 28735 \frac{FT^3}{HR}$

SINCE THE ENGINE SPEED IS CONSTANT,

$\dot{V}_{a2} = \dot{V}_{a1} = 28735 \frac{FT^3}{HR}$

STEP 8: $\dot{W}_{a2} = (28735)(.0634) = 1821.8 \frac{LBM}{HR}$

ASSUME THE METERED FUEL INJECTION VOLUME IS THE SAME.

$\dot{W}_{f2} = \dot{W}_{f1} = 96 \frac{LBM}{HR}$

$R_{a/f,2} = \frac{1821.8}{96} =$ **18.98**

$BSFC_2 = \frac{96}{168} =$ **$.57 \frac{LBM}{HP\text{-}HR}$**

7 FIRST WORK WITH THE ORIGINAL SYSTEM TO FIND THE STEAM FLOW.

$h_1 = 1227.6$ BTU/LBM (SUPERHEATED)

$h_2 = 38.04$ (SUBCOOLED)

$h_3 = 147.92$ (SUBCOOLED)

LET X = FRACTION OF STEAM IN MIXTURE

$147.92 = X(1227.6) + (1-X)(38.04)$

$X = .0924$

SO, THE STEAM FLOW IS

$(.0924)(2000) = 184.8 \frac{LBM}{HR}$

SINCE THE PRESSURE DROP ACROSS THE HEATER IS 5 PSI,

$P_4 = P_6 + 5 = 20 + 5 = 25$ PSIA

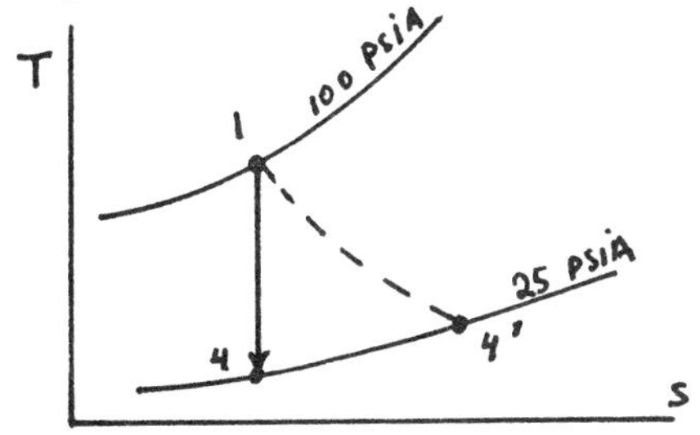


FROM THE MOLLIER DIAGRAM,

$h_4 \approx 1116$ (LIQUID-VAPOR MIXTURE)

$h_4' = 1227.6 - (.60)(1227.6 - 1116)$

$= 1160.6$ BTU/LBM

$W_{out} = (.96)(184.8)\frac{LBM}{HR}(1227.6 - 1160.6)\frac{BTU}{LBM}$

$=$ **$11{,}886.3 \frac{BTU}{HR}$**

LET X = FRACTION OF STEAM IN.

$147.92 = X(1160.6) + (1-X)(38.04)$

$X = .0979$

$\dot{m}_6 = \left(\frac{184.8}{.0979}\right) =$ **$1887.6 \frac{LBM}{HR}$**

8

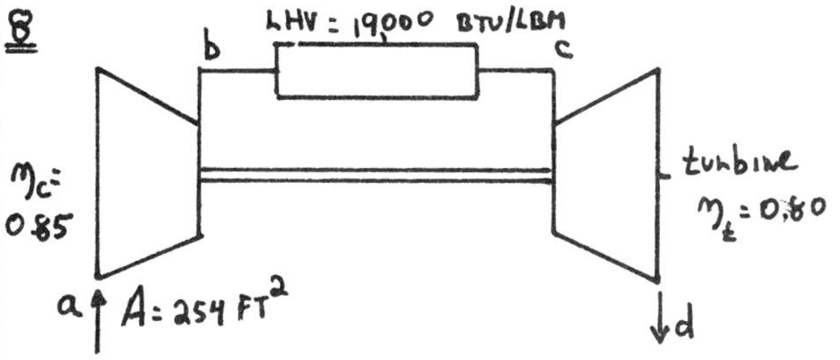


INITIALLY

$Q = 50{,}000$ CFM

BHP = 6000

BSCF = 0.609

at a:

$P_a = 12$ psia

$T_a = 35°F = 495°R$

$h_a = 118.28$ (FROM AIR TABLE, OR INTERPOLATE FROM PAGE 6-35.)

$P_{r,a} = 1.0224$

AIR DENSITY $= \frac{P}{RT} = \frac{(12)(144)}{(53.3)(35+460)} = 0.0655$ LBM/FT$^3$

AIR MASS FLOW RATE : $\dot{m} = Q\rho = (50{,}000)(0.0655)$

$= 3274.8$ LBM/MIN

at b:

$P_b = 8P_a = (8)(12) = 96$ psia

ASSUMING ISENTROPIC COMPRESSION,

$P_{r,b} = 8 \times P_{r,a} = (8)(1.0224) = 8.1792$

THIS $P_{r,b}$ CORRESPONDS TO

$T_b = 893°R$

$h_b = 214.54$

BUT COMPRESSION IS NOT ISENTROPIC

$h'_b = 118.28 + \dfrac{214.54 - 118.28}{0.85} = 231.52$

$T'_b = 962°R$

$W_{compression} = 231.52 - 118.28 = 113.24$ BTU/LBM

at c:

$T_c = 1800°F = 2260°R$ (NO CHANGE IF MOVED)

$P_c = 96$ PSIA {IN REALITY, SOME PRESSURE DROP WOULD BE EXPECTED ACROSS COMBUSTOR}

$h_c = 577.51$

$P_{r,c} = 286.6$

THE ENERGY REQUIREMENTS ARE

$\dot{m}\Delta h = 3274.8\,(577.51 - 231.52) = 1.133$ EE6 BTU/MIN

THE IDEAL FUEL RATE IS

$\dfrac{(1.133 \text{ EE6}) \text{ BTU/MIN } (60) \text{ MIN/HR}}{(19{,}000) \text{ BTU/LBM}}$

$= 3577.9$ LBM/HR

THE IDEAL BSFC IS

$\dfrac{3577.9}{6000} = 0.596$ LBM/HP-HR

THE ACTUAL BSFC IS GIVEN, THE COMBUSTOR EFFICIENCY IS

$\dfrac{0.596}{0.609} = 0.979$ (97.9%)

at d:

$P_d = 12$ PSIA {DETERMINED BY ATMOSPHERIC CONDITIONS}

IF ISENTROPIC EXPANSION, EQUATION 6.28 WOULD HOLD:

$P_{r,d} = \dfrac{286.6}{8} = 35.825$

$T_d = 1335°R$

$h_d = 325.99$

$h'_d = 577.51 - (0.80)(577.51 - 325.99)$

$= 376.29$ BTU/LBM

$W_{turbine} = 577.51 - 376.29 = 201.22$

THE THEORETICAL NET HORSEPOWER, CORRECTED FOR NON-ISENTROPIC EXPANSION, AND SUBTRACTING WORK OF COMPRESSION, IS

$IHP = \dfrac{\dot{m}\,\Delta h\, J}{\text{CONVERSION TO HP}}$

$= \dfrac{(3274.8)\frac{\text{LBM}}{\text{MIN}}\,(201.22 - 113.24)\frac{\text{BTU}}{\text{LBM}}\,(778)\frac{\text{FT-LBF}}{\text{BTU}}}{33{,}000\ \frac{\text{FT-LBF}}{\text{HP-MIN}}}$

$= 6793$ HP

THE FRICTION HORSEPOWER IS

$FHP = IHP - BHP = 6793 - 6000 = 793$ HP

AT 0 ALTITUDE

AT a:

$P = 14.7$ PSIA

$T = 70°F = 530°R$

$h_a = 126.6$

$P_{r,a} = 1.2983$

$\rho = \dfrac{(14.7)(144)}{(53.3)(530)} = 0.0749$

$\dot{m}_{AIR} = (50{,}000)(0.0749) = 3745$ LBM/MIN

AT b: $P_1$

$P_b = 8 \times 14.7 = 117.6$

$P_{r,b} = 8 \times 1.2983 = 10.3864$

$T_b = 954°R$

$h_b = 229.57$

$h'_b = 126.6 + \dfrac{229.57 - 126.6}{.85} = 247.74$

$T'_b = 1027°R$

$W_{compression} = 247.74 - 126.6 = 121.14$ BTU/LBM

AT C:

$P_c = 117.6$ (ASSUMING NO PRESSURE DROP ACROSS THE COMBUSTOR)

$T_c = 2260°R$

$h_c = 577.51$

$P_{r,c} = 286.6$

ENERGY REQUIREMENTS ARE

$\dot{m}\Delta h = (3745)(577.51 - 247.74) = 1.235$ EE6 BTU/MIN.

ASSUMING A CONSTANT COMBUSTION EFFICIENCY OF 97.9%,

$\text{FUEL RATE} = \dfrac{(1.235 \text{ EE6})(60)}{(19{,}000)(.979)} = 3983.6$ LBM/HR

AT D:

$P_d = 14.7$ psi

$P_{r,d} = \dfrac{286.6}{8} = 35.825$ (NO CHANGE)

$T_d = 1335°R$

$h_d = 325.99$

$h'_d = 376.29$ (NO CHANGE)

$W'_{turb} = 201.22$ (NO CHANGE)

THE HORSEPOWER CORRECTED FOR NON-ISENTROPIC EXPANSION IS

$IHP = \dfrac{(3745)(201.22 - 121.14)(778)}{33{,}000} = 7070$

ASSUMING FRICTIONAL HORSEPOWER IS CONSTANT,

$BHP = IHP - FHP$

$= 7070 - 793 = \boxed{6277 \text{ HP}}$

$BSFC = \dfrac{3983.6}{6277} = \boxed{0.635 \ \frac{\text{LBM}}{\text{HP-HR}}}$

# Compressible Fluid Dynamics

WARM-UPS

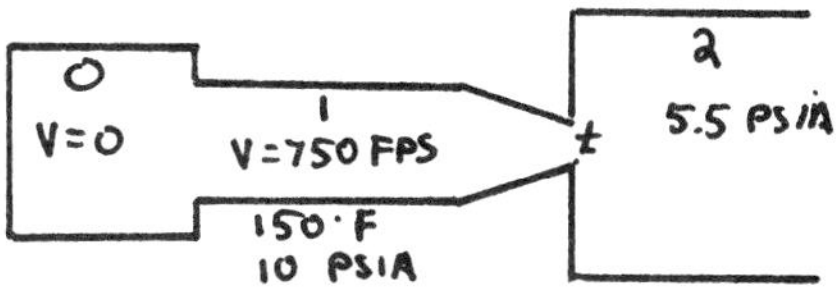


1 $T_1 = 460 + 150 = 610°R$

THE VELOCITY OF SOUND AT POINT 1 IS

$$C_1 = \sqrt{kg_cRT} = \sqrt{(1.4)(32.2)(53.3)610} = 1210.7 \text{ FT/SEC}$$

THE ENTRANCE MACH NUMBER IS

$$M_1 = \frac{V_1}{C_1} = \frac{750}{1210.7} = .619 \ \{\text{SAY } .62\}$$

SO, FROM PAGE 8-17,

$$\frac{T_1}{T_0} = \frac{T}{TT} = .9286$$

$$\frac{P_1}{P_0} = \frac{P}{TP} = .7716$$

$$\text{SO } T_0 = \frac{610}{.9286} = 656.9 °R$$

$$P_0 = \frac{10}{.7716} = 12.96 \text{ PSIA}$$

$$\frac{P_2}{P_0} = \frac{5.5}{12.96} = .424$$

FROM PAGE 8-5, THE CRITICAL PRESSURE RATIO IS .5283. SINCE $\frac{P_2}{P_0} < .5283$, THE FLOW IS CHOKED AND $M_t = 1$. FROM PAGE 8-18 FOR $M = 1$

$$\frac{P}{P_0} = \frac{P}{TP} = .5283$$

$$\frac{T}{T_0} = \frac{T}{TT} = .8333$$

$$\text{SO } P = (.5283)(12.96) = 6.85 \text{ PSIA}$$

$$T = (.8333)(656.9) = 547.4 °R$$

2 ASSUME STP IS 14.7 PSIA AND 32°F.

THUS, $T_x = 460 + 32 = 492°R$

$P_x = 14.7$ PSIA

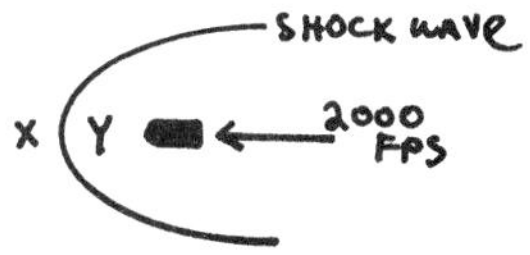


$$C = \sqrt{(1.4)(32.2)(53.3)(492)} = 1087.3$$

$$M = \frac{2000}{1087.3} = 1.84$$

ANALYZE THIS AS THOUGH THE BULLET WAS STATIONARY AND THE STP AIR WAS MOVING AT $M = 1.84$. FROM THE COMPRESSIBLE FLOW TABLES {CFT} FOR $M = 1.84$

$$\frac{P_x}{TP_y} = .2060$$

$$\frac{T_x}{TT_x} = .5963$$

SO

$$TP_y = \frac{14.7}{.2060} = 71.36 \text{ PSIA}$$

SINCE A SHOCK WAVE IS ADIABATIC,

$$TT_x = TT_y = \frac{492}{.5963} = 825.1$$

FROM PAGE 6-35,

$$h \approx 197.9 \ \frac{\text{BTU}}{\text{LBM}}$$

3 ASSUME THE PRESSURE AND TEMPERATURE GIVEN ARE THE CHAMBER PROPERTIES, SO

$TT = 240°F = 700°R$

$TP = 160$ PSIA

a) $\frac{20}{160} < .5283$, SO THE NOZZLE IS SUPERSONIC AND THE THROAT FLOW IS SONIC.

b) THE TOTAL DENSITY IS

$$TD = \frac{(TP)}{R(TT)} = \frac{(160)(144)}{(53.3)(700)} = .6175 \text{ LBM/FT}^3$$

AT THE THROAT,

$M = 1$

$$\frac{D}{TD} = .6339$$

$$\frac{T}{TT} = .8333$$

$$\text{SO } T^* = (700)(.8333) = 583.3 °R$$

$$D^* = (.6175)(.6339) = .3914 \text{ LBM/FT}^3$$

$$C^* = \sqrt{(1.4)(53.3)(32.2)(583.3)} = 1183.9 \text{ FT/SEC}$$

$$A^* = \frac{\dot{M}}{D^*C^*} = \frac{(4.5) \text{ LBM/SEC}}{(.3914)\frac{\text{LBM}}{\text{FT}^3}(1183.9)\frac{\text{FT}}{\text{SEC}}} = .00971 \text{ FT}^2$$

c) AT EXIT, $\frac{P}{TP} = \frac{20}{160} = .125$

SEARCHING THE CFT GIVES

$M \approx 2.01$

d) AT $M = 2.01$

$$\frac{A}{A^*} = 1.7016$$

$$\text{SO } A_e = (1.7016)(.00971) = .01652 \text{ FT}^2$$

4 $C = \sqrt{(1.4)(32.2)(53.3)(460+60)} = 1117.8$ FT/SEC

$$\text{SO } M = \frac{2700}{1117.8} = 2.415$$

FROM TABLE 8.3 OR FIGURE 8.6(b),

$\theta_{MAX} \approx 45°$

THE ACTUAL VALUE MAY BE SMALLER.

5 $TT = 460 + 80 = 540°R$

$TP = 100$ PSIA

FROM THE CFT AT $M = 2$,

$$\frac{P}{TP} = .1278; \quad \frac{T}{TT} = .5556; \quad \frac{A}{A^*} = 1.6875 \ \{\text{MORE}\}$$

WARM-UP #5 CONTINUED

so, $T = (.5556)(540) = 300°R$

$A = (1.6875)(1) = 1.6875$

$p = (.1278)(100) = 12.78$ psia

$$c = \sqrt{(1.4)(32.2)(53.3)(300)} = 849 \text{ ft/sec}$$

$$V = Mc = (2)(849) = 1698 \text{ ft/sec}$$

$$D = \rho = p/RT = \frac{(12.78)(144)}{(53.3)(300)} = .1151 \text{ lbm/ft}^3$$

$$\dot{m} = VA\rho = (1698)\frac{1.6875}{144}(.1151) = 2.29 \text{ lbm/sec}$$

6 FROM PAGE 8-18 OF THE CFT AT $M_x = 2$

$M_y = .5744$

$\frac{T_y}{T_x} = 1.687$

so $T_y = (1.687)(500) = 843.5°R$

$$c_y = \sqrt{(1.4)(32.2)(53.3)(843.5)} = 1423.6 \text{ ft/sec}$$

$$V_y = Mc = (.5744)(1423.6) = 822 \text{ ft/sec}$$

7 TP = 100 psia

TT = (460 + 70) = 530°R

SEARCHING THE CFT FOR $\left(\frac{A}{A^*}\right) = 1.555$,

$M = 1.9$

$\frac{T}{TT} = .5807$

$\frac{p}{TP} = .1492$

so $T = (.5807)(530) = 307.8°R$

$p = (.1492)(100) = 14.92$ psia

8 SINCE V IS UNKNOWN, ASSUME THAT THE STATIC TEMPERATURE IS $(40+460) = 500°R$.

THEN,

$$\rho = \frac{p}{RT} = \frac{(10)(144)}{(53.3)(500)} = .054 \text{ lbm/ft}^3$$

$$\text{so } V = \frac{\dot{m}}{A\rho} = \frac{(20) \text{ lbm/sec}}{(1)\text{ ft}^2(.054)\frac{\text{lbm}}{\text{ft}^3}} = 370.4 \text{ ft/sec}$$

AT 500°F,

$$c = \sqrt{(1.4)(32.2)(53.3)(500)} = 1096.1 \text{ ft/sec}$$

$$\text{so } M = \frac{370.4}{1096.1} = .338 \text{ (say .34)}$$

AT $M = .34$, $\frac{T}{TT} = .9774$

so, A CLOSER APPROXIMATION TO T WOULD BE

$T = (.9774)(500) = 488.7°R$

$$\rho = \frac{p}{RT} = \frac{(10)(144)}{(53.3)(488.7)} = .0553 \text{ lbm/ft}^3$$

$$V = \frac{\dot{m}}{A\rho} = \frac{20}{(1)(.0553)} = 361.7 \text{ ft/sec}$$

$$c = \sqrt{(1.4)(32.2)(53.3)(488.7)} = 1083.6 \text{ ft/sec}$$

$$M = \frac{361.7}{1083.6} = .334 \text{ (say .33)}$$

AT $M = .33$,

$\frac{A}{A^*} = 1.8707$, so

$$A_{\text{smallest}} = \frac{1}{1.8707} = .535$$

9 $\frac{p_x}{TP_y} = \frac{1.38}{20} = .069$

SEARCHING THE CFT, $M = 3.3$

10 $h_1 = 1428.9$ BTU/lbm

USING THE MOLLIER DIAGRAM ASSUMING ISENTROPIC EXPANSION,

$h_2 = 1362$

FROM EQN 6.92 OR 8.62, ASSUMING $V_1 = 0$,

$$V = \sqrt{(2)(32.2)(778)(1428.9 - 1362)}$$

$$= 1830.8 \text{ ft/sec}$$

CONCENTRATES

1 AT THAT POINT,

$$c = \sqrt{(1.4)(32.2)(53.3)(1000)} = 1550.1 \text{ ft/sec}$$

$$\text{so } M = \frac{600}{1550.1} = .387$$

AT $M = .39$,

$\frac{T}{TT} = .9705$

$\frac{p}{TP} = .9004$

$\frac{A}{A^*} = 1.6234$

so

$$TT = \frac{1000}{.9705} = 1030.4$$

$$TP = \frac{50}{.9004} = 55.5$$

$$A^* = \frac{.1}{1.6234} = .0616$$

AT $M = 1$,

$\frac{p}{TP} = .5283$

$\frac{T}{TT} = .8333$

so $p^* = (55.5)(.5283) = 29.32$ psia

$T^* = (1030.4)(.8333) = 858.6°R$

2 THE EXPANSION IS NOT ISENTROPIC AND $V_i \neq 0$, SO EQN 8.62 SHOULD BE USED.

$h_1 = 1322.1$

FROM THE MOLLIER DIAGRAM, ASSUMING ISENTROPIC EXPANSION,

$h_2 = 1228$

(MORE)

CONCENTRATES # 2 CONTINUED

$$h_2' = 1322.1 - .85(1322.1 - 1228)$$
$$= 1242.1 \text{ BTU/LBM}$$

KNOWING $h = 1242.1$ AND $p = 80$ psi ESTABLISHES

$T_2' = 420°\text{F}$, $v_2' = 6.383$ } FROM DETAILED SUPERHEAT TABLES

SO $\rho_2 = 1/v_2 = 1/6.383 = .1567 \text{ LBM/FT}^3$

FROM EQUATION 8.62,

$$V_2' = \sqrt{(2)(32.2)(778)(1322.1 - 1242.1) + (300)^2}$$
$$= 2024.4 \text{ FT/SEC AT EXIT}$$

$$A_e = \frac{\dot{m}}{V\rho} = \frac{3 \text{ LBM/SEC}}{\left(2024.4 \frac{\text{FT}}{\text{SEC}}\right)\left(.1567 \frac{\text{LBM}}{\text{FT}^3}\right)} = .009457$$

FROM FIGURE 8.13 FOR $420 + 460 = 880°\text{R}$ STEAM, $k_{\text{STEAM}} = 1.31$

FROM TABLE 6.4, $R_{\text{STEAM}} = 85.8$

$$c = \sqrt{(1.31)(32.2)(85.8)(880)} = 1784.6 \text{ FT/SEC}$$

SO $M = \frac{2024.4}{1784.6} = 1.13$

FROM THE CFT AT $M = 1.13$ AND $k = 1.30$

$\frac{A}{A^*} = 1.0139$ {OR, GET EXACT VALUE FROM EQN 8.22}

SO $A^* = \frac{A_e}{1.0139} = \frac{.009457}{1.0139} = .00933 \text{ FT}^2$

TIMED

1 THIS IS A FANNO FLOW PROBLEM. IT'S NOT CLEAR WHAT THE PROBLEM WANTS, BUT ASSUME WE ARE TO CHECK FOR CHOKED FLOW.

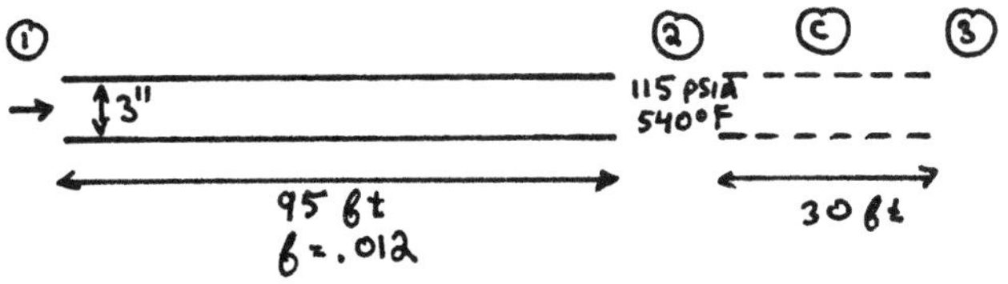


SINCE $P$ IS DECREASING, THE FLOW IS INITIALLY SUB-SONIC.

METHOD 1

AT POINT 2, FROM THE SUPERHEAT TABLES,
$v_2 = 5.066 \text{ FT}^3/\text{LBM}$

$$V = \frac{\dot{m}}{A\rho} = \frac{\dot{m}v}{A}$$
$$= \frac{(35200)(5.066)}{(3600)\left(\frac{\pi}{4}\right)\left(\frac{3}{12}\right)^2} = 1009 \text{ FT/SEC}$$

ASSUME $k = 1.29$ (SEE FIGURE 8.13 AT 1000°R)

$$c = \sqrt{k g_c R T} = \sqrt{(1.29)(32.2)(85.8)(540 + 460)}$$
$$= 1888 \text{ FT/SEC}$$

SO, THE MACH NUMBER AT POINT 2 IS

$$M_2 = \frac{1009}{1888} = 0.534$$

FROM EQUATION 8.38, THE DISTANCE FROM POINT 2 TO WHERE THE FLOW BECOMES CHOKED IS

$$x_{\text{MAX}} = \frac{3/12}{(4)(.012)}\left[\frac{1 - (.534)^2}{(1.29)(.534)^2} + \frac{1.29 + 1}{2(1.29)} \ln\left(\frac{(1.29 + 1)(.534)^2}{2\left[1 + \left(\frac{1.29 - 1}{2}\right)(.534)^2\right]}\right)\right]$$
$$= 5.208\left[1.943 + .888 \ln\left(\frac{0.6530}{2.083}\right)\right]$$
$$= 4.75 \text{ FT}$$

SO THE FLOW WILL BE CHOKED IN LESS THAN 30 FT

METHOD 2

USE FANNO FLOW TABLE FOR $k = 1.3$

AT $M_2 = .53$, $\frac{4fL_{\max}}{D} = .949$

AT $M_c = 1$, $\frac{4fL_{\max}}{D} = 0$

$$L_{\max} = \frac{(.949 - 0)\left(3/12\right)}{(4)(.012)} = 4.94 \text{ FT}$$

2 SINCE THE PRESSURE DROPS TO 14.7 PSIA, THIS MUST BE A SUPERSONIC NOZZLE.

ASSUME THE GIVEN PROPERTIES ARE TOTAL PROPERTIES.

AT THE THROAT, $M = 1$, AND

$$\frac{T}{TT} = .8333$$
$$\frac{P}{TP} = .5283$$

SO

$$T_{\text{throat}} = (.8333)(660) = 550°\text{R}$$
$$P_{\text{throat}} = (.5283)(160) = 84.53 \text{ psia}$$

$$c^* = \sqrt{kgRT} = \sqrt{(1.4)(32.2)(53.3)(550)}$$
$$= 1150 \text{ ft/sec}$$

{MORE}

SINCE AIR IS AN IDEAL GAS,

$$\rho_{throat} = \frac{P}{RT} = \frac{(84.53)(144)}{(53.3)(550)}$$

$$= .415\ \text{LBM/FT}^3$$

OVERALL NOZZLE EFFICIENCY IS $C_D = 0.90$.

FROM $\dot{m} = C_D A v \rho_1$,

$$A^* = \frac{\dot{m}}{C_D v \rho} = \frac{3600}{(3600)(1150)(.415)(.90)}$$

$$= \boxed{.002328\ \text{FT}^2}$$

$$D_{throat} = \sqrt{\frac{4A}{\pi}} = \sqrt{\frac{(4)(.002328)}{\pi}}$$

$$= 0.0544\ \text{FT}$$

AT THE EXIT, $P = 14.7$

$$\frac{P}{TP} = \frac{14.7}{160} = .0919$$

FINDING THIS IN THE ISENTROPIC FLOW TABLES YIELDS

$$M = 2.22$$

$$\frac{A}{A^*} = 2.041$$

SO,

$$A_{exit} = (2.041)(.002328) = .004751\ \text{FT}^2$$

$$D_{exit} = \sqrt{\frac{4(.004751)}{\pi}} = .0778\ \text{FT}$$

THE LONGITUDINAL DISTANCE FROM THROAT TO EXIT IS

$$x = \frac{.0778 - .0544}{(2)(\tan 3°)} = 0.223\ \text{FT}$$

THE ENTRANCE VELOCITY IS NOT KNOWN, SO THE ENTRANCE AREA CANNOT BE FOUND. HOWEVER, THE LONGITUDINAL DISTANCE FROM ENTRANCE TO THROAT IS

$$.05(.223) = 0.0112\ \text{FT}$$

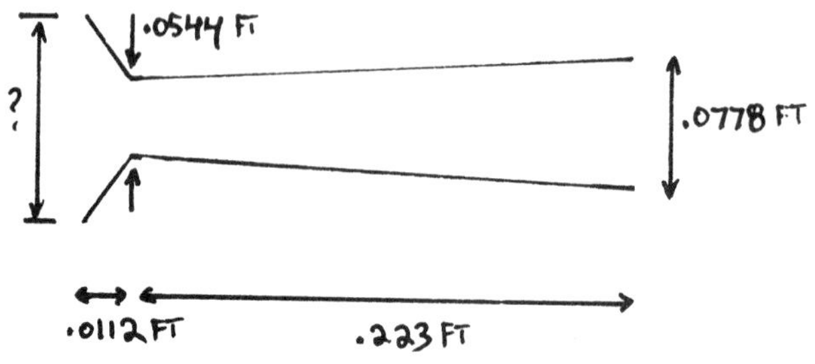

# Combustion

1

| | G | AW | G/AW |
|---|---|---|---|
| C | 40.0 | 12 | 3.33 |
| H | 6.7 | 1 | 6.7 |
| O | 53.3 | 16 | 3.33 |

$\rightarrow CH_2O$

2 ONE MOLE OF ANY GAS OCCUPIES A VOLUME OF 22.4 LITERS.

$\#$ gMOLES OF $CH_4 = \frac{200}{22.4} = 8.929$

1 MOLE OF METHANE HAS A MOLECULAR WEIGHT OF 16 g.

SO, THE METHANE HAS A MASS OF

$(16)(8.929) = 142.86$ g

IN POUNDS,

$(142.86\ g)(.0022046\ LB/g) = .315$ LB

THE AVAILABLE HEAT IS

$(.315)(24{,}000\ BTU/LB) = 7560$ BTU

$\Delta T = (95-15)\left(\frac{9}{5}\right) = 144°F$

$q = mc_p\Delta T/m$

$$M = \frac{(.5)(7560)(.4536\ kg/lbm)}{(1)(144)} = 11.91\ kg$$

3 $C_3H_8 + 5O_2 \rightarrow 3CO_2 + 4H_2O$

(MW) 44 + 160 → 132 + 72

$$\frac{CO_2}{C_3H_8} = \frac{132}{44} = \frac{X}{15}$$

$X = 45$ LBM/HR

$$V = \frac{WRT}{P} = \frac{(45)(35.1)(530)°R}{(14.7)(144)} = 395.5\ FT^3/HR$$

4 $CH_4 + 2O_2 \rightarrow CO_2 + 2H_2O$

(volumes) 1 2 1 2

SO, IDEAL OXYGEN VOLUME IS

$2(4000) = 8000$ CFH

ACTUAL VOLUME $= (1.3)(8000) = 10400$ CFH

ACTUAL OXYGEN WEIGHT

$$= \frac{PV}{RT} = \frac{(15)(144)(10400)}{(48.3)(460+100)} = 830.5$$

BUT AIR IS .2315 OXYGEN BY WEIGHT, SO NITROGEN WEIGHT IS

$$\left(\frac{1-.2315}{.2315}\right)(830.5) = 2757\ LBM/HR$$

5 $C + O_2 \rightarrow CO_2$

12 + 32 → 48 (44)

SO $\frac{32}{12} = 2.67$ LBM OXYGEN REQ'D PER POUND CARBON

$2H_2 + O_2 \rightarrow 2H_2O$

4 + 32 → 36

SO $\frac{32}{4} = 8$ $\frac{LBM\ O_2}{LBM\ H_2}$

$S + O_2 \rightarrow SO_2$

32.1 + 32 → 64.1

SO $\frac{32}{32.1} = 1$ $\frac{LBM\ O_2}{LBM\ S}$

NITROGEN DOES NOT BURN

$(.84)(2.67) + (.153)(8) + (.003)(1) = 3.47$ $\frac{LBM\ O_2}{LBM\ FUEL}$

BUT AIR IS .2315 OXYGEN

$$AIR = \frac{3.47}{.2315} = 15.0\ \frac{LBM\ AIR}{LBM\ FUEL}$$

6 IDEALLY,

$C_3H_8 + 5O_2 \rightarrow 3CO_2 + 4H_2O$

44 + 160 → 132 + 72

THE EXCESS OXYGEN IS $(160)(.2) = 32$

SINCE AIR IS .2315 OXYGEN BY WEIGHT, THE NITROGEN IS

$$\frac{(1-.2315)}{.2315}(160+32) = 637.4$$

% $CO_2$ BY WEIGHT =

$$\frac{132}{132+72+32+637.4} = .151\ (WET)$$

CONCENTRATES

1 STEP 1 FIND THE WEIGHT OF OXYGEN IN THE STACK GASES. ASSUME THE STACK GASES ARE AT 60°F AND 14.7 PSIA WHEN SAMPLED. FROM TABLE 6.4

$R_{CO_2} = 35.1$ $R_{CO} = 55.2$ $R_{O_2} = 48.3$

SO $\rho_{CO_2} = \frac{P}{RT} = \frac{(14.7)(144)}{(35.1)(460+60)} = .1160$

$$\rho_{CO} = \frac{(14.7)(144)}{(55.2)(520)} = .07374$$

$$\rho_{O_2} = \frac{(14.7)(144)}{(48.3)(520)} = .08428$$

$CO_2$ IS $\frac{32}{44} = .7273$ OXYGEN

$CO$ IS $\frac{16}{28} = .5714$ OXYGEN

$O_2$ IS ALL OXYGEN

IN 100 $FT^3$ OF STACK GASES, THE TOTAL OXYGEN WEIGHT IS

$$(.12)(100)(.7273)(.1160) + (.01)(100)(.07374)(.5714) + (.07)(100)(.08428)(1.00) = 1.644 \ \frac{LBM\ O_2}{100\ FT^3}$$

STEP 2 SINCE AIR IS 23.15% OXYGEN BY WEIGHT, THE AIR PER 100 $FT^3$ IS

$$\frac{1.644}{.2315} = 7.102 \ \frac{LBM\ AIR}{100\ FT^3}$$

STEP 3 FIND THE WEIGHT OF CARBON IN THE STACK GASES.

$CO_2$ IS $\frac{12}{44} = .2727$ CARBON

$CO$ IS $\frac{12}{28} = .4286$ CARBON

THEN

$$(.12)(100)(.1160)(.2727) + (.01)(100)(.07374)(.4286) = .4112 \ \frac{LBM\ CARBON}{100\ FT^3}$$

STEP 4 THE COAL IS 80% CARBON, SO THE AIR PER POUND OF COAL IS

$$\frac{(.80)\ \frac{LBM\ CARBON}{LBM\ COAL}}{(.4112)\ \frac{LBM\ CARBON}{100\ FT^3}} \left(7.102\right) \frac{LBM\ AIR}{100\ FT^3} = 13.82 \ \frac{LBM\ AIR}{LBM\ COAL}$$

THIS DOES NOT INCLUDE AIR TO BURN THE HYDROGEN, SINCE ORSAT IS A DRY ANALYSIS.

STEP 5. FROM EQN 9.3

THE THEORETICAL AIR FOR THE HYDROGEN IS

$$34.34\left(.04 - \frac{.02}{8}\right) = 1.288 \ \frac{LBM\ AIR}{LBM\ COAL}$$

STEP 6 IGNORING ANY EXCESS AIR FOR THE HYDROGEN, THE AIR PER POUND COAL IS

$$13.82 + 1.29 = 15.11 \ \frac{LBM\ AIR}{LBM\ COAL}$$

ALTERNATE SOLUTION

USE EQN 9.13.

$$\frac{LBM\ AIR}{LBM\ FUEL} = \frac{(3.04)(N_2)[C]}{(CO_2)+(CO)}$$

$$= \frac{(3.04)(80\%)(0.80)}{12\% + 1\%} = 15 \ \frac{LBM\ AIR}{LBM\ FUEL}$$

2 STEP 1 FROM TABLE 9.22

THE HEATING VALUE PER POUND OF COAL IS

$$(.75)(14093) + \left(.05 - \frac{.03}{8}\right)(60958) = 13389 \ \frac{BTU}{LBM}$$

STEP 2 THE GRAVIMETRIC ANALYSIS OF 1 LBM OF FUEL IS

CARBON: ............ .75 LBM

FREE HYDROGEN $\left(.05 - \frac{.03}{8}\right)$ = .......... .0463

WATER $9(.05 - .0463)$ .......... .0333

NITROGEN .......... .02

STEP 3 THE THEORETICAL STACK GASES PER POUND OF COAL (TABLE 9.11)

FOR .75 LBM COAL ARE

$CO_2$: $(.75)(3.67) = 2.753$

$N_2$: $(.75)(8.78) = 6.585$

ALL PRODUCTS ARE SUMMARIZED IN THE FOLLOWING TABLE:

| | $CO_2$ | $N_2$ | $H_2O$ |
|---|---|---|---|
| FROM CARBON | 2.753 | 6.583 | |
| FROM $H_2$ | | 1.218 | .417 |
| FROM $H_2O$ | | | .0333 |
| FROM $O_2$ | SHOWS UP IN $CO_2$, $H_2O$ | | |
| FROM $N_2$ | | .02 | |
| TOTALS: | 2.753 | 7.821 | .4503 |

STEP 4 ASSUME THE STACK GASES LEAVE AT 1000°F. THEN, $T_{AVE} = \frac{1}{2}\left((60+460)+(1000+460)\right) = 990$ {SAY 1000°R}

FROM TABLE 9.14

$\bar{c}_p\ (CO_2) = .251$

$\bar{c}_p\ (N_2) = .255$

$\bar{c}_p\ (H_2O) = .475$

THE HEAT REQUIRED TO RAISE THE COMBUSTION PRODUCTS FROM 1 LBM OF COAL 1°F IS

$$(2.753)(.251) + (7.821)(.255) + (.4503)(.475) = 2.9 \text{ BTU}$$

STEP 5 ASSUMING ALL COMBUSTION HEAT GOES INTO THE STACK GASES, THE FINAL TEMPERATURE IS

$$T_2 = T_1 + \frac{HHV}{2.9}$$

$$= 60 + \frac{13389}{2.9} = 4677°F$$

STEP 6 IN REALITY, THERE WILL BE APPROXIMATELY 40% EXCESS AIR, AND 75% OF THE HEAT WILL BE ABSORBED BY THE BOILER

$$\text{EXCESS AIR} = \frac{(40)(7.821)}{.7685} = 4.071 \text{ LBM}$$

FROM TABLE 9.14, $\overline{C}_p$ FOR AIR AT 1000°R IS

$$(.7685)(.255) + (.2315)(.236) = .251$$

$$\text{THEN, } T_2 = 60 + \frac{13389(1-.75)}{2.9 + (4.071)(.251)} = 913.5°$$

3 STEP 1 USE TABLE 9.11 TO FIND THE STOICHIOMETRIC OXYGEN REQUIRED PER POUND OF FUEL OIL

| | | |
|---|---|---|
| $C \rightarrow CO_2$: | $(.8543)(2.67)$ = | 2.2810 |
| $H \rightarrow H_2O$: | $(.1131)(8)$ = | .9048 |
| $S \rightarrow SO_2$: | $(.0034)(1)$ = | .0034 |
| LESS OXYGEN IN FUEL | | −.0270 |
| | | 3.1622 LBM OXY / LBM FUEL |

STEP 2 THE THEORETICAL NITROGEN IS $\frac{3.1622}{.2315}(.7685) = 10.497$

THE ACTUAL NITROGEN WITH EXCESS AIR AND THE NITROGEN IN THE FUEL IS

$$.0022 + (1.6)(10.497) = 16.8$$

THIS IS A VOLUME OF

$$V = \frac{WRT}{P} = \frac{(16.8)(55.2)(460+60)}{(14.7)(144)} = 227.81 \text{ FT}^3$$

STEP 3 THE EXCESS OXYGEN

$$(.6)(3.1622) = 1.897 \ \frac{\text{LBM AIR}}{\text{LBM FUEL}}$$

THIS IS A VOLUME OF

$$V = \frac{WRT}{P} = \frac{(1.897)(48.3)(460+60)}{(14.7)(144)} = 22.51 \text{ FT}^3$$

STEP 4 FROM TABLE 9.11, THE 60°F COMBUSTION PRODUCT VOLUMES PER POUND OF FUEL WILL BE

| | | |
|---|---|---|
| $CO_2$ : | $(.8543)(31.6)$ = | 27.0 $FT^3$ |
| $H_2O$ : | $(.1131)(189.5)$ = | 21.43 |
| $SO_2$ : | $(.0034)(11.84)$ = | .04 |
| $N_2$ : | FROM STEP 2 = | 227.81 |
| $O_2$ : | FROM STEP 3 | 22.51 |
| | | 298.79 $FT^3$ |

STEP 5 AT 600°F, THE WET VOLUME WILL BE

$$V_{WET} = \left(\frac{460+600}{460+60}\right)(298.79) = 609 \text{ FT}^3$$

AT 600°F, THE DRY VOLUME WILL BE

$$V_{DRY} = \left(\frac{460+600}{460+60}\right)(298.79 - 21.43) = 565.4 \text{ FT}^3$$

STEP 6 % $CO_2$ BY VOLUME (DRY)

$$= \frac{27}{27 + .04 + 227.81 + 22.51} = .097$$

4 NOTE THAT THE HYDROGEN AND OXYGEN LISTED MUST BOTH BE FREE SINCE THE WATER VAPOR IS LISTED SEPARATELY.

STEP 1 THE USEABLE % OF CARBON PER POUND OF FUEL IS

$$.5145 - \frac{2816(.209)}{15395} = .4763$$

STEP 2 THE THEORETICAL OXYGEN REQUIRED PER POUND OF FUEL IS

| | | |
|---|---|---|
| $C \rightarrow CO_2$ : | $(.4763)(2.67)$ = | 1.2717 |
| $H_2 \rightarrow H_2O$ : | $(.0402)(8.0)$ = | .3216 |
| $S \rightarrow SO_2$ : | $(.0392)(1.0)$ = | .0392 |
| MINUS $O_2$ IN FUEL | | −.0728 |
| | | 1.5597 |

STEP 3 THE THEORETICAL AIR IS

$$\frac{1.5597}{.2315} = 6.737 \ \frac{\text{LBM AIR}}{\text{LBM FUEL}}$$

STEP 4 IGNORING FLY ASH, THE THEORETICAL DRY PRODUCTS ARE

| | | |
|---|---|---|
| $CO_2$ : | $(.4763)(3.67)$ | = 1.748 |
| $SO_2$ : | $(.0392)(2.0)$ | = .0784 |
| $N_2$ : | $(.0093 + (.7685)(6.737))$ | = 5.187 |
| | | 7.013 |

STEP 5 THE EXCESS AIR IS

$$13.3 - 7.013 = 6.287$$

STEP 6 THE TOTAL AIR SUPPLIED WAS

$$6.287 + 6.737 = 13.024 \ \frac{\text{LBM AIR}}{\text{LBM FUEL}}$$

5 STEP 1 THE DENSITY OF CARBON DIOXIDE AT 60°F IS

$$\rho = \frac{p}{RT} = \frac{(14.7)(144)}{(35.1)(460+60)} = .1160$$

THE CARBON DIOXIDE IS $\frac{12}{44} = .2727$ CARBON, SO THE WEIGHT OF CARBON PER 100 $FT^3$ OF STACK GASES IS $[(.2727)(.1160)(100)(.095)] = .3005$

STEP 2 THE DENSITY OF NITROGEN IS

$$\rho = \frac{(14.7)(144)}{(55.2)(460+60)} = .07375 \text{ LBM/FT}^3$$

SO, THE MASS OF NITROGEN IN 100 $FT^3$ IS

$$(.815)(100)(.07375) = 6.0106 \text{ LBM}$$

STEP 3 THE ACTUAL NITROGEN PER POUND OF COAL IS

$$\frac{(.65)(1-.03)\frac{\text{LBM CARBON}}{\text{LBM FUEL}}}{(.3005)\frac{\text{LBM CARBON}}{100 \text{ FT}^3}}(6.0106)\frac{\text{LBM N}_2}{100 \text{ FT}^3}$$

$$= 12.611 \frac{\text{LBM N}_2}{\text{LBM FUEL}}$$

STEP 4 ASSUMING THE ASH PIT LOSS COAL HAS THE SAME COMPOSITION AS THE UNBURNED COAL, THE THEORETICAL NITROGEN PER POUND OF FUEL BURNED IS

$$(.7685)(9.45)(1-.03) = 7.0444 \text{ LBM/LBM}$$

STEP 5 THE % EXCESS AIR IS

$$\frac{12.611 - 7.0444}{7.0444} = .790$$

6 FROM EQN 9.20,

$$[C] = (1-.03)(.6734) = .6532$$

$$q_4 = \frac{(10{,}150)[.6532](1.6)}{15.5 + 1.6} = 620 \frac{\text{BTU}}{\text{LBM}}$$

7 FROM TABLE 6.5, DENSITY IS VOLUMETRICALLY WEIGHTED.

METHANE

$B = .93$

$$\rho = \frac{p}{RT} = \frac{(14.7)(144)}{(96.4)(460+60)} = .0422$$

$$\frac{\text{FT}^3 \text{ AIR}}{\text{FT}^3 \text{ FUEL}} = 9.52 \text{ \{TABLE 9.11\}}$$

PRODUCTS: 1 $FT^3$ $CO_2$, 2 $FT^3$ $H_2O$

HHV = 1013 {TABLE 23.15}

THE RESULTS FOR ALL THE FUEL COMPONENTS IS TABULATED IN THE FOLLOWING TABLE

| | | | | | VOLUMES OF PRODUCTS | | |
|---|---|---|---|---|---|---|---|
| GAS | B | $\rho$ | $FT^3$ AIR | HHV | $CO_2$ | $H_2O$ | OTHER |
| $CH_4$ | .93 | .0422 | 9.52 | 1013. | 1 | 2 | |
| $N_2$ | .034 | .0737 | — | — | — | — | 1 $N_2$ |
| CO | .0045 | .0737 | 2.38 | 322 | 1 | — | — |
| $H_2$ | .0182 | .0053 | 2.39 | 325 | — | 1 | — |
| $C_2H_4$ | .0025 | .0739 | 14.29 | 1614 | 2 | 2 | — |
| $H_2S$ | .0018 | .0900 | 7.15 | 647 | — | 1 | 1 $SO_2$ |
| $O_2$ | .0035 | .0843 | — | — | — | — | — |
| $CO_2$ | .0022 | .1160 | — | — | 1 | — | — |

a) THE COMPOSITE DENSITY IS

$$\sum B\rho = .0431$$

b) THE THEORETICAL AIR REQUIREMENTS ARE

$$\sum B_i(\text{FT}^3 \text{ AIR}) - \frac{\text{OXYGEN IN FUEL}}{.209}$$

$$= 8.9564 - \frac{.0035}{.209} = 8.94 \frac{\text{FT}^3 \text{ AIR}}{\text{FT}^3 \text{ FUEL}}$$

c) THE AIR IS 20.9% OXYGEN BY VOLUME, SO THE THEORETICAL OXYGEN WILL BE

$$(8.94)(.209) = 1.868 \text{ FT}^3/\text{FT}^3$$

THE EXCESS OXYGEN WILL BE

$$(.4)(1.868) = .747 \text{ FT}^3/\text{FT}^3$$

SIMILARLY, THE TOTAL NITROGEN IN THE STACK GASES IS

$$(1.4)(.791)(8.94) + .034 = 9.934$$

THE STACK GASES PER $FT^3$ OF FUEL ARE

EXCESS $O_2$: = .747

NITROGEN: = 9.934

$SO_2$: = .0018

$CO_2$: $(.93)(1)+(.0045)(1)+(.0025)(2)+(.0022)(1) = .9417$

$H_2O$: $(.93)(2)+(.0182)(1)+(.0025)(2)+(.0018)(1) = 1.885$

THE TOTAL WET VOLUME IS 13.51

THE TOTAL DRY VOLUME IS 11.62

THE VOLUMETRIC ANALYSES ARE

| | $O_2$ | $N_2$ | $SO_2$ | $CO_2$ | $H_2O$ |
|---|---|---|---|---|---|
| WET | .055 | .735 | — | .070 | .140 |
| DRY | .064 | .855 | — | .081 | — |

8 REFER TO THE PROCEDURE STARTING ON PAGE 9-15 {WHICH IS LIBERALLY INTERPRETED}

STEP 1 HEAT ABSORBED IN THE BOILER IS

$$(11.12) \text{ LBM WATER } (970.3) \frac{\text{BTU}}{\text{LBM}} = 10789.7 \frac{\text{BTU}}{\text{LBM}}$$

STEP 2 THE LOSSES ARE:

a) HEATING STACK GASES: THE BURNED CARBON PER POUND OF FUEL IS

$$.7842 - (.315)(.0703) = .7621 \text{ LBM/LBM}$$

FROM EQUATION 9.16, THE WEIGHT OF DRY FUEL GAS IS

$$\frac{\{11(14.0) + 8(5.5) + 7(.42 + 80.08)\}\{.7621 + \frac{.01}{1.833}\}}{3(14 + .42)}$$

$$= 13.51 \frac{\text{LBM STACK GASES}}{\text{LBM FUEL}}$$

ASSUME $\bar{c}_p = .245$ BTU/LBM-°F

ASSUME THE AIR IS 73°F INITIALLY.

THEN

$q_1 = (13.51)(.245)(575-73) = 16616$ BTU/LBM

b) THE COAL ANALYSIS GIVEN ADDS TO 100% WITHOUT THE 1.91% MOISTURE, WE MUST ASSUME THAT THE $H_2$ AND $O_2$ ARE FREE GASES. WITH LITTLE ERROR, THE AMOUNT OF MOISTURE CAN BE TAKEN AS

$.0191 \ \frac{\text{LBM}}{\text{LBM COAL}}$

b1) TO HEAT, EVAPORATE, AND SUPERHEAT THE WATER FORMED FROM COMBUSTION:

FROM FIGURE 8.13, ASSUME $\bar{C}_p = .46$ FOR SUPERHEATED STEAM

$$(.0556)\text{ LBM HYDROGEN}\left(9\right)\frac{\text{LBM WATER}}{\text{LBM HYDROGEN}} \times$$

$$\times\left((1)(212-73) + (970.3) + .46(575-212)\right)$$

$= 638.7$ BTU/LBM

b2) TO EVAPORATE THE MOISTURE IN THE COAL

$$(.0191)\left((1)(212-73) + (970.3) + .46(575-212)\right)$$

$= 24.4$ BTU/LBM

c) FROM THE PSYCHROMETRIC CHART,

$\omega = 90 \ \frac{\text{GRAINS}}{\text{LBM AIR}}$

$$\frac{90}{7000} = .0129 \ \frac{\text{LBM WATER}}{\text{LBM AIR}}$$

SINCE THE CARBON BURNED PER POUND OF FUEL IS

$.7842 - (.315)(.0703) = .7621$

EQN 9.13 {MODIFIED FOR SULFUR CONTENT} IS

$$\frac{\text{LBM AIR}}{\text{LBM FUEL}} = \frac{3.04(80.08)\left[.7621 + \frac{.01}{1.833}\right]}{(14 + .42)}$$

$= 12.96$

THE ENERGY TO SUPERHEAT THE MOISTURE IN THE AIR IS APPROXIMATELY

$$(.0129)\frac{\text{LBM WATER}}{\text{LBM AIR}}(12.96)\frac{\text{LBM}}{\text{AIR}}(.46)\frac{\text{BTU}}{\text{LBM-°F}}$$

$$\times (575-73)\text{°F}$$

$= 38.6$ BTU/LBM

d) IN INCOMPLETE COMBUSTION OF CARBON {EQN 9.20}

$$q_4 = \frac{10,143[.7621](.42)}{14 + .42} = 225.1$$

e) IN UNBURNED CARBON {SEE EQN 9.21}

$$q_5 = \frac{(14,093)(.0703)(31.5}{100} = 312.1$$

f) RADIATION AND UNACCOUNTED FOR:

$14000 - 10789.7 - 1661.6 - 638.7 - 24.4 -$

$- 38.6 - 225.1 - 312.1$

$= 309.8$

9
===

(a) THE HARDNESS IS CALCULATED FROM FACTORS IN APPENDIX D.

| ION | AMOUNT AS SUBSTANCE | | FACTOR | | AMOUNT AS $CaCO_3$ |
|---|---|---|---|---|---|
| $Ca^{++}$ | 80.2 | x | 2.5 | = | 200.5 |
| $Mg^{++}$ | 24.3 | x | 4.1 | = | 99.63 |
| $Fe^{++}$ | 1.0 | x | 1.79 | = | 1.79 |
| $Al^{+++}$ | 0.5 | x | 5.56 | = | 2.78 |

HARDNESS = **304.7 mg/ℓ**

(b) TO REMOVE THE CARBONATE HARDNESS FIRST REMOVE THE $CO_2$. AGAIN USING APPENDIX D,

$CO_2$ 19 x 2.27 = 43.13 mg/ℓ as $CaCO_3$

THEN, ADD LIME TO REMOVE THE CARBONATE HARDNESS. WE DON'T CARE IF THE $HCO_3^-$ COMES FROM $Ca^{++}$, $Mg^{++}$, $Fe^{++}$, OR $Al^{+++}$. ADDING LIME WILL REMOVE IT.

THERE MIGHT BE EXTRA $Mg^{++}$, $Ca^{++}$, $Fe^{++}$, OR $Al^{+++}$ IONS LEFT OVER IN THE FORM OF NON-CARBONATE HARDNESS, BUT THE CARBONATE HARDNESS IS

$HCO_3^-$ 185 x 0.82 = 151.7 mg/ℓ as $CaCO_3$

THE TOTAL EQUIVALENTS TO BE NEUTRALIZED ARE

43.13 + 151.7 = 194.83 mg/ℓ

CONVERT THESE $CaCO_3$ EQUIVALENTS BACK TO $Ca(OH)_2$ USING APPENDIX D:

$$\frac{194.83}{1.35} = \boxed{1443 \text{ mg/ℓ of } Ca(OH)_2}$$

NO SODA ASH IS REQUIRED, SINCE IT IS USED TO REMOVE NON-CARBONATE HARDNESS

10 $Ca(HCO_3)_2$ AND $MgSO_4$ BOTH CONTRIBUTE TO HARDNESS. SINCE 100 mg/l OF HARDNESS IS THE GOAL, LEAVE ALL $MgSO_4$ IN THE WATER. SO, NO SODA ASH IS REQUIRED. PLAN ON TAKING OUT $(137+72-100) = 109$ mg/l OF $Ca(HCO_3)_2$.

INCLUDING THE EXCESS, THE LIME REQUIREMENT IS

$$\text{pure } Ca(OH)_2 = 30 + \frac{109}{1.35} = 110.74 \text{ mg/l}$$

$$(110.74)(8.345) = 924 \text{ LB/million gallons}$$

(b) HARDNESS REMOVED = $(137+72-100) = 109$ mg/l

THE CONVERSION FROM mg/l TO LB/MG IS 8.345.

$$(109) \text{ mg/l } (8.345) = 909.6 \ \frac{\text{LB HARDNESS}}{\text{MILLION GALLONS}}$$

$$\left(\frac{.5}{1000}\right)\frac{\text{LB}}{\text{GRAIN}}(909.6)\frac{\text{LB}}{\text{MIL GAL}}(7000)\frac{\text{GRAIN}}{\text{LB}} =$$

$$= 3.18 \text{ EE} \ \text{LB/million gallons}$$

THE REMAINING COMBUSTION POWER IS

$$2.09 \text{ EE}8 - 1.73 \text{ EE}6 = 2.07 \text{ EE}8 \text{ BTUH}$$

LOSSES IN THE STEAM GENERATOR AND ELECTRICAL GENERATOR REDUCE THIS FURTHER.

$$(.86)(.95)(2.07 \text{ EE}8) = 1.7 \text{ EE}8 \text{ BTUH}$$

WITH AN ELECTRICAL OUTPUT OF 17,000 KW, THE REMAINING THERMAL ENERGY WOULD BE REMOVED BY THE COOLING WATER.

$$\Delta T = \frac{Q}{\dot{m}C_p} = \frac{1.7\text{EE}8 - (17{,}000)(3413)\frac{\text{BTUH}}{\text{KW}}}{[(225)(62.4)](3600)(1.0)}$$

$$= 2.21^\circ\text{F}$$

d) THE ALLOWABLE EMISSION RATE IS

$$\frac{(1)\frac{\text{LBM}}{\text{MBTU}}(15300)\frac{\text{LBM}}{\text{HR}}(13653)\frac{\text{BTU}}{\text{LBM}}}{1{,}000{,}000\frac{\text{BTU}}{\text{MBTU}}} = 20.89 \ \frac{\text{LBM}}{\text{HR}}$$

$$\eta = \frac{1709.7 - 20.89}{1709.7} = .988$$

TIMED

1a) SILICON IN ASH IS $SiO_2$, WITH A MOLECULAR WEIGHT OF

$$28.09 + 2(16) = 60.09.$$

THE OXYGEN TIED UP IS

$$\frac{2(16)}{28.09}(.061) = .0695$$

THE SILICON ASH PER HOUR IS

$$(15300)(.061 + .0695) = 1996.7 \text{ LBM/HR}$$

THE REFUSE SILICON IS

$$410(1-.30) = 287 \text{ LBM/HR}$$

THE EMISSION RATE IS

$$1996.7 - 287 = 1709.7 \text{ LBM/HR}$$

b) FROM TABLE 9.11 EACH POUND OF SULFUR PRODUCES 2 LBM $SO_2$

$$(15300)(.0244)(2) = 746.6 \text{ LBM/HR}$$

(c) FROM EQUATION 9.12, THE HEATING VALUE OF THE FUEL IS

$$14{,}093(.7656) + 60{,}958\left(.055 - \frac{.077}{8}\right) + 3983(.0244)$$

$$= 13{,}653 \text{ BTU/LBM}$$

THE GROSS AVAILABLE COMBUSTION POWER IS

$$(15{,}300)\frac{\text{LBM}}{\text{HR}}(13{,}653)\frac{\text{BTU}}{\text{LBM}} = 2.09 \text{ EE}8 \text{ BTUH}$$

THE CARBON CONTENT OF THE ASH REDUCES THIS SLIGHTLY.

$$(410)(0.30)(14{,}093) = 1.73 \text{ EE}6 \text{ BTUH}$$

2 FOR PROPANE

$$C_3H_8 + 5O_2 \longrightarrow 3CO_2 + 4H_2O$$

$$\text{MW: } 44.09 + 160 \longrightarrow 132.03 + 72.06$$

OXYGEN IS

$$(160)(1.4) = 224$$

EXCESS: $224 - 160 = 64$

THE WEIGHT RATIO OF NITROGEN TO OXYGEN IS

$$\frac{G_N}{G_O} = \frac{B_N R_O}{B_O R_N} = \frac{(.40)(48.3)}{(.60)(55.2)} = .583$$

ACCOMPANYING NITROGEN

$$(224)(.583) = 130.6$$

SO, THE MASS BALANCE PER MOLE OF PROPANE IS

$$C_3H_8 + O_2 + N_2 \longrightarrow CO_2 + H_2O + O_2 + N_2$$

$$44.09 + 224 + 130.6 \longrightarrow 132.03 + 72.06 + 64 + 130.6$$

AT STANDARD INDUSTRIAL CONDITIONS {60°F, 1 ATMOS} THE PROPANE DENSITY IS

$$\rho = \frac{P}{RT} = \frac{(14.7)(144)}{(35.0)(460+60)} = .1163 \ \frac{\text{LBM}}{\text{FT}^3}$$

SINCE 250 SCFM FLOW, THE MASS FLOW RATE OF THE PROPANE IS

$$(250)(.1163) = 29.075 \ \frac{\text{LBM}}{\text{MIN}}$$

SCALING THE OTHER MASS BALANCE FACTORS DOWNWARD,

$$C_3H_8 + O_2 + N_2 \longrightarrow CO_2 + H_2O + O_2 + N_2$$

$$29.08 + 147.72 + 86.1 \longrightarrow 87.07 + 47.52 + 42.20 + 86.1$$

THE OXYGEN FLOW IS

$$\boxed{147.72 \ \frac{\text{LBM}}{\text{MIN}}}$$

THE SPECIFIC VOLUMES OF THE REACTANTS ARE

$$v_{C_3H_8} = \frac{RT}{P} = \frac{(35.0)(460+80)}{(14.7)(144)} = 8.929 \ \frac{\text{FT}^3}{\text{LBM}}$$

$$v_{O_2} = \frac{(48.3)(540)}{(14.7)(144)} = 12.321$$

$$v_{N_2} = \frac{(55.2)(540)}{(14.7)(144)} = 14.082$$

THE TOTAL INCOMING VOLUME IS

$$(29.08)(8.929) + (147.72)(12.321) + (86.1)(14.082) = 3292 \text{ CFM}$$

SINCE VELOCITY MUST BE KEPT BELOW 400 FPM,

$$A = \frac{Q}{V} = \frac{3292}{400} = \boxed{8.23 \text{ FT}^2}$$

SIMILARLY, THE SPECIFIC VOLUMES OF THE PRODUCTS ARE

$$v_{CO_2} = \frac{(35.1)(460+460)}{(8)(144)} = 28.03 \ \frac{\text{FT}^3}{\text{LBM}}$$

$$v_{H_2O} = \frac{(85.8)(920)}{1152} = 68.52$$

$$v_{O_2} = \frac{(48.3)(920)}{1152} = 38.57$$

$$v_{N_2} = \frac{(55.2)(920)}{1152} = 44.08$$

THE TOTAL EXHAUST VOLUME IS

$$(87.07)(28.03) + (47.52)(68.52) + (42.20)(38.57) + (86.1)(44.08) = \boxed{11{,}120 \text{ CFM}}$$

$$A_{\text{stack}} = \frac{11{,}120}{800} = \boxed{13.9 \text{ FT}^2}$$

FOR IDEAL GASES, THE PARTIAL PRESSURE IS VOLUMETRICALLY WEIGHTED.

THE WATER VAPOR PARTIAL PRESSURE IS

$$8\left[\frac{(47.52)(68.52)}{11{,}120}\right] = 2.34 \text{ PSIA}$$

THE SATURATION TEMPERATURE CORRESPONDING TO 2.34 PSIA IS

$$\boxed{132°\text{F} = T_{\text{dewpoint}}}$$

**3**

(a) ASSUME THE NITROGEN AND OXYGEN CAN BE VARIED INDEPENDENTLY.

$$C_3H_8 + 5O_2 \longrightarrow 3CO_2 + 4H_2O$$

| | $C_3H_8$ | $5O_2$ | $3CO_2$ | $4H_2O$ |
|---|---|---|---|---|
| WTS: | 44.09 | 160 | 132.03 | 72.06 |
| MOLES: | (1) | (5) | (3) | (4) |

ASSUME 0% EXCESS OXYGEN. (JUSTIFIED SINCE ENTHALPY INCREASE INFORMATION IS NOT GIVEN FOR OXYGEN.)

USE THE ENTHALPY OF FORMATION DATA TO CALCULATE THE HEAT OF REACTION (MAXIMUM FLAME TEMPERATURE).

OXYGEN AND NITROGEN HAVE $\Delta H_f = 0$. SO,

$$3(-169{,}300) + 4(-104{,}040) - (28{,}800) = -952{,}860 \ \frac{\text{BTU}}{\text{MOLE}}$$

THE NEGATIVE SIGN SIGN INDICATES EXOTHERMIC REACTION.

LET X BE THE # OF MOLES OF NITROGEN PER MOLE OF PROPANE. USE THE NITROGEN TO COOL THE COMBUSTION.

$$952{,}860 = 3(39{,}791) + 4(31{,}658) + X(24{,}471)$$

$$X = 28.89 \text{ moles}$$

$$M_{N_2} = (28.89)(28.016) = \boxed{809.4 \ \frac{\text{LBM}}{\text{MOLE}}}$$

$$M_{O_2} = \boxed{160 \ \frac{\text{LBM}}{\text{MOLE}}}$$

EACH MOLE OF PROPANE PRODUCES 4 MOLES OF $H_2O$

$$M_{H_2O} = (4)(18) = 72 \ \text{LBM}$$

ALL FLOWS ARE PER MOLE PROPANE

(b) PARTIAL PRESSURE IS VOLUMETRICALLY WEIGHTED, THIS IS THE SAME MOLE WEIGHTING.

| PRODUCT | MOLES | VOLUMETRIC FRACTION |
|---|---|---|
| $CO_2$ | 3 | 0.0836 |
| $H_2O$ | 4 | 0.1115 |
| $N_2$ | 28.89 | 0.8049 |
| $O_2$ | 0 | 0 |
| | 35.89 | 1.000 |

THE PARTIAL PRESSURE OF THE WATER VAPOR IS

$$P_{H_2O} = 0.1115 \times 14.7 = 1.64 \text{ PSIA}$$

THIS CORRESPONDS (APPROXIMATELY) TO 118°F. SINCE $T_{\text{STACK}} = 100$°F, THE MAXIMUM VAPOR PRESSURE IS 0.9492 PSIA.

LET n BE THE NUMBER OF MOLES OF WATER IN THE STACK GAS.

$$\frac{0.9492}{14.7} = \frac{n}{35.89}$$

$$n = 2.317 \text{ moles}$$

THE LIQUID WATER REMOVED IS

$$4 - 2.317 = \boxed{1.683 \text{ MOLES OF } H_2O \text{ PER MOLE } C_3H_8}$$

EXPRESSING THIS RESULT IN POUNDS,

$$(1.683)\left(18 \ \frac{\text{LBM}}{\text{MOLE}}\right) = \boxed{30.29 \ \frac{\text{LBM } H_2O}{\text{MOLE } C_3H_8}}$$

# Heat Transfer

## WARM-UPS

**1** USE THE VALUE OF K AT THE AVERAGE TEMPERATURE $\frac{1}{2}(150+350) = 250.$

$$K = .030(1 + .0015(250)) = .04125$$

**2** USE EQUATION 10.3:

$$q = \frac{KA\Delta T}{L}$$

$$= \frac{(.038)\ \frac{\text{BTU-FT}}{\text{FT}^2\text{-HR-}^\circ\text{R}}\ (350^\circ)\text{F}}{(1)\ \text{FT}} = 13.3\ \frac{\text{BTU}}{\text{HR-FT}^2}$$

**3** $\Delta T_A = 190^\circ\text{F} - 85^\circ\text{F} = 105^\circ\text{F}$

$\Delta T_B = 160 - 85 = 75^\circ\text{F}$

FROM EQUATION 10.66,

$$\Delta T_M = \frac{105 - 75}{\ln\left(\frac{105}{75}\right)} = 89.16^\circ\text{F}$$

**4** a) h SHOULD BE EVALUATED AT $\frac{1}{2}(T_{PIPE} + T_{FLUID})$. THE MIDPOINT PIPE TEMPERATURE IS $\frac{1}{2}(160 + 190) = 175^\circ\text{F}$. AT THE START OF THE HEATING PROCESS,

$$T_{FILM} = \frac{1}{2}(85 + 175) = 130^\circ\text{F}$$

b) THE INITIAL FILM COEFFICIENT SHOULD BE EVALUATED AT

$$T = \frac{1}{2}(190^\circ\text{F} + 160^\circ\text{F}) = \boxed{175^\circ\text{F}}$$

**5** FROM PAGE 10-14, ASSUME COUNTER-FLOW OPERATION TO FIND $\Delta T_M$.

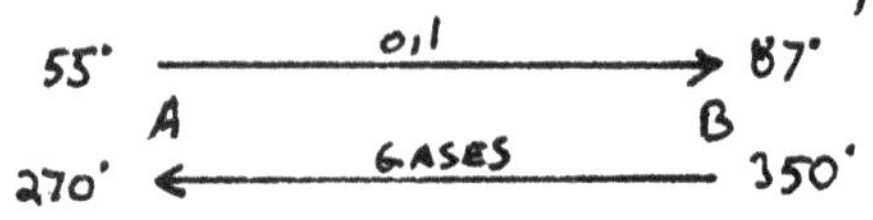


$\Delta T_A = 270 - 55 = 215$

$\Delta T_B = 350 - 87 = 263$

FROM EQUATION 10.66,

$$\Delta T_M = \frac{215 - 263}{\ln\left(\frac{215}{263}\right)} = 238.19^\circ\text{F}$$

**6** IF THE STEAM IS 87% WET, THE QUALITY IS $x = .13$. FROM EQUATION 6.34,

$$v = .01727 + .13(8515 - .01727)$$

$$= 1.122\ \text{FT}^3/\text{LBM}$$

$$\rho = 1/v = \frac{1}{1.122} = .891\ \frac{\text{LBM}}{\text{FT}^3}$$

**7** FROM PAGE 10-23,

$$\mu = (.458\ \text{EE-3})\ \frac{\text{LBM}}{\text{FT-SEC}}\ (3600)\ \frac{\text{SEC}}{\text{HR}}$$

$$= 1.6488\ \frac{\text{LBM}}{\text{FT-HR}}$$

**8**

ASSUMING THAT THE WALLS ARE RERADIATING, NON-CONDUCTING, AND VARY IN TEMPERATURE FROM 2200°F AT THE INSIDE TO 70°F AT THE OUTSIDE, FIGURE 10.15, CURVE 6 CAN BE USED TO FIND $F_{1\text{-}2}$ USING $x = 3''/6'' = 0.5$, $F_{1\text{-}2} = 0.38$

$$q = AE$$

$$= \left[\frac{3\ \text{IN}}{12\ \frac{\text{IN}}{\text{FT}}}\right]^2 (.1713\ \text{EE-8})\frac{\text{BTU}}{\text{HR-FT}^2\text{-}^\circ\text{R}}\ (0.38)$$

$$\times \left[(2200 + 460)^4 - (70 + 460)^4\right] {}^\circ\text{R}^4$$

$$= 2033.6\ \text{BTU/HR}$$

**9**

$$U = \frac{1}{\sum\left(\frac{L_i}{K_i}\right) + \sum\frac{1}{h_j}}$$

$$L = \frac{4\ \text{IN}}{12\ \text{IN/FT}} = .333\ \text{FT}$$

$K = 13.9\ \text{BTU-FT/FT}^2\text{-HR-}^\circ\text{F}$

$h_{INSIDE} \approx 1.65$ FROM TABLE 10.2

$h_{OUTSIDE}$ IS MORE DIFFICULT. ASSUMING LINEARITY BETWEEN 0 AND 15 MPH,

$$h_{OUTSIDE} \approx 1.65 + \frac{10}{15}(6.00 - 1.65) = 4.55$$

$$U = \frac{1}{\left(\frac{.333}{13.9}\right) + \frac{1}{1.65} + \frac{1}{4.55}} = 1.18\ \frac{\text{BTU}}{\text{FT}^2\text{-HR-}^\circ\text{F}}$$

10 FROM EQUATION 3.62

$$N_{Re} = \frac{Dv}{\nu}$$

$$D = \frac{.6 \text{ IN}}{12 \frac{\text{IN}}{\text{FT}}} = .05 \text{ FT}$$

$v = 2$ FT/SEC

ASSUMING SAE 10W AT 100°F

$\nu = 45$ CENTISTOKES {PAGE 4-27, 15TH EDITION, CAMERON HYDRAULIC DATA}

FROM TABLE 3.2,

$$(45)\text{ CENTISTOKES}\left(\frac{1}{100}\right)\frac{\text{STOKES}}{\text{CENTISTOKE}}\left(\frac{1}{929}\right)\frac{\text{FT}^2}{\text{SEC-STOKE}}$$

$$= 4.84 \text{ EE-4 } \text{FT}^2/\text{SEC}$$

$$N_{Re} = \frac{(.05)\text{ FT}(2)\text{ FT/SEC}}{(4.84 \text{ EE-4})\text{ FT}^2/\text{SEC}} = 206.6$$

## CONCENTRATES

1 SINCE THE WALL TEMPERATURES ARE GIVEN {NOT ENVIRONMENT TEMPERATURES} IT IS NOT NECESSARY TO CONSIDER FILMS. USING EQUATION 10.6 ON A 1 SQ. FT BASIS:

$$q = \frac{(1)\text{ FT}^2(1000-200)\text{°F}}{\frac{\left(\frac{3}{12}\right)}{.06} + \frac{\left(\frac{5}{12}\right)}{.5} + \frac{\left(\frac{6}{12}\right)}{.8}} = 142.2 \text{ BTU/HR}$$

NOW, SOLVING FOR $\Delta T$ FROM EQN. 10.6 UP TO THE FIRST INTERFACE

$$\Delta T = \frac{(142.2)\frac{\text{BTU}}{\text{HR}}\left(\frac{(3/12)\text{ FT}}{.06 \text{ BTU-FT/FT}^2\text{-HR-°F}}\right)}{(1)\text{ FT}^2} = 592.5\text{°F}$$

THE TEMPERATURE AT THE FIRST INTERFACE IS

$$T = 1000 - 592.5 = 407.5\text{°F}$$

SIMILARLY FOR THE OTHER INTERFACE,

$$\Delta T = \frac{(142.)\frac{(6/12)}{.8}}{(1)} = 88.9$$

$$T = 200 + 88.9 = 288.9\text{°F}$$

(2) REFERRING TO FIG. 10.2,

$$r_a = \frac{d}{2} = \frac{3.5 \text{ IN}}{(2)\left(12 \frac{\text{IN}}{\text{FT}}\right)} = 0.1458 \text{ FT}$$

$$r_b = \frac{4.0}{(2)(12)} = 0.1667 \text{ FT}$$

$$r_c = r_b + t_{\text{insulation}} = 0.1667 + \frac{2}{12} = 0.3334 \text{ FT}$$

INTERPOLATING FROM PAGE 10-21, FOR STEEL PIPE, $k \approx 25.4$.

INITIALLY ASSUME $h_o = 1.5$ BTU/HR-FT²-°F

EITHER THE SEIDER-TATE OR GRAETZ CORRELATION CAN BE USED TO FIND $h_i$. FOR FULLY-DEVELOPED LAMINAR FLOW (I.E., A LONG DISTANCE DOWN A LONG PIPE), THE NUSSELT NUMBER IS

$$N_{Nu} = \frac{h_i D}{k} \approx 3.66$$

AT 350°F, $k_{AIR} \approx 0.0203$ BTU/HR-FT²-°F

$$h_i = \frac{k N_{Nu}}{D} = \frac{(0.0203)(3.66)}{(2)(0.1458)} = 0.255$$

NEGLECT THERMAL RESISTANCE BETWEEN PIPE AND INSULATION, FROM EQN. 10.11,

$$q = \frac{(2\pi)(100 \text{ FT})(350\text{°F} - 50\text{°F})}{\frac{1}{(0.1458)(.255)} + \frac{\ln\left(\frac{.1667}{.1458}\right)}{25.4} + \frac{\ln\left(\frac{.3334}{.1667}\right)}{0.05} + \frac{1}{(.3334)(1.5)}}$$

$$= \frac{188{,}496}{26.90 + 0.00527 + 13.863 + 2.00}$$

$$= \boxed{4407 \text{ BTU/HR}}$$

GREATER ACCURACY REQUIRES FINDING $h_o$. SOLVE EQN 10.11 FOR $\Delta T$ UP TO THE OUTER FILM TO FIND THE INSULATION SURFACE TEMPERATURE,

$$\Delta T = \frac{4407 \frac{\text{BTU}}{\text{HR}}}{(2\pi \text{ FT})(100 \text{ FT})}\left[26.90 + 0.00527 + 13.863\right]\frac{\text{HR}^2\text{-FT-°F}}{\text{BTU}}$$

$$= 286\text{°F}$$

$$T_{SURFACE} \approx 350\text{°F} - 286\text{°F} = 64\text{°F}$$

THE OUTER FILM SHOULD BE EVALUATED AT

$$T_{FILM} = \tfrac{1}{2}(T_s + T_\infty) = \tfrac{1}{2}(64\text{°F} + 50\text{°F}) \approx 57\text{°F}$$

FOR 57°F AIR,

$$N_{Pr} = 0.72$$

$$\frac{g\beta\rho^2}{\mu^2} \approx 2.64 \times 10^6$$

FROM EQN 10.57 WHERE $L = 2r_c$ IS THE OUTSIDE DIAMETER IN FEET,

$$N_{Gr} = L^3 \Delta T\left(\frac{g\beta\rho^2}{\mu^2}\right)$$

$$= (0.6667)^3(64-50)(2.64 \times 10^6)$$

$$= 1.1 \times 10^7$$

THEN, $N_{Pr}N_{Gr} = (0.72)(1.1 \times 10^7)$

$= 7.9 \times 10^6$

FROM TABLE 10.7 IN THIS RANGE,

$$h_o \approx 0.27\left(\frac{\Delta T}{L}\right)^{0.25}$$

$$= 0.27\left(\frac{64-50}{0.6667}\right)^{0.25} = 0.578$$

AT THE SECOND ITERATION, THE HEAT FLOW IS

$$q = \frac{188{,}496}{26.90 + 0.00527 + 13.863 + \frac{1}{(0.3334)(0.578)}}$$

$= \boxed{4102 \text{ BTU/HR}}$

ADDITIONAL ITERATIONS WILL IMPROVE THE ACCURACY FURTHER.

**3** SINCE NO INFORMATION WAS GIVEN ABOUT THE PIPE TYPE, MATERIAL, OR THICKNESS, IT CAN BE ASSUMED THAT THE PIPE RESISTANCE IS NEGLIGIBLE,

AND SO $T_{PIPE} = T_{SAT}$.

FOR 300 PSIA STEAM, $T_{SAT} = 417.33°F$

WHEN A VAPOR CONDENSES, THE VAPOR AND CONDENSED LIQUID ARE AT THE SAME TEMPERATURE. THEREFORE, THE ENTIRE PIPE IS ASSUMED TO BE AT 417.33°F.

THE OUTSIDE FILM SHOULD BE EVALUATED AT $\frac{1}{2}(417.33 + 70) = 243.7°F$

INTERPOLATING FROM PAGE 10-24,

$N_{Pr} = .715$

$$\frac{g\beta\rho^2}{\mu^2} = .673 \text{ EE}6$$

$$N_{Gr} = \frac{L^3 g\beta\rho^2 \Delta T}{\mu^2} = \left(\frac{4}{12}\right)^3(.673\text{EE}6)(417.33-70)$$

$= 8.66 \text{ EE}6$

SO $N_{Pr}N_{Gr} = (.715)(8.66 \text{ EE}6) = 6.2 \text{ EE}6$

FROM TABLE 10.7,

$$h_o \approx .27\left(\frac{\Delta T}{L}\right)^{.25} = .27\left(\frac{417.33-70}{\frac{4}{12}}\right)^{.25} = 1.53$$

FROM EQN. 10.11,

$$q = \frac{2\pi(50)(417.33-70)}{\frac{1}{\left(\frac{2}{12}\right)(1.53)}} = 2.782 \text{ EE}4 \text{ BTU/HR}$$

TO DETERMINE THE RADIATION LOSS, ASSUME OXIDIZED STEEL PIPE, SURROUNDED COMPLETELY.

$F_a = 1$

$T_\infty = 70°F = 530°R$

$\epsilon_{steel} = 0.80$

FROM EQN 10.85 WITH $F_e = \epsilon_{steel}$ AND $F_a = 1$

$$q = \text{EA}$$

$$= (.1713\text{EE-}8)(.80)(1)\left[(\pi)\left(\tfrac{4}{12}\right)(50)\right] \times \left[(417.33+460)^4 - (530)^4\right]$$

$= 36{,}850$ BTU/HR

THE TOTAL HEAT LOSS IS

$27{,}820 + 36{,}850 = 64{,}670$ BTU/HR

THE ENTHALPY DECREASE PER POUND IS

$$\frac{64{,}670 \text{ BTU/HR}}{5000 \text{ LBM/HR}} = 12.93 \text{ BTU/LBM}$$

THIS IS A QUALITY LOSS OF

$$\Delta x = \frac{\Delta h}{h_{fg}} = \frac{12.93}{809} = 0.016$$

OR 1.6%

**4** SOLUTION FOR 3RD AND PRIOR PRINTINGS

THE HEAT LOSS IS

$$\left(2 \frac{W}{FT}\right)\left(3.413 \frac{BTU}{HR\text{-}W}\right) = 6.83 \text{ BTU/HR-FT}$$

FROM PAGE 10-24 FOR 100°F FILM,

$N_{Pr} = .72$

$$\frac{g\beta\rho^2}{\mu^2} = 1.76 \text{ EE}6$$

FROM EQN 10.57,

$$N_{Gr} = L^3 \Delta T\left(\frac{g\beta\rho^2}{\mu^2}\right)$$

$\Delta T = (T_{WIRE} - 60°F)$

BUT $T_{WIRE}$ IS UNKNOWN, SO ASSUME 150°F

$$N_{Gr} = \left(\frac{.6}{12}\right)^3(150-60)(1.76 \text{ EE}6) = 1.98 \text{ EE}4$$

$N_{Pr}N_{Gr} = (.72)(1.98 \text{ EE}4) = 1.42 \text{ EE}4$

FROM TABLE 10.7,

$$h_o \approx .27\left(\frac{\Delta T}{L}\right)^{.25} = .27\left(\frac{150-60}{\frac{.6}{12}}\right)^{.25} = 1.76$$

THEN FROM EQN. 10.38,

$$T_{WIRE} = \frac{q}{hA} + T_\infty =$$

$$= \frac{6.83}{(1.76)\left(\frac{.6}{12}\right)(\pi)(1)} + 60 = 84.7°F$$

THIS CANNOT BE SINCE $T_{FILM}$ CANNOT EXCEED $T_{WIRE}$. THE LOWEST $T_{WIRE}$ CAN BE IS 100°F.

CONCENTRATES #4, CONTINUED

SO, TRY $T_{wire} = 100°F$.

$$N_{gr} = \left(\frac{.6}{12}\right)^3(100-60)(1.76\ EE6) = 8.8\ EE3$$

$$N_{Pr}N_{gr} = (.72)(8.8\ EE3) = 6.34\ EE3$$

FROM TABLE 10.7,

$$h_o \approx .27\left(\frac{\Delta T}{L}\right)^{.25} = .27\left(\frac{100-60}{\frac{.6}{12}}\right)^{.25} = 1.44$$

FROM EQN. 10.38,

$$T_{wire} = \frac{6.83}{(1.44)(.6/12)(\pi)(1)} + 60 = 90.2°F$$

SINCE THE WIRE TEMPERATURE IS STILL LESS THAN THE FILM TEMPERATURE, THE PROBLEM STATEMENT MUST CONTAIN INCONSISTENT DATA. THIS CONCLUSION IS VERIFIED BY THE OBSCURE METHOD PRESENTED IN 1948 BY W. ELENBAAS IN THE JOURNAL OF APPLIED PHYSICS.

SOLUTION FOR 4TH AND SUBSEQUENT PRINTINGS

THE HEAT LOSS IS

$$\left(8\ \frac{W}{FT}\right)\left(3.413\ \frac{BTU}{HR\text{-}W}\right) = 27.3\ BTU/HR\text{-}FT$$

FROM PAGE 10-24 FOR 100°F FILM, $N_{Pr} = 0.72$.

$$\frac{g\beta\rho^2}{\mu^2} = 1.76\ EE6$$

FROM EQN. 10.57,

$$N_{Gr} = L^3\Delta T\left(\frac{g\beta\rho^2}{\mu^2}\right)$$

$$\Delta T = T_{wire} - 60°F$$

BUT $T_{wire}$ IS UNKNOWN, SO ASSUME $T_{wire} = 150°F$

$$N_{Gr} = \left(\frac{0.6}{12}\right)^3(150-60)(1.76\ EE6) = 1.98\ EE4$$

$$N_{Pr}N_{Gr} = (0.72)(1.98\ EE4) = 1.42\ EE4$$

FROM TABLE 10.7,

$$h_o \approx 0.27\left(\frac{\Delta T}{L}\right)^{0.25} = 0.27\left(\frac{150-60}{\frac{0.6}{12}}\right)^{0.25} = 1.76$$

THEN, FROM EQN 10.38,

$$T_{wire} = \frac{q}{hA} + T_\infty = \frac{27.3}{(1.76)\left(\frac{0.6}{12}\right)(\pi)(1)} + 60$$

$$= 158.7\ °F$$

TRY $T_{wire} = 158\ °F$

$$N_{Gr} = \left(\frac{0.6}{12}\right)^3(158-60)(1.76\ EE6) = 21{,}560$$

$$N_{Pr}N_{Gr} = (0.72)(21{,}560) = 15{,}523$$

FROM TABLE 10.7,

$$h_o = 0.27\left(\frac{\Delta T}{L}\right)^{0.25} = 0.27\left(\frac{158-60}{\frac{0.6}{12}}\right)^{0.25} = 1.80$$

FROM EQN 10.38,

$$T_{wire} = \frac{27.3}{(1.80)\left(\frac{0.6}{12}\right)(\pi)(1)} + 60$$

$$= \boxed{156.6\ °F\ \ (\text{SAY } 157\ °F)}$$

5 THE BULK TEMPERATURE OF THE WATER IS $\frac{1}{2}(70° + 190°) = 130°$

FROM PAGE 10-23 FOR 130° WATER,

$$C_p = .999$$
$$\nu = .582\ EE\text{-}5$$
$$N_{Pr} = 3.45$$
$$K = .376$$

THE REQUIRED HEAT IS

$$q = MC_p\Delta T = (2940)(.999)(190-70)$$

$$= 352{,}447\ BTU/HR$$

FROM EQUATION 10.40,

$$N_{Re} = \frac{VD}{\nu} = \frac{(3)\frac{FT}{SEC}\left(\frac{.9}{12}\right)FT}{.582\ EE\text{-}5} = 3.87\ EE4$$

THEN FROM EQUATION 10.39,

$$h_i = \frac{(.376)(.0225)(3.87EE4)^{.8}(3.45)^{.4}}{\left(\frac{.9}{12}\right)}$$

$$= 866$$

THE PROCEDURE ON PAGE 10-3 MUST BE USED TO FIND $h_o$. 134 PSIA STEAM HAS $T_{SAT} = 350°F$. ASSUME $T_{sv} - T_s = 20$.

SO, THE WALL IS AT $350 - 20 = 330°F$ AND THE FILM SHOULD BE EVALUATED AT $\frac{1}{2}(330 + 350) = 340°F$.

FROM PAGE 6-29, FOR 340°F STEAM,

$$\rho_\ell = 1/v_\ell = 1/.01787 = 55.96\ LBM/FT^3$$

$$\rho_v = 1/3.788 = .26\ LBM/FT^3$$

FROM PAGE 6-29 FOR 350°F STEAM,

$$h_{fg} = 870.7\ BTU/LBM$$

FROM PAGE 10-23, FOR 340°F WATER,

$$K \approx .392$$

$$\mu_\ell = (.109\ EE\text{-}3)\frac{LBM}{FT\text{-}SEC}(3600)\frac{SEC}{HR}$$

$$= .392\ \frac{LBM}{FT\text{-}HR}$$

USING 4.17 EE8 TO CONVERT ft/SEC² TO ft/HR² AND EQUATION 10.59,

$$h_o = .725\left[\frac{55.96(55.96-.26)(4.17\text{EE}8)(870.7)(.392)^3}{\left(\frac{1}{12}\right)(.392)(350-330)}\right]^{.25}$$

$= 2317$ BTU/HR-FT²-°F

USING EQN 10.71,

$$r_o = \frac{1}{(2)(12)} = .0417$$

$$r_i = \frac{.9}{(2)(12)} = .0375$$

$K_{copper} = 216$ {FROM PAGE 10-21, ASSUMING PURE COPPER PIPES}

$$U_o = \frac{1}{\frac{1}{2317} + \frac{.0417}{216}\ln\left(\frac{.0417}{.0375}\right) + \frac{.0417}{(.0375)(866)}}$$

$= 576.0$ BTU/HR-FT²-°F

FOR CROSS-FLOW OPERATION,

70°F
190°F
A
B
350°F
350°F

$\Delta T_A = 350-70 = 280$°F

$\Delta T_B = 350-190 = 160$°F

$$\Delta T_M = \frac{280-160}{\ln\left(\frac{280}{160}\right)} = 214.4\text{°F}$$

FROM EQN. 10.67,

$$A_o = \frac{q}{U_o \Delta T_M} = \frac{352{,}447}{(576.0)(214.4)}$$

$= 2.85$ FT²

6 THIS IS A TRANSIENT PROBLEM,

$C_e = 100{,}000$ BTU/°F {EQN. 10.21}

$R_e = 1/6500 = .0001538$ BTU/HR-°F

FROM EQUATION 10.24,

$$T_{8\,HRS} = 40 + (70-40)\,\text{EXP}\left[\frac{-8}{(100{,}000)(.0001538)}\right]$$

$= 57.8$°F

7 SINCE WE KNOW THE DUCT OUTSIDE TEMPERATURE, WE DON'T NEED $K_{DUCT}$ AND $h_i$. EVALUATE $h_o$ AT $\frac{1}{2}(80+200) = 140$°F.

FROM PAGE 10-24,

$N_{Pr} = .72$

$\frac{g\beta\rho^2}{\mu^2} = 1.396$ EE6

$$N_{Gr} = L^3 \Delta T \frac{g\beta\rho^2}{\mu^2} = \left(\frac{9}{12}\right)^3(200-80)(1.396\text{ EE}6)$$

$= 7.07$ EE7

SO $N_{Pr}N_{Gr} = (.72)(7.07\text{ EE}7) = 5.09\text{EE}7$

FROM TABLE 10.7,

$$h = .27\left(\frac{\Delta T}{L}\right)^{.25} = .27\left(\frac{200-80}{\frac{9}{12}}\right)^{.25} = .96$$

THE DUCT AREA PER FOOT OF LENGTH IS

$A = pL = \left(\frac{9}{12}\right)\pi(1) = 2.356$ FT²

THE CONVECTION LOSSES ARE

$q = hA\Delta T = (.96)(2.356)(200-80) = 271.4$ BTU/HR-FT

THE RADIATION LOSSES ARE GIVEN BY EQN. 10.85 ASSUME $\epsilon_{DUCT} \approx 0.97$ {FROM PAGE 10-26}. THEN $F_e = \epsilon_{DUCT} = 0.97$. $F_A = 1$ SINCE THE DUCT IS ENCLOSED.

$$q = (2.356)(.1713\text{ EE}-8)(.97)\left[(200+460)^4 - (70+460)^4\right]$$

$= 433.9$ BTU/HR-FT

$q_{total} = 271.4 + 433.9 = 705.3$ BTU/HR-FT LENGTH

8 1 FOOT OF ROD HAS THE VOLUME

$V = \left(\frac{.4}{12}\right)^2 \frac{\pi}{4}(1) = 8.727$ EE-4 FT³/FT

THE HEAT OUTPUT PER FOOT OF ROD IS

$q = (8.727\text{ EE}-4)\,\frac{\text{FT}^3}{\text{FT}}\,(4\text{ EE}7)\,\frac{\text{BTU}}{\text{HR-FT}^3}$

$= 3.491$ EE4 BTU/HR-FT

THE SURFACE TEMPERATURE OF THE CLADDING IS FOUND FROM EQN 10.38

$$T_s = \frac{q}{hA} + T_\infty$$

$A = \pi\left(\frac{.44}{12}\right)(1) = .1152$

$$T_s = \frac{3.491\text{ EE}4}{(10000)(.1152)} + 500 = 530.3$$

FOR THE CLADDING,

$$r_o = \frac{.2+.02}{12} = .01833 \text{ FT}$$

$$r_i = \frac{.2}{12} = .01667$$

$K \approx 10.9$

FROM EQN. 10.9,

$$\Delta T = \frac{q \ln\left(\frac{r_o}{r_i}\right)}{2\pi k L} = \frac{(3.491\text{ EE}4)\ln\left(\frac{0.01833}{0.01667}\right)}{(2\pi)(10.9)(1)}$$

$= 48.4\,°F$

$T_{\text{INSIDE CLADDING}} = T_{\text{OUTSIDE FUEL ROD}}$

$= 530.3 + 48.4 = 578.7\,°F$

FROM EQN 10.29,

$$T_{\text{CENTER}} = T_o + \frac{r_o^2 q^*}{4k}$$

$$= 578.7 + \frac{\left(\frac{0.2}{12}\right)^2(4\text{ EE}7)}{(4)(1.1)} = 3104\,°F$$

**9** CONSIDER THIS AN INFINITE CYLINDRICAL FIN WITH

$T_s = 450°F$ $\quad T_\infty = 80°F$ $\quad h = 3$

$$p = \frac{\pi\left(\frac{1}{16}\right)}{12} = .01636$$

$K \approx 215$

$$A = \frac{\pi}{4}\left(\frac{1}{(16)(12)}\right)^2 = 2.131\text{ EE-}5$$

THEN FROM EQN 10.33 WITH 2 FINS JOINED AT THE MIDDLE,

$$q = 2\sqrt{(3)(.01636)(215)(2.131\text{ EE-}5)}\,(450-80)$$

$= 11.1$ BTU/HR THIS IGNORES RADIATION

**10** THIS IS A SINGLE TUBE IN CROSS FLOW. REFER TO PAGE 10-10.

$$\frac{hD}{K} = C\left(N_{Re}\right)^n$$

THE FILM IS EVALUATED AT

$\frac{1}{2}(100+150) = 125$

FOR AIR AT 125°F

$K = .0159$

$\nu = .195\text{ EE-}3$

$$N_{Re} = \frac{VD}{\nu} = \frac{(100)(.35/12)}{.195\text{ EE-}3} = 14957$$

BASED ON $N_{Re}$, C AND n CAN BE FOUND

$C = .174$ AND $n = .618$

$$h = \frac{(.0159)(.174)(14957)^{.618}}{\left(\frac{.35}{12}\right)} = 36.07 \text{ BTU/HR-FT}^2\text{-}°F$$

**TIMED**

**1** ASSUME THAT THE EXPOSED WALL TEMPERATURES ARE EQUAL TO THE RESPECTIVE AMBIENT TEMPERATURES.

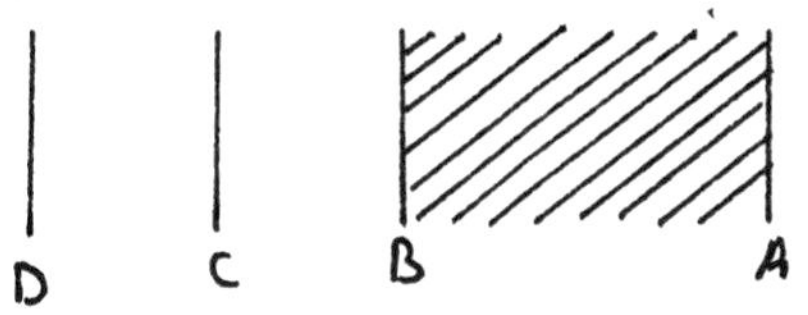


$T_A = 80°F = 540°R$

$$\frac{q_{A-B}}{A} = \frac{k\Delta T}{L} = \frac{(0.025)(540-T_B)}{\left(\frac{4\text{ in}}{12\text{ in/ft}}\right)} = 40.5 - 0.75T_B \quad (1)$$

SINCE THE AIR SPACES ARE EVACUATED, ONLY RADIATION SHOULD BE CONSIDERED FOR B-C AND C-D

$$E_{B-C} = \frac{q_{B-C}}{A} = \sigma F_e F_A\left[T_B^4 - T_C^4\right]$$

$F_A = 1$ SINCE THE FREEZER IS ASSUMED LARGE

$$F_e = \frac{(.5)(.1)}{(.1)+(1-.1)(.5)} = .0909 \quad \{\text{FROM TABLE } 10.10\}$$

$$E_{BC} = (.1713\text{ EE-}8)(.0909)(1)\left[T_B^4 - T_C^4\right]$$

$$= (1.56\text{ EE-}10)T_B^4 - (1.56\text{ EE-}10)T_C^4 \quad (2)$$

SIMILARLY

$$E_{C-D} = (.1713\text{ EE-}8)(.0909)(1)\left[T_C^4 - (460-60)^4\right]$$

$$= (1.56\text{ EE-}10)T_C^4 - 3.99 \quad (3)$$

BUT (2) = (3)

$$(1.56\text{ EE-}10)T_B^4 - (1.56\text{ EE-}10)T_C^4 = (1.56\text{ EE-}10)T_C^4 - 3.99$$

$$T_B^4 - T_C^4 = T_C^4 - (2.56\text{ EE}10)$$

$$T_B^4 = 2T_C^4 - (2.56\text{ EE}10)$$

$$T_C^4 = \tfrac{1}{2}T_B^4 + (1.28\text{ EE}10) \quad (4)$$

AND (1) = (2)

$$40.5 - .075\,T_B = (1.56\text{EE-}10)T_B^4 - (1.56\text{ EE-}10)T_C^4$$

$$T_C^4 = T_B^4 + (4.81\text{ EE}8)T_B - 2.6\text{EE}11 \quad (5)$$

BUT (4) = (5)

$$\tfrac{1}{2}T_B^4 + 1.28\text{ EE10} = T_B^4 + (4.81\text{ EE8})T_B - 2.6\text{ EE11}$$

OR $T_B^4 + (9.62\text{ EE8})T_B = 5.456\text{ EE11}$

BY TRIAL + ERROR, $T_B = 501.4°R$

THEN FROM EQN (1)

$$\frac{q}{A} = 40.5 - (.075)(501.4) = 2.895\ \text{BTU/FT}^2\text{-HR}$$

FROM EQN. (4)

$$T_C = \sqrt[4]{\left(\tfrac{1}{2}\right)(501.4)^4 + (1.28\text{ EE10})} = 459°R$$

NOW, CHECK THE INITIAL ASSUMPTION ABOUT WALL TEMPERATURE. ASSUME A FILM EXISTS OF, SAY, 2 BTU/FT²-HR-°F.

USING EQN 10.38

$$(T_S - T_\infty) = \frac{q}{Ah} = \frac{2.9}{(1)2} = 1.5°F$$

SINCE THIS IS SMALL, OUR ASSUMPTION IS VALID.

---

2 THE EXPOSED DUCT AREA IS

$$\frac{(18+18+12+12)\text{ IN}}{12\ \text{IN/FT}}(50)\text{ FT} = 250\text{ FT}^2$$

THE DUCT IS NON-CIRCULAR.
THE INSIDE FILM COEFFICIENT CANNOT BE EVALUATED WITHOUT KNOWING THE EQUIVALENT DIAMETER OF THE DUCT. FROM TABLE 3.6,

$$D_e = \frac{(2)(18)(12)}{(18+12)\ 12\ \text{IN/FT}} = 1.2\text{ FT}$$

CHECK THE REYNOLDS NUMBER. FROM PAGE 10-24 FOR 100°F AIR,

$\nu = .180\text{ EE-3}$

$$N_{Re} = \frac{VD}{\nu} = \frac{\frac{800\text{ FPM}}{60\text{ SEC/MIN}}(1.2)\text{ FT}}{.180\text{ EE-3}\ \frac{\text{FT}^2}{\text{SEC}}} = 8.9\text{ EE4}$$

SINCE THIS IS TURBULENT, EQN 10.43 MAY BE USED. AT 100°F, $\rho = .071$ LBM/FT³

$$G = V\rho = (800)(.071)/60 = .947\ \text{LBM/SEC-FT}^2$$

$$C = .00351 + .000001583(100) = .003668$$

$$h_i = \frac{(.003668)[(3600)(.947)]^{.8}}{(1.2)^{.2}} = 2.37$$

IGNORING THE DUCT THERMAL RESISTANCE, THE TOTAL HEAT TRANSFER COEFFICIENT IS

$$\frac{1}{U} = \frac{1}{h_i} + \frac{1}{h_o} = \frac{1}{2.37} + \frac{1}{2} = .922$$

$$U = 1/.922 = 1.08$$

$$q = UA(T_{ave} - T_\infty) = UA\left(\tfrac{1}{2}(T_{IN} + T_{OUT}) - T_\infty\right)$$

SINCE $T_{OUT}$ IS UNKNOWN, ASSUME $T_{OUT} \approx 95°F$

$$q \approx \left(1.08\ \frac{\text{BTU}}{\text{HR-FT}^2\text{-°F}}\right)(250\text{ FT})\left(\frac{100+95°F}{2} - 70°F\right)$$

$= 7425$ BTU/HR

NOTICE THAT $\Delta T$ (NOT $\Delta T_M$) IS USED. THIS IS IDIOSYNCRATIC TO THE HVAC INDUSTRY.

b) THE FLOW PER HOUR IS

$$3600\,GA = (3600)(.947)\left(\frac{(12)(18)}{144}\right)$$

$$= 5113.8\ \text{LBM/HR}$$

$C_p = .240$

$$\Delta T = \frac{q}{\dot m c_p} = \frac{7425}{(5113.8)(.240)} = 6.05°F$$

$T_{OUT} = 100 - 6.0 = 94°F$ (CLOSE ENOUGH)

c) USE PAGE 5-7 FOR CLEAN GALVANIZED DUCTWORK WITH $\epsilon = .0005$ AND ABOUT 40 JOINTS PER 100 FEET.

THE EQUIVALENT DIAMETER FOR THIS CHART IS GIVEN BY EQN 5.30

$$D_e = (1.3)\frac{[(12)(18)]^{.625}}{[(12)+(18)]^{.25}} = 16\text{ INCHES}$$

THEN, $\Delta p = .057\ \frac{\text{IN. WATER}}{100\text{ FEET}}$

SO, THIS DUCT SYSTEM HAS A LOSS OF

$$\left(\frac{50}{100}\right)(.057) = .029\text{" W.G.}$$

---

3 THIS IS A TRANSIENT PROBLEM (SEE PAGE 10-5). CHECK THE BIOT MODULUS TO SEE IF THE LUMPED PARAMETER METHOD CAN BE USED.

$$L = \frac{V}{A_s} = \frac{\frac{4}{3}\pi r^3}{4\pi r^2} = \frac{r}{3}$$

FOR THE LARGEST BALL,

$$L = \frac{\frac{1.5}{(2)(12)}}{3} = .0208\text{ FT}$$

EVALUATE K FOR STEEL AT

$\frac{1}{2}(1800+250) \approx 1000°F$

FROM PAGE 10-21 FOR MILD STEEL, $K \approx 22$. {ACTUALLY K FOR A HIGH ALLOY STEEL IS ABOUT 12-15}

$$N_{bi} = \frac{hL}{k} = \frac{(56)(.0208)}{22} = .053$$

FOR SMALLER BALLS, $N_{bi}$ WILL BE EVEN SMALLER.

SINCE $N_{bi} < .10$, THE LUMPED PARAMETER METHOD CAN BE USED. THE ASSUMPTIONS ARE:

- HOMOGENEOUS BODY TEMPERATURE
- MINIMAL RADIATION LOSSES
- OIL BATH REMAINS AT 110°F
- CONSTANT h {NO VAPORIZATION OF OIL}

FROM EQN. 10.21 AND 10.22,

$$R_eC_e = \frac{c_p\rho V}{hA_s} = \frac{c_p\rho r}{3h}$$

USE $\rho = 490$ LBM/FT$^3$ AND $C = .11$ EVEN THOSE ARE FOR 32°F. {ACTUALLY, $C \approx .16$ AT 900°F}

$$R_eC_e = \frac{(.11)(490)r}{(3)(56)} = .3208\, r_{FEET}$$

$$= .01337\, D_{INCHES}$$

TAKING ln OF EQN 10.24,

$$\ln(T_t - T_\infty) = \ln\left(\Delta T e^{-t/R_eC_e}\right)$$

$$\ln(T_t - T_\infty) = \ln \Delta T + \ln\left(e^{-t/R_eC_e}\right)$$

$$\ln(T_t - T_\infty) = \ln \Delta T - \frac{t}{R_eC_e}$$

BUT $T_t = 250$ AND $T_\infty = 110$

AND $\Delta T = 1800 - 110 = 1690$

$$\ln(250-110) = \ln(1690) - \frac{t}{R_eC_e}$$

$$4.942 = 7.432 - \frac{t}{R_eC_e}$$

$$t = R_eC_e(2.49)$$

$$= (.01337)D(2.49)$$

$$= .0333\, D_{INCHES}$$

b) FROM EQUATION 10.23, THE TIME CONSTANT IS

$$\frac{c_p\rho L}{h} = \frac{(.11)(490)\frac{D}{(2)(3)(12)}}{56} = \frac{D}{74.81}\ (\text{hr})$$

4

USING STEAM TABLES,

$h_1 = 167.99$ BTU/LBM

$h_2 = 364.17$

$h_3 = 1201.0$

$h_4 = 374.97$

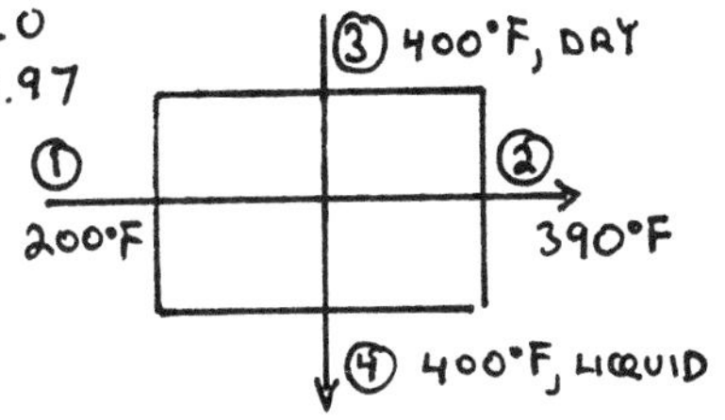


SO, THE HEAT TRANSFER IS

$$q = m\Delta h = (500{,}000)(364.17 - 167.99)$$

$$= 9.809 \text{ EE7 BTUH}$$

THE FLOW AREA PER PIPE IS

$$A = \frac{\pi}{4}D^2 = \frac{\left(\frac{\pi}{4}\right)(.875 - 2\times.0625)^2}{144} = .003068 \text{ FT}^2$$

AT 390°F THE WATER VOLUME IS LARGEST. FROM THE STEAM TABLES,

$$\rho_2 = \frac{1}{.01850} = 54.05$$

THE WATER VOLUME IS

$$Q = \frac{m}{\rho} = \frac{500{,}000}{54.05} = 9250.7 \text{ FT}^3/\text{HR}$$

THE REQUIRED NUMBER OF TUBES IS

$$\# = \frac{Q}{A_{tube}v} = \frac{9250.7}{(.003068)(5)(3600)}$$

$$= 167.5 \quad (\text{SAY } 168)$$

(NORMALLY, 20% WOULD BE ADDED)

NOW, USE EQUATION 10.68

$$q = F_cUA\Delta T_M$$

CALCULATE $\Delta T_M$ AS IF IN COUNTERFLOW.

$$\Delta T_A = 400 - 390 = 10$$

$$\Delta T_B = 400 - 200 = 200$$

$$\Delta T_m = \frac{10 - 200}{\ln\left(\frac{10}{200}\right)} = 63.4\ °F$$

$F_c$ IS NOT REQUIRED EVEN THOUGH THIS IS A 2-PASS HEAT EXCHANGER BECAUSE ONE OF THE FLUIDS HAS A CONSTANT TEMPERATURE.

THE OUTSIDE AREA OF ONE FOOT OF 168 TUBES IS

$$A = \frac{(168)(1)(\pi)(.875)}{12} = 38.48 \text{ FT}^2$$

THE REQUIRED AREA IS

$$A = \frac{q}{U\Delta T_M} = \frac{9.809 \text{ EE7}}{(700)(63.4)} = 2210 \text{ FT}^2$$

THE EFFECTIVE TUBE LENGTH IS

$$L = \frac{2210}{38.48} = 57.4 \text{ FT}$$

**5**

$$h_{probe} = \frac{(.024)(3480)^{.8}}{\left(\frac{.5}{12}\right)^{.4}} = 58.3 \frac{\text{BTU}}{\text{HR-FT}^2\text{-°R}}$$

$$T_{walls} = 600°\text{F} = 1060°\text{R}$$

THE THERMOCOUPLE GAINS HEAT BY RADIATION FROM THE WALLS. IT ALSO LOSES HEAT BY CONVECTION TO THE GAS. NEGLECT CONDUCTION AND THE INSIGNIFICANT KINETIC ENERGY.

$$E_{LOSS,CONVECTION} = E_{GAIN,RADIATION}$$

$$E = \frac{q}{A_{probe}} = h(T_{probe} - T_{GAS}) = \sigma \epsilon (T^4_{walls} - T^4_{probe})$$

(a) 300°F = 760°R,

$$58.3(T_{probe} - 760) = (.1713\ \text{EE-8})(.8)[(1060)^4 - (T_{probe})^4]$$

$$58.3\, T_{probe} - 44308 = 1730 - (1.37\ \text{EE-9})\, T^4_{probe}$$

$$1.37\ \text{EE-9}\, T^4_{probe} + 58.3\, T_{probe} = 46038$$

BY TRIAL AND ERROR, $\boxed{T_{probe} = 781°\text{R}}$

(b) IF $T_{probe} = 300°\text{F} = 760°\text{R}$, THEN

$$h(T_{probe} - T_{gas}) = \sigma \epsilon (T^4_{walls} - T^4_{probe})$$

$$T_{gas} = 760 - \frac{(0.1713\ \text{EE-8})(0.8)[(1060)^4 - (760)^4]}{58.3}$$

$$= \boxed{738.2\ °\text{R}}$$

**6** CALCULATE THE DUCT CHARACTERISTICS. THE DUCT DIAMETER IS

$$d = \frac{12\ \text{IN}}{12\ \text{IN/ft}} = 1.0\ \text{FT}$$

THE OUTSIDE DUCT AREA IS

$$A_{SURFACE} = \pi d L = (\pi)(1\ \text{FT})(50\ \text{FT}) = 157.1\ \text{FT}^2$$

THE CROSS SECTIONAL AREA IN FLOW IS

$$A_{FLOW} = \frac{\pi}{4} d^2 = \left(\frac{\pi}{4}\right)(1\ \text{FT})^2 = 0.7854\ \text{FT}^2$$

THE ENTERING AIR DENSITY IS

$$\rho = \frac{P}{RT} = \frac{\left(14.7 \frac{\text{LBF}}{\text{IN}^2}\right)\left(144 \frac{\text{IN}^2}{\text{FT}^2}\right)}{\left(53.3 \frac{\text{FT-LBF}}{\text{LBM-°R}}\right)(45° + 460°\text{R})} = 0.07864 \frac{\text{LBM}}{\text{FT}^3}$$

THE MASS FLOW RATE IS

$$\dot{m} = \rho Q = \left(0.07864 \frac{\text{LBM}}{\text{FT}^3}\right)\left(500 \frac{\text{FT}^3}{\text{MIN}}\right)\left(60 \frac{\text{MIN}}{\text{HR}}\right) = 2359.2\ \text{LBM/HR}$$

THE MASS FLOW RATE PER UNIT AREA IS

$$G = \frac{\dot{m}}{A_{flow}} = \frac{2359.2\ \text{LBM/HR}}{(0.7854\ \text{FT}^3)\left(3600 \frac{\text{SEC}}{\text{HR}}\right)} = 0.8344\ \text{LBM/FT}^2\text{-SEC}$$

SINCE THE DUCT MATERIAL AND THICKNESS WERE NOT GIVEN, DISREGARD THE THERMAL RESISTANCE OF THE DUCT.

ESTIMATE TEMPERATURES TO GET INITIAL FILM COEFFICIENTS, USING THE GIVEN FINAL AIR TEMPERATURE.

$$T_{BULK,AIR} = \frac{1}{2}(T_{AIR,IN} + T_{AIR,OUT}) = \frac{1}{2}(45°\text{F} + 50°\text{F}) = 47.5°\text{F}\quad [\text{ESTIMATE}]$$

$$T_{SURFACE,DUCT} = 70°\text{F}\quad \text{AT MIDLENGTH [ESTIMATE]}$$

THE FILM COEFFICIENTS ARE NOT HIGHLY SENSITIVE TO SMALL TEMPERATURE DIFFERENCES, SO CALCULATE THE FILM COEFFICIENTS BASED ON THESE ESTIMATES.

INSIDE, $h_i$

SINCE THIS IS FORCED CONVECTION, USE EQNS 10.44 AND 10.43

$$C = 0.00351 + 0.000001583\, T_{BULK}$$
$$= 0.00351 + (0.000001583)(47.5°\text{F})$$
$$= 0.003585$$

$$h_i = \frac{C(3600G)^{0.8}}{D^{0.2}} = \frac{(0.003585)[(3600)(0.8344)]^{0.8}}{(1\ \text{FT})^{0.2}}$$

$$= 2.17\ \text{BTU/HR-FT}^2\text{-°R}$$

OUTSIDE, $h_o$

THIS IS NATURAL CONVECTION. ESTIMATE $h_o$ AT MID-LENGTH

$$T_{FILM} = \frac{1}{2}(T_{SURFACE} + T_\infty) = \frac{1}{2}(70°\text{F} + 80°\text{F}) = 75°\text{F}$$

AT 75°F, $N_{Pr} = 0.72$ AND

$$\frac{g\beta\rho^2}{\mu^2} \approx 2.27 \times 10^6\ \frac{1}{\text{ft}^3\text{-°F}}$$

$$N_{Gr} = L^3\left(\frac{\rho^2\beta g}{\mu}\right)(T_\infty - T_{SURFACE})$$

$$= (1\ \text{FT})^3\left(2.27 \times 10^6 \frac{1}{\text{FT}^3\text{-°F}}\right)(80°\text{F} - 70°\text{F})$$

$$= 2.27 \times 10^7$$

$$N_{Pr} N_{Gr} = (2.27 \times 10^7)(0.72) = 1.63 \times 10^7$$

FROM TABLE 10.7, THE APPROXIMATE FILM COEFFICIENT FOR A HORIZONTAL CYLINDER IS

$$h_o = 0.27\left(\frac{T_\infty - T_s}{L}\right)^{0.25} = (0.27)\left(\frac{80°\text{F} - 70°\text{F}}{1\ \text{FT}}\right)^{0.25}$$

$$= 0.48\ \text{BTU/HR-FT}^2\text{-°F}$$

THE OVERALL HEAT TRANSFER COEFFICIENT IS

$$\frac{1}{U} = \frac{1}{h_o} + \frac{1}{h_i} = \frac{1}{2.17 \frac{\text{BTU}}{\text{HR-FT}^2\text{-°F}}} + \frac{1}{0.48 \frac{\text{BTU}}{\text{HR-FT}^2\text{-°F}}}$$

$$= 2.544\ \frac{\text{HR-FT}^2\text{-°F}}{\text{BTU}}$$

$$U = \frac{1}{2.544 \frac{\text{HR-FT}^2\text{-°F}}{\text{BTU}}} = 0.393\ \text{BTU/HR-FT}^2\text{-°F}$$

THE HEAT TRANSFER DUE TO CONVECTION IS

$$q_{convection} = UA(T_\infty - T_{bulk,air})$$
$$= \left(0.393 \frac{BTU}{HR\text{-}FT^2\text{-}°F}\right)(157.1\ FT^2)(80°F - 47.5°F)$$
$$= 2006.6\ BTU/HR$$

ASSUME THE ROOM AND DUCT "SEE" EACH OTHER COMPLETELY. THEN $F_e = \epsilon$ AND $F_A = 1.0$. THE HEAT EXCHANGE DUE TO RADIATION IS

$$q_{radiation} = \sigma F_e F_A A_{surface}\left[T_\infty^4 - T_{surface}^4\right]$$
$$= \left(0.1713 \times 10^{-8} \frac{BTU}{HR\text{-}FT^2\text{-}°R}\right)(0.28)(1.0)(157.1\ FT^2)$$
$$\times \left[(80 + 460°R)^4 - (70 + 460°R)^4\right]$$
$$= 461.6\ BTU/HR$$

THE TOTAL HEAT TRANSFER TO THE AIR IS

$$q_{total} = q_{convection} + q_{radiation}$$
$$= 2006.6 \frac{BTU}{HR} + 461.6 \frac{BTU}{HR}$$
$$= 2468.2\ BTU/HR$$

AT 47.5°F, THE SPECIFIC HEAT OF AIR IS APPROXIMATELY 0.240 BTU/LBM-°F. THE LEAVING AIR TEMPERATURE IS

$$T_{out} = T_{in} + \Delta T = T_{in} + \frac{q_{total}}{c_p \dot{m}}$$
$$= 45°F + \frac{2468.2\ BTU/HR}{\left(0.240 \frac{BTU}{LBM\text{-}°F}\right)\left(2359.2 \frac{LBM}{HR}\right)}$$
$$= 45°F + 4.36°F = 49.36°F$$

THIS AGREES WITH THE ENGINEER'S ESTIMATE.

---

**7** (a) $D_o = \frac{4.25}{12} = .354$

$$A_{pipe} = (.354)(\pi)(35') = 38.9\ FT^2$$

THE ENTERING DENSITY IS

$$\rho = \frac{(25)(144)}{(53.3)(460 + 500)} = .0704$$
$$\dot{m} = (200)(.0704)(60) = 844.8\ LBM/HR$$

AT LOW PRESSURES, THE AIR ENTHALPY IS FOUND FROM PAGE 6-35

$$h_1 = 231.06\ BTU/LBM \quad AT\ 960°R$$
$$h_2 = 194.25\ BTU/LBM \quad AT\ 810°R$$
$$q_{loss} = \dot{m}\Delta h = 844.8(231.06 - 194.25)$$
$$= 31097\ BTU/HR$$

ASSUMING THE MID-POINT PIPE SURFACE IS AT

$$\tfrac{1}{2}(500 + 350) = 425°$$

THEN,

$$U = \frac{q}{A\Delta T} = \frac{31097}{(38.9)(425 - 70)}$$
$$= \boxed{2.25 \frac{BTU}{FT^2\text{-}HR\text{-}°F}}$$

(b) IGNORE THE PIPE THERMAL RESISTANCE. IGNORE THE INSIDE FILM WHICH WILL NOT BE A LARGE FACTOR COMPARED TO THE OUTSIDE FILM AND RADIATION.

WORK WITH THE MIDPOINT PIPE TEMPERATURE OF 425°F.

RADIATION

ASSUME $F_a = 1$

$F_e = e \approx .9$ FOR 500°F ENAMEL PAINT OF ANY COLOR.

$$E = (.1713\ EE{-}8)(.9)\left[(425 + 460)^4 - (70 + 460)^4\right]$$
$$= 824.1\ BTU/HR\text{-}FT^2$$

NOTICE THE OMISSION OF $A$ WHICH WOULD BE DIVIDED OUT IN THE NEXT STEP.

THE RADIANT HEAT TRANSFER COEFFICIENT IS

$$h_r = \frac{E}{\Delta T} = \frac{824.1}{425 - 70} = 2.32 \frac{BTU}{HR\text{-}FT^2\text{-}°F}$$

OUTSIDE COEFFICIENT

EVALUATE THE FILM AT THE MIDPOINT, SO THE FILM TEMPERATURE IS

$$\tfrac{1}{2}(425 + 70) = 247.5 \quad (\text{SAY } 250°F)$$

$$N_{Pr} = .72$$
$$N_{Gr} = (.354)^3(425 - 70)(.647\ EE6) = 1.02\ EE7$$

SINCE $N_{Pr} N_{Gr} < EE9$, THE SIMPLIFIED COEFFICIENT IS

$$\bar{h} = (.27)\left(\frac{425 - 70}{.354}\right)^{.25} = 1.52$$

THE OVERALL FILM COEFFICIENT IS

$$h_t = h_r + \bar{h} = 2.32 + 1.52 = \boxed{3.84}$$

THIS IS CONSIDERABLY MORE THAN ACTUAL.

(c)
- LOWER EMISSIVITY DUE TO DIRTY OUTSIDE OF DUCT.
- $h_r$ AND $\bar{h}$ ARE NOT REALLY ADDITIVE.
- MID-POINT CALCULATIONS SHOULD BE REPLACED WITH INTEGRATION ALONG THE LENGTH
- THE INTERNAL FILM RESISTANCE WAS IGNORED
- CONDUCTIVITY OF STEEL PIPE IS DISREGARDED.

**8**

$100 \text{ gpm} = (100)(.1337)(60)(62.4) = 50057 \frac{\text{LBM}}{\text{HR}}$

$q_1 = M C_p \Delta T = (50057)(1)(140-70)$

$= 350.4 \text{ EE4 } \frac{\text{BTU}}{\text{HR}}$

$\Delta T_A = 230 - 70 = 160$

$\Delta T_B = 230 - 140 = 90$

$$\Delta T_M = \frac{160-90}{\ln\left(\frac{160}{90}\right)} = 121.66^\circ$$

$$U_1 = \frac{q}{A \Delta T_M} = \frac{350.4 \text{ EE4}}{(50)(121.66)} = 576 \frac{\text{BTU}}{\text{HR-FT}^2\text{-}^\circ\text{F}}$$

$q_2 = (50057)(1)(122-70) = 260.3 \text{ EE4}$

AFTER FOULING,

$\Delta T_{B,2} = 230 - 122 = 108$

$$\Delta T_M = \frac{160-108}{\ln\left(\frac{160}{108}\right)} = 132.3^\circ$$

$$U_2 = \frac{260.3 \text{ EE4}}{(50)(132.3)} = 393.5$$

FROM EQUATION 10.73

$$\frac{1}{U_2} = \frac{1}{U_1} + R_f$$

$$R_f = \frac{1}{393.5} - \frac{1}{576}$$

$$= \boxed{.000805 \frac{\text{HR-FT}^2\text{-}^\circ\text{F}}{\text{BTU}}}$$

**9** Heat is lost from the top and sides, by radiation and convection.

$A_{sides} = \frac{(1.5)(.75)}{144}\pi = 0.0245 \text{ ft}^2$

$A_{top} = \frac{\pi}{4}\left(\frac{.75}{12}\right)^2 = 0.003068 \text{ ft}^2$

$A_{total} = 0.0276 \text{ ft}^2$

Use EQUATION 10.89 and EQUATION 10.85

$$\dot{q} = h_t A_1 (T_1 - T_3) = h_{sides} A_{sides} (T_s - T_\infty) + h_{top} A_{top} (T_s - T_\infty) + \sigma F_e F_a A (T_s - T_\infty)$$

For first guess at $T_s$, use $h \sim 1.65$ (PAGE 10-3, table 10.2)

$\dot{q} = 5 \text{ W} (3.413 \frac{\text{BTU}}{\text{hr}}/\text{W}) = 17.065 \text{ BTU/hr}$

(APPENDIX G, PAGE 10-27)

$$17.065 = 1.65(0.0245)(T_s - 535) + 1.65(0.003068)(T_s - 535) + (0.65)(0.0276)(0.1713 \text{ EE}-8)(T_s^4 - 535^4)$$

Guess

$T_s \sim 800\,^\circ\text{R}$, $Q = 22.12$,

$\sim 600\,^\circ\text{R}$, $= 4.4$,

$\sim 700\,^\circ\text{R}$, $= 12.4$,

$\sim 750\,^\circ\text{R}$, $= 17.0$

Use $T_s \sim 750\,^\circ\text{R}$ as first approximation.

$T_f = \frac{1}{2}(535 + 750) = 642.5\,^\circ\text{R} \sim 640\,^\circ\text{R}$

At $T_f = 640 - 460 = 180^\circ\text{F}$, PAGE 10-29,

$N_{Pr} = 0.72$, $\frac{g\beta\rho^2}{\mu^2} = 1.03 \text{ EE6}$

Using EQUATIONS 10.57 and 10.58,

$$N_{Gr} = \left(\frac{g\beta\rho^2}{\mu^2}\right) L^3 (T_s - T_\infty)$$

$(N_{Gr} N_{Pr})_{sides} = (1.03 \text{ EE6})\left(\frac{1.5}{12}\right)^3 (750-535)(.72) = 3.11 \text{ EE5}$

$(N_{Gr} N_{Pr})_{top} = (1.03 \text{ EE6})\left(\frac{0.75}{12}\right)^3 (750-535)(.72) = 3.89 \text{ EE4}$

$h_{top} = 0.27\left(\frac{T_s - 535}{0.75/12}\right)^{0.25} = 0.54(750-535)^{0.25} = 2.068$

$h_{sides} = 0.29\left(\frac{T_s - 535}{1.5/12}\right)^{0.25} = 0.488(750-535)^{0.25} = 1.869$

Then

$$17.065 = 1.869(0.0245)(T_s - 535) + 2.07(0.003068)(T_s - 535) + (0.65)(0.0276)(0.1713 \text{ EE}-8)(T_s^4 - 535^4)$$

By trial and error, $\boxed{T_s \sim 736\,^\circ\text{R} = 276\,^\circ\text{F}}$

(b)

$$q_{convection} = \frac{1.869(0.0245)(736-535) + 2.068(.003068)(736-535)}{17.065}$$

$$= \frac{10.479}{17.065} = 0.614 = \boxed{61.4\% \text{ by convection}}$$

$$q_{radiation} = \frac{(.65)(.0276)(.1713 \text{ EE}-8)(736^4 - 535^4)}{17.065}$$

$$= \frac{6.5}{17.065} = 0.381 = \boxed{38.1\% \text{ by radiation}}$$

SOME ROUNDING ERRORS KEEP THE TOTAL FROM EQUALING 100%

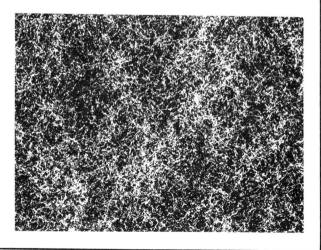

# HVAC

## WARM-UPS

**1** THE TOTAL PERIMETER INCLUDING THE CENTER-LINE IS

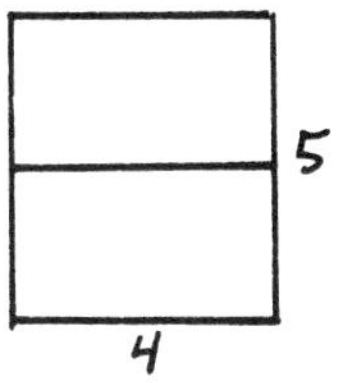


$3(4) + 10 = 22$ FT

ASSUME NO WEATHER-STRIPPING. FROM TABLE 11.4,

$$B = 80 \ \frac{FT^3}{HR\text{-}FT}$$

SO

$$Q = BL = (80)\frac{FT^3}{HR\text{-}FT}(22)\ FT = 1760\ FT^3/HR$$

**2** THE OFFICE VOLUME IS

$$60 \times 95 \times 10 = 57000\ FT^3$$

FROM TABLE 11.5, CHOOSE 6 AS THE NUMBER OF AIR CHANGES PER HOUR. THEN,

$$Q = \frac{(57000)\ FT^3\ (6)\ 1/HR}{(60)\ \frac{MIN}{HR}} = 5700\ CFM$$

b) FROM PAGE 11-6, THE PREFERRED VENTILATION RATE IS

$$Q = \tfrac{1}{2}(45)(15) + \tfrac{1}{2}(45)(40) = 1238\ CFM$$

**3** USE EQN. 11.18 AND TABLE 11.8

FOR THE METHANOL:

FROM STEP 2,

| | | |
|---|---|---|
| ADD | 2 | BECAUSE TLV = 200 |
| ADD | 1 | ASSUMING UNIFORM EVOLUTION |
| ADD | 3 | ASSUMING POOR BOOTH VENTILATION |
| | K = 6 | |

$$CFH = \frac{(4\ EE8)(.792)(2)(6)}{(32.04)(200)} = 6.0\ EE5$$

FOR THE METHYLENE CHLORIDE

$$CFH = \frac{(4\ EE8)(1.336)(2)(6)}{(84.94)(500)} = 1.5\ EE5$$

$$TOTAL = (6.0 + 1.5)\ EE5 = 7.5\ EE5$$

**4** A PSYCHROMETRIC CHART IS NEEDED FOR THIS.

$\omega$ = 79 GRAINS/LBM DRY AIR

$$= \frac{79\ GRAINS}{7000\ \frac{GRAINS}{LBM}} = .01129\ \frac{LBM}{LBM\ DRY\ AIR}$$

$h$ = 31.5 BTU/LBM DRY AIR

FROM TABLE 6.5, $\bar{c}_p$ IS GRAVIMETRICALLY WEIGHTED. ASSUME $c_p$ = .241 BTU/LBM-°F FOR AIR.

$$G_{AIR} = \frac{1}{1 + .01129} = .989$$

$$G_{WATER} = \frac{.01129}{1 + .01129} = .011$$

USING .4 AS THE $c_p$ OF SUPERHEATED STEAM,

$$\bar{c}_p = (.989)(.241) + (.011)(.4) = .243\ \frac{BTU}{LBM\text{-}°F}$$

**5**

$$q = \frac{(12000)\ W\ (3413)\ \frac{BTU}{KW\text{-}HR}\ (1.2)}{(1000)\ \frac{W}{KW}} + \frac{(12)(.8)(10)\ HP\ (2545)\ \frac{BTU}{HP\text{-}HR}}{.9}$$

$$= 3.206\ EE5\ BTU/HR$$

**6** FROM PAGE 9-9, TABLE 9.6, BUNKER C (#6) OIL,

$$HV = \tfrac{1}{2}(151{,}300 + 155{,}900) = 153{,}600\ BTU/GAL$$

FROM EQN 11-15, ASSUMING $T_i = 70°$, $\eta = .70$

$$\text{FUEL CONSUMPTION} = \frac{(24)\ \frac{HR}{DAY}\ (3.5\ EE6)\ \frac{BTU}{HR}\ (4772)\ °F\text{-}DAYS}{(70-0)\ °F\ (.70)(153{,}600)\ \frac{BTU}{GAL}}$$

$$= 53{,}260\ GAL$$

$$COST = (53{,}260)\ GAL\ (.15)\ \$/GAL = \$\ 7989$$

**7**

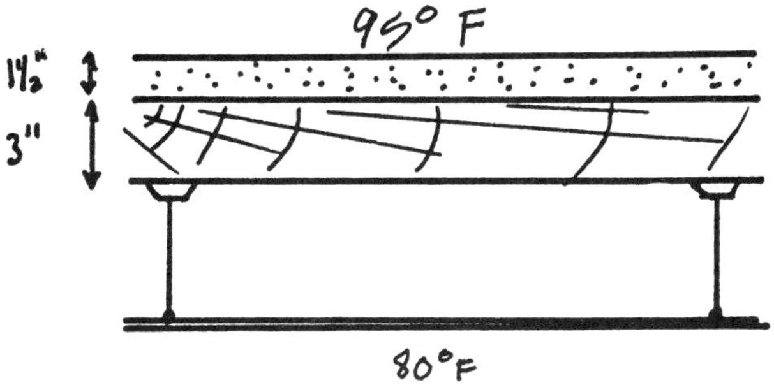


IT IS NORMAL TO ASSUME A 15 MPH OUTSIDE WIND. $h$ = 6.00 {PAGE 11-2} SO THE FILM THERMAL RESISTANCE IS

$$R_1 = \frac{1}{6} = .17$$

FROM PAGE 11-25 FOR ROOF INSULATION USED ABOVE DECKING,

$$R_2 = (1.5)\ IN\ (2.78)\ 1/IN = 4.17$$

ASSUME A PINE (SOFTWOOD) DECK. FROM PAGE 11-26

$$R_3 = (3)(1.25) = 3.75$$

WARM-UP PROBLEM #7 CONTINUED

FROM PAGE 10-26,

$\epsilon_{wood} = .93$

$\epsilon_{paper} = .95$ {USED FOR THE ACCOUSTICAL TILE}

FROM EQN 11.2,

$$E = \frac{1}{\frac{1}{.93} + \frac{1}{.95} - 1} = .89$$

FROM TABLE 11.2 {ASSUMING 4" SEPARATION},

$C \approx .81$

SO $R_4 = 1/.81 = 1.23$

FOR 3/4" ACCOUSTICAL TILE {FROM PAGE 11-25}

$R_5 = 1.78$

FOR THE INSIDE FILM, $h = 1.65$ {PAGE 11-2}

$$R_6 = \frac{1}{1.65} = .61$$

THE TOTAL RESISTANCE IS

$$R_t = R_1 + R_2 + R_3 + R_4 + R_5 + R_6$$
$$= .17 + 4.17 + 3.75 + 1.23 + 1.78 + .61$$
$$= 11.71$$

$$U = 1/R_t = .0854 \ \frac{\text{BTU}}{\text{HR-FT}^2\text{-}^\circ\text{F}}$$

8

$$BF = \left(\frac{1}{3}\right)^4 = .0123$$

9 USE THE SLAB EDGE METHOD.

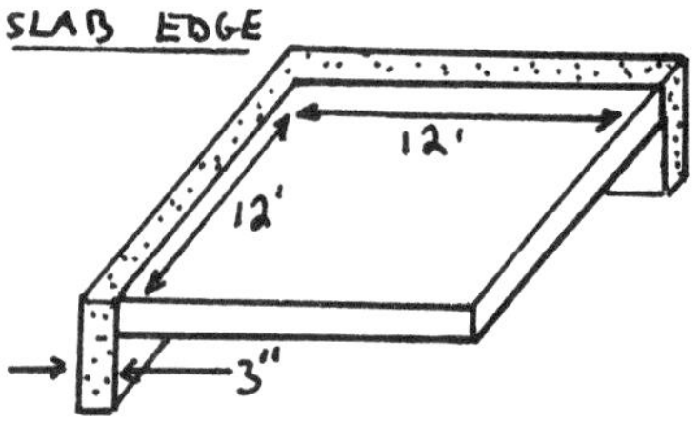


$P = 12 + 12 = 24$ FT.

CHOOSE $F = .55$ {PAGE 11-3}

ASSUME $T_i = 70°F$

FROM EQN 11.4,

$$q = (24)\text{ FT}(.55)\frac{\text{BTU}}{\text{HR-FT-}^\circ\text{F}}(70 - (-10))$$
$$= 1056 \text{ BTU/HR}$$

10 a) THE TOTAL VENTILATION IS

$$Q = (60 \ \text{CFM/PERSON})(4500 \text{ PEOPLE})\left(60 \ \frac{\text{MIN}}{\text{HR}}\right)$$
$$= 1.62 \text{ EE7 CFH}$$

BECAUSE $T_0 = 0°F$ {LESS THAN FREEZING} THERE WILL BE NO MOISTURE IN THE AIR. THE OUTSIDE DENSITY IS

$$\rho = \frac{p}{RT} = \frac{(14.6 \text{ PSI})(144)}{(53.3)(460 + 0°)} = .08575 \ \frac{\text{LBM}}{\text{FT}^3}$$

SO, $\dot{w} = Q\rho = (1.62 \text{ EE7 CFH})\left(.08575 \ \frac{\text{LBM}}{\text{FT}^3}\right)$

$$= 1.389 \text{ EE6 LBM/HR}$$

ASSUME $C_p = .241$ BTU/LBM-°F.

THE INCREASE IN AIR TEMPERATURE ONCE IT ENTERS THE AUDITORIUM IS DUE TO THE BODY HEAT LESS ANY CONDUCTIVE LOSSES THROUGH THE WALLS, ETC. FROM TABLE 11.13 THE OCCUPANT LOAD IS 225 BTU/HR-PERSON.

SO,

$$q_{\text{IN FROM PEOPLE}} = (225)(4500) = 1.01 \text{ EE6 BTU/HR}$$

SINCE THE AIR LEAVES AT 70°F, FROM

$$q = wC_p\Delta T$$

$$(1.01 \text{ EE6}) = (1.389 \text{ EE6})(.241)(70 - T_{IN})$$

OR $T_{IN} = 66.98°F$

b) THE COIL HEAT NEEDED IS

$$q = wC_p\Delta T$$
$$= (1.389 \text{ EE6})(.241)(66.98 - 0)$$
$$= 2.24 \times 10^7 \text{ BTU/HR}$$

c) SINCE $2.24 \times 10^7 < 1.25 \times 10^7$, THE FURNACE IS TOO SMALL.

CONCENTRATES

1

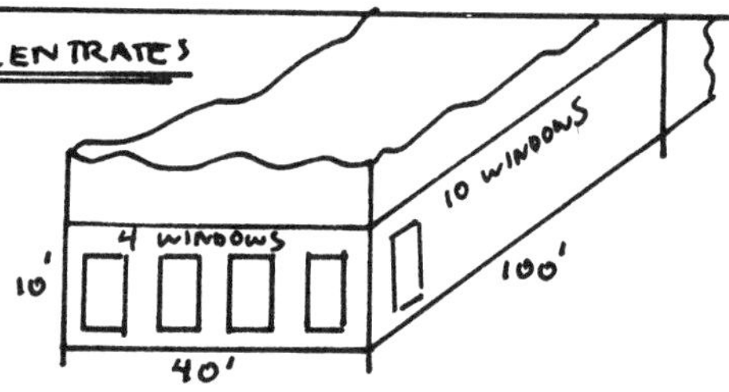


THE TOTAL EXPOSED (GLASS AND WALL) AREA IS

$$10(100 + 100 + 40) = 2400 \text{ FT}^2$$

THE WINDOW AREA IS

$$(10 + 10 + 4) \text{ WINDOWS}(4 \times 6)\text{ FT}^2 = 576 \text{ FT}^2$$

THE INSIDE AND OUTSIDE FILM COEFFICIENTS ARE 1.65 AND 6.00 $\frac{\text{BTU}}{\text{HR-FT}^2\text{-F}}$ RESPECTIVELY (PAGE 11-2)

$$U_{wall} = \frac{1}{\frac{1}{1.65} + \frac{1}{.2} + \frac{1}{6.00}} = .173$$

SO, THE HEAT LOSS FOR THE WALLS IS

$$q_1 = UA\Delta T = (.173)(2400 - 576)(75 - (-10))$$
$$= 26822 \text{ BTU/HR}$$

FROM PAGE 11-26, FOR THE WINDOWS ASSUMING A 1/4" AIR SPACE,

$$U_{windows} = 1/1.63 = .61$$

SO, THE HEAT LOSS FOR THE WINDOWS IS

$$q_2 = (.61)(576)(75 - (-10))$$
$$= 29866 \text{ BTU/HR}$$

FOR THE SAKE OF COMPLETENESS, CALCULATE THE AIR INFILTRATION. ASSUME DOUBLE-HUNG, UNLOCKED

{MORE}

CONCENTRATE #1 CONTINUED

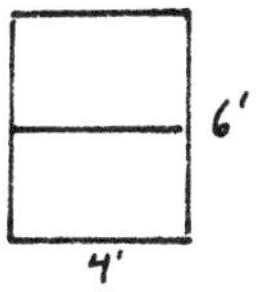


FROM TABLE 11.4, $B = 32 \frac{CFH}{FT}$

$L = 4+4+4+6+6 = 24$ FT

SO, $Q = (14)_{\text{windows}}(32)(24)$
$= 10752$ CFH

{USE WINDOWS ON 2 SIDES FOR 3 EXPOSED walls}

FROM EQN 11.5

$q_3 = (.018)(10752)(75-(-10)) = 16451$

SO, $q_{total} = q_1 + q_2 + q_3$
$= 26,822 + 29,866 + 16,451$
$= 73,139$ BTU/HR

**2** SEE PAGE 11-11 (PARAGRAPH C). LOCATE POINTS A + B ON THE PSYCHROMETRIC CHART.

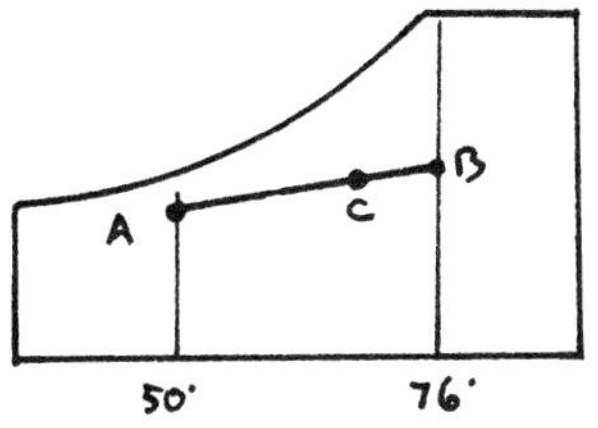


THE LENGTH OF THE LINE (A-B) ON MY CHART IS 6.6 CM. READING FROM THE CHART,

$v_A = 13.0$ FT³/LBM

$v_B = 13.69$ FT³/LBM

SO $\rho_A = 1/13.0 = .0769$ LBM/FT³

$\rho_B = 1/13.69 = .073$

$M_A = \rho Q_A = (.0769)(1000) = 76.9$

$M_B = (.073)(1500) = 109.5$

THE GRAVIMETRIC FRACTION OF FLOW A IS

$$\frac{76.9}{76.9 + 109.5} = .41$$

SO, POINT C IS LOCATED $(.41)(6.6) = 2.7$ CM FROM POINT B. THIS DETERMINES

$T_{dry\ bulb} = 65.5$ °F

$\omega = 57$ GRAINS/LBM

$T_{dp} = 51$ °F

**3** a) BF $= 1 - .70 = .30$

b) FROM EQN 11.59

$$.70 = \frac{60 - T_{OUT}}{60 - 45}$$

$T_{OUT} = 49.5$ °F

**4** FROM THE PSYCHROMETRIC CHART

AT 1 $\omega_1 = 99$ GRAINS/LBM
$h_1 = 38.3$ BTU/LBM
$v_1 = 14.29$ FT³/LBM

AT 2 $\omega_2 = 165.8$
$h_2 = 46.4$

SO, THE MOISTURE ADDED IS

$$\frac{(165.8 - 99)\ \frac{\text{GRAINS}}{\text{LBM AIR}}}{(7000)\ \frac{\text{GRAINS}}{\text{LBM WATER}}\ (14.29)\ \frac{\text{FT}^3}{\text{LBM AIR}}} = 6.68\ \text{EE-4}\ \frac{\text{LBM}}{\text{FT}^3}$$

THE ENTHALPY CHANGE IS

$$\frac{(46.4 - 38.3)}{14.29} = .567\ \text{BTU/FT}^3\ \text{AIR}$$

**5** Follow THE PROCEDURE STARTING ON PAGE 11-15

$$SHR = \frac{200,000}{200,000 + 50,000} = .8$$

DRAW THE CONDITION LINE:

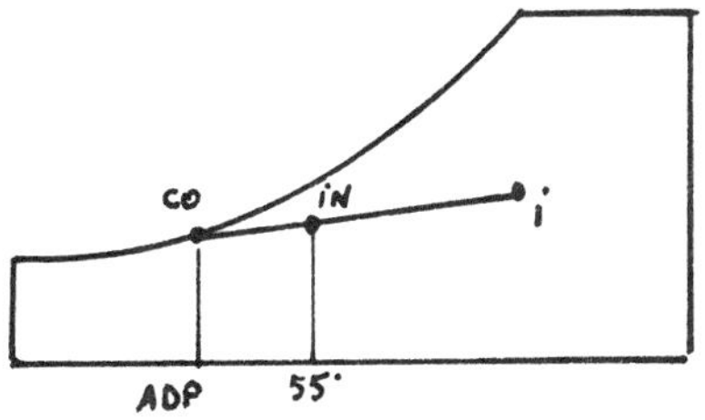


ADP = 50.8 °F.

$T_{co}$ = ADP

CHOOSE 20° AS A TEMPERATURE GAIN,

$T_{db,IN} = (75-20) = 55$ °F

FROM EQN 11.66 (IN CFM NOT CFH)

$$CFM_{IN} = \frac{200,000}{\left(\frac{60}{55.3}\right)(75-55)} = 9091\ \text{CFM}$$

THIS IS A MIXING PROBLEM

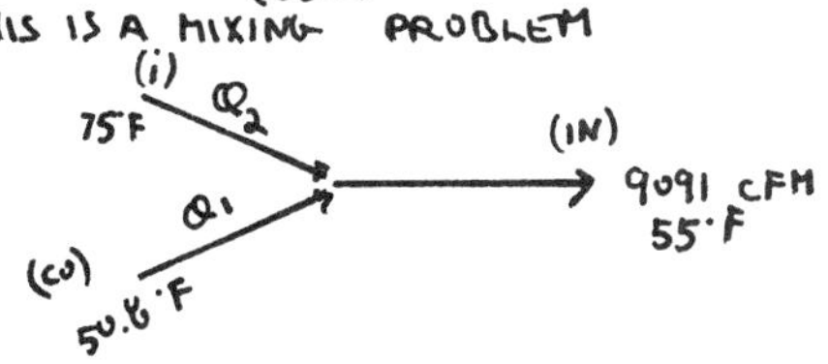


THE LINES (CO-i) AND (IN-i) HAVE LENGTHS OF 6.15 CM AND 5.05 CM RESPECTIVELY {ON MY CHART}

$$\frac{5.05}{6.15} = .821$$

$Q_1 = (.821)(9091) = 7464$ CFM

**6** ASSUME HEATING LOAD IS BASED ON $T_{id} = 70$°F. THE HEAT LOSS PER DEGREE IS

$$\frac{q}{\Delta T} = \frac{650,000}{70 - 0} = 9285.7\ \frac{\text{BTU}}{\text{HR-°F}}$$

THE AVERAGE OUTDOOR TEMPERATURE DURING THE HEATING SEASON IS FOUND FROM

$5252 = 245(65 - \bar{\bar{T}}_o)$ OR $\bar{\bar{T}}_o = 43.56$ °F

{MORE}

CONCENTRATES # 6 CONTINUED

8:30 A.M. TO 5:30 P.M. IS 9 HOURS. THE TOTAL WINTER HEAT LOSS IS

$$(245)(9285.7)\left[(9)(70-43.56)+(15)(50-43.56)\right]$$
$$= 7.61 \text{ EE8 BTU}$$

$$\text{FUEL CONSUMPTION} = \frac{7.61 \text{ EE8}}{(13{,}000)(.70)} = 83{,}626 \text{ LB/YR}$$

7 THIS IS A MIXING PROBLEM. THE PROCESS CANNOT BE REPRESENTED BY A CONSTANT ENTHALPY LINE BECAUSE $T_{WATER} < T_{wb,\ INCOMING}$

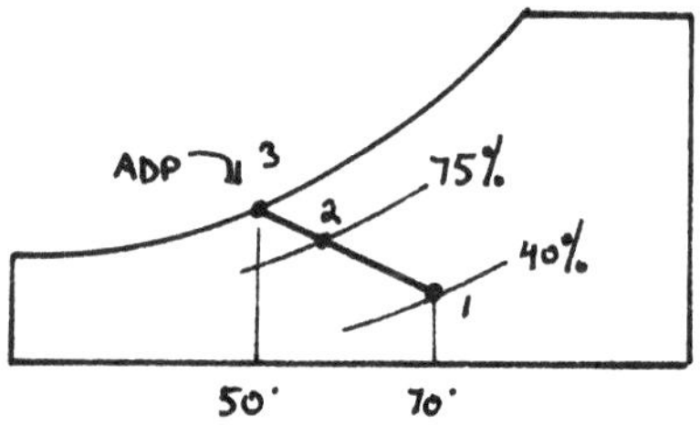


AT 1
$h_1 = 23.6$ BTU/LBM
$\omega_1 = 44$ GRAINS/LBM
$v_1 = 13.48$ FT³/LBM

SO THE MASS OF INCOMING AIR IS

$$\dot{m}_1 = \frac{(1800)\text{ CFM}}{(13.48)\ \frac{\text{FT}^3}{\text{LBM}}} = 133.53\ \frac{\text{LBM}}{\text{MIN}}$$

AT 2
$h_2 = 21.4$
$\omega_2 = 51$
$T_{db,2} = 56.3$
$T_{wb} = 51.9$

THE ADDED WATER CONTENT IS

$$\frac{(133.5)\ \frac{\text{LBM AIR}}{\text{MIN}}\ (51-44)\ \frac{\text{GRAINS WATER}}{\text{LBM AIR}}}{(7000)\ \frac{\text{GRAINS WATER}}{\text{LBM WATER}}} = .134\ \frac{\text{LBM}}{\text{MIN}}$$

8 THIS REQUIRES USING THE ENERGY AND WEIGHT BALANCES {EQUATIONS 11.44 – 11.46}

FROM PROBLEM #7:
$\omega_1 = 44$ GRAINS/LBM = .006286 LBM/LBM
$h_1 = 23.6$ BTU/LBM
$\dot{M} = 133.53$ LBM/MIN

FOR 1 ATMOSPHERE STEAM, $h = 1150.4$ BTU/LBM

WEIGHT BALANCE
$$133.53(1+.006286) + M_{STEAM} = 133.53\left(1+\frac{\omega_2}{7000}\right)$$

ENERGY BALANCE
$$133.53(23.6) + M_{STEAM}(1150.4) = (133.53)\,h_2$$

SINCE NO SINGLE RELATIONSHIP EXISTS BETWEEN $\omega_2$, $M_{STEAM}$, AND $h_2$, A TRIAL AND ERROR SOLUTION IS REQUIRED. ONCE $M_{STEAM}$ IS SELECTED, $\omega_2$ AND $h_2$ CAN BE FOUND FROM THE ABOVE 2 EQUATIONS:

$$\omega_2 = 52.42(m_{STEAM}) + 44$$
$$h_2 = 8.615(m_{STEAM}) + 23.6$$

KNOWING $\omega_2$ AND $h_2$ DETERMINES THE RELATIVE HUMIDITY. IF THE RELATIVE HUMIDITY IS 75%, THEN $W_{STEAM}$ WAS CHOSEN CORRECTLY. HERE ARE MY TRIAL + ERROR ITERATIONS

| $M_{STEAM}$ | $\omega_2$ | $h_2$ | RH |
|---|---|---|---|
| .3 | 59.7 | 26.2 | 53% |
| .5 | 70.2 | 27.9 | 61% |
| .7 | 80.7 | 29.6 | 72% |
| .8 | 86 | 30.5 | 75% |

SO $\omega_2 = 86$ GRAINS/LBM
$h_2 = 30.5$ BTU/LBM
$T_{db} = 71.1$ °F
$T_{wb} = 65.6$ °F

9 FOR THE INCOMING AIR,
$v_1 = 13.37$ FT³/LBM
$\omega_1 = 51$ GRAINS/LBM

FROM EQN 11.5 USING SENSIBLE HEATING AS THE LIMITING FACTOR,

$$\text{CFH} = \frac{500{,}000}{(.018)(75-65)} = 2.765 \text{ EE6 CFH}$$

$$\text{OR } \dot{M} = \frac{2.765 \text{ EE6}}{13.37} = 2.068 \text{ EE5 LBM/HR}$$

ASSUME THAT THIS AIR ABSORBS ALL THE MOISTURE. THEN, THE FINAL HUMIDITY RATIO WILL BE

$$\omega_2 = 51 + \frac{(175)\ \frac{\text{LBM}}{\text{HR}}\ (13.37)\ \frac{\text{FT}^3}{\text{LBM}}(7000)\ \text{GR/LBM}}{(2.765 \text{ EE6})\ \frac{\text{FT}^3}{\text{HR}}}$$
$$= 56.9 \text{ GR/LBM}$$

THE FINAL CONDITIONS ARE
$T_{db} = 75$ °F
$\omega_2 = 56.9$ GR/LBM
RH = 43%.

THIS IS BELOW RH = 60%

10 LOCATE POINT (OUT) AND (CO) ON THE PSYCHROMETRIC CHART.
$v_{OUT} = 13.94$ FT³/LBM
$h_{OUT} = 36.2$ BTU/LBM

{MORE}

CONCENTRATES #10 CONTINUED

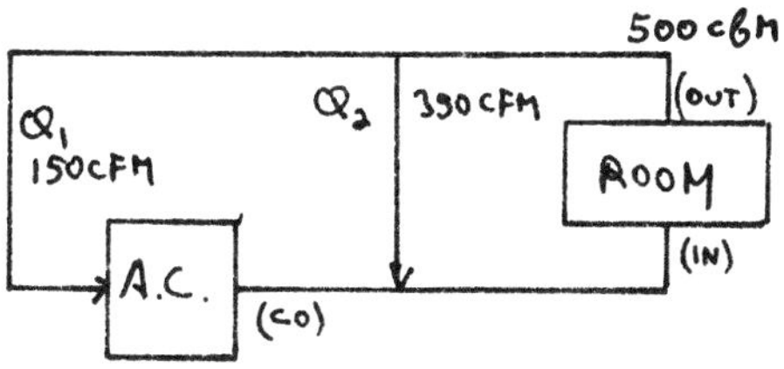


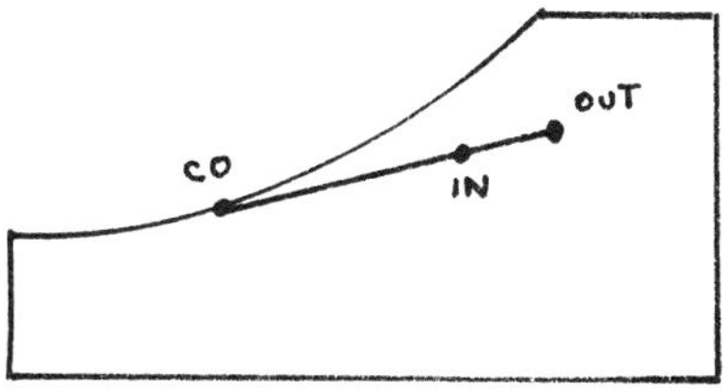


$h_{co} = 20.2$

so $\dot{w}_1 = \dfrac{(150)\ \frac{FT^3}{MIN}}{(13.94)\ \frac{FT^3}{LBM}} = 10.76\ LBM/MIN$

$\dot{w}_2 = \dfrac{350}{13.94} = 25.11$

THE % BYPASS IS

$x = \dfrac{25.11}{25.11 + 10.76} = .70$

ON MY CHART, THE LENGTH OF LINE (CO-OUT) IS 8.5 CM, POINT (IN) IS LOCATED $(.70)(8.5) = 6$ CM FROM POINT CO.

a) AT THAT POINT,

$T_{db,IN} = 71.2\ °F$

$\omega_{IN} = 92\ gR/LBM$

$RH_{,IN} = 80\%$

b) THE CONDITIONER CAPACITY IS

$$\frac{(10.76)\ \frac{LBM}{MIN}\ (36.2 - 20.2)\ \frac{BTU}{LBM}}{(200)\ \frac{BTU}{MIN\text{-}TON}} = .86\ TON$$

THE WATER DECREASE IS

$$\frac{(343.6)\ \frac{LBM\ AIR}{MIN}\ (177 - 73)\ \frac{gR}{LBM\text{-}AIR}}{(7000)\ \frac{gR}{LBM\ WATER}} = 5.1\ \frac{LBM}{MIN}$$

c) $(343.6)\ \frac{LBM}{MIN}\ (50.7 - 25.8)\ \frac{BTU}{LBM} = 8555.6\ \frac{BTU}{MIN}$

d) IT'S NOT CLEAR FROM THE PROBLEM WHETHER THIS IS A WET OR DRY CYCLE. ASSUME WET SINCE THE PROBLEM SAYS 'SATURATED' AT 100°F

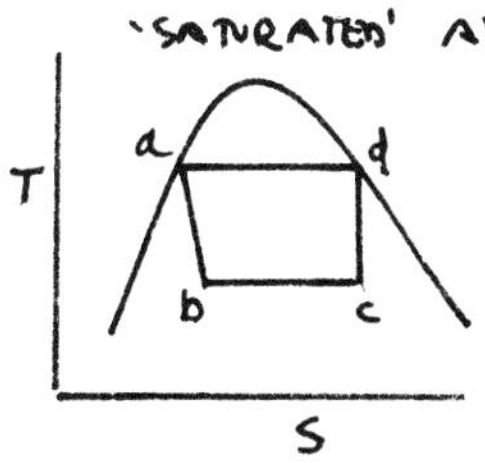


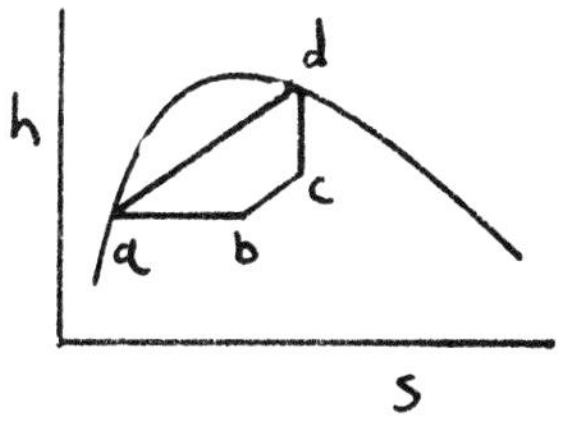


e) FROM FIG. 7.44,

AT a
$T = 100°F$
$p = 131.6$ PSIA
$h = 31.16$ BTU/LBM
$s = .06316$ BTU/LBM-°F
$v = .0127$ FT³/LBM

AT b
$T = 50$
$p = 61.39$
$h = h_a = 31.16$
$x = \dfrac{31.16 - 19.27}{64.51} = .184$

$s = .04126 + .184(.12659) = .06455$
$v = .0118 + .184(.673 - .0118) = .1335$

AT d
$T = 100°F$
$p = 131.6$
$h = 88.62$
$s = .16584$
$v = .319$

AT c
$T = 50°F$
$p = 61.39$ PSIA
$s = s_d = .16584$
$x = \dfrac{.16584 - .04126}{.12659} = .984$

$h = 19.27 + .984(64.57) = 82.75$

$v = .0118 + .984(.673 - .0118) = .662$

TIMED

1 AT 1

$h_1 = 50.7$ BTU/LBM

$v_1 = 14.55$ FT³/LBM

$\omega_1 = 177$ gR/LBM

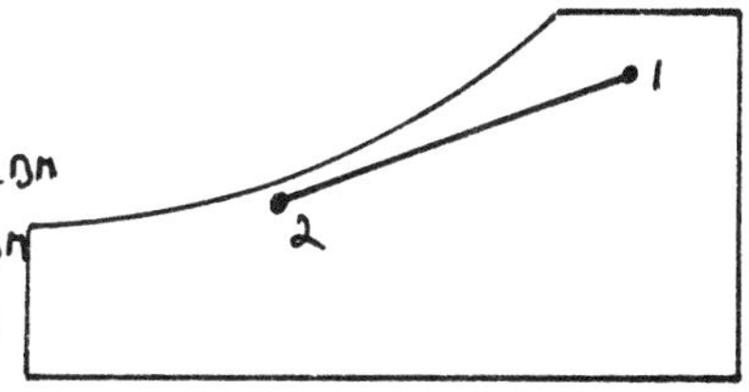


AT 2

$h_2 = 25.8$

$\omega_2 = 73$

b) THE AIR WEIGHT IS

$$\frac{(5000)\ \frac{FT^3}{MIN}}{(14.55)\ \frac{FT^3}{LBM}} = 343.6\ LBM/MIN$$

2 THE HEAT LOSS PER °F IS

$$\frac{q}{\Delta T} = \frac{200{,}000}{75 - 0} = 2666.7\ \frac{BTU}{HR\text{-}°F}$$

THE AVERAGE OUTDOOR TEMPERATURE DURING THE HEATING SEASON IS FOUND FROM THE FOLLOWING ANALYSIS.

$$DD = 4200 = \sum_{i=1}^{210} (65 - \overline{T}_o) = 210(65 - \overline{\overline{T}}_o)$$

$$\overline{\overline{T}}_o = 45°F$$

$\overline{T}_o$ AND $\overline{\overline{T}}_o$ ARE NOT THE SAME. $\overline{\overline{T}}_o$ IS THE AVERAGE TEMPERATURE OVER ALL HOURS IN THE ENTIRE 210 DAY HEATING SEASON

THE ORIGINAL HEAT LOSS DURING THE HEATING SEASON WAS

$$q = (24)\ \frac{HR}{DAY}\ (210)\ DAYS\ (2666.7)\ \frac{BTU}{HR\text{-}°F}\ (75 - 45)\ °F$$
$$= 4.03\ EE8\ BTU$$

(MORE)

TIMED #2, CONTINUED

THE REDUCED HEAT LOSS IS

$$q = (24)(210)(2666.7)(68-45) = 3.09\text{ EE8 BTU}$$

THE REDUCTION IS

$$\frac{4.03-3.09}{4.03} = .233 \qquad \boxed{23.3\%}$$

3 THIS IS JUST A STANDARD BYPASS PROBLEM.

THE INDOOR CONDITIONS ARE GIVEN:

$T_i = 75°F$ $\phi_i = 50\%$

THE OUTDOOR CONDITIONS ARE GIVEN:

$T_o = 90°F$ db $T_{o,wb} = 76°F$

THE VENTILATION IS GIVEN AS 2000 CFM.

THE LOADS ARE GIVEN:

$q_s = 200{,}000$ BTUH

$q_l = 450{,}000$ gr/HR

WE WANT TO FIND THE SENSIBLE HEAT RATIO. BUT, $q_l$ MUST BE EXPRESSED IN BTUH.

$\omega \approx .0095$ FOR THE ROOM CONDITIONS. (ACTUALLY, IT SHOULD BE A LITTLE LESS SINCE MOISTURE IS REMOVED BETWEEN $T_{IN}$ AND $T_i$)

ASSUME P = 14.7 PSIA. THEN, FROM EQN 11.26,

$$.0095 = \frac{.622\,P_W}{14.7 - P_W} \approx \frac{.622\,P_W}{14.7}$$

SO, $P_W = .22$ psia

FOR P = .22 psia, FROM THE STEAM TABLES,

$h_{fg} \approx 1060$ BTU/LBM

EACH POUND OF WATER EVAPORATED REQUIRES APPROXIMATELY 1060 BTU

$$q_l = \frac{(450{,}000)\frac{\text{gr}}{\text{HR}}(1060)\frac{\text{BTU}}{\text{LB}}}{(7000)\frac{\text{gr}}{\text{LB}}} = 68143\ \text{BTU/HR}$$

FROM EQUATION 11.64,

$$SHR = \frac{200{,}000}{200{,}000 + 68143} = .75$$

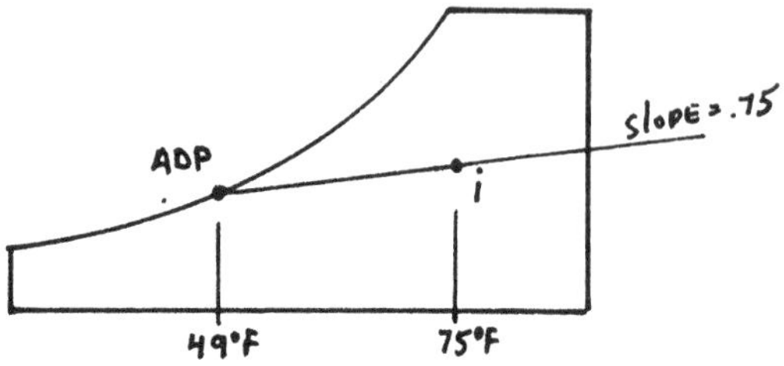


USING $T_i$ AND $\phi_i$, LOCATE POINT i ON THE PSYCH CHART AND DRAW THE CONDITION LINE WITH SLOPE = .75 THROUGH IT.

THE LEFT-HAND INTERSECTION SHOWS ADP = 49°F. SINCE THE AIR LEAVES THE CONDITIONER SATURATED, WE KNOW

$BF_{coil} = 0$

$\boxed{T_{co} = 49°F}$

NOW CALCULATE THE AIR FLOW THROUGH THE ROOM.

$$(CFM)_{IN} = \frac{200{,}000}{\left(\frac{60}{55.3}\right)(75-58)} = \boxed{10{,}843}$$

$$Q_{IN} = 60(10{,}843) = 650{,}600\ \text{CFH}$$

SINCE $T_{IN} = 58°F$ IS GIVEN, LOCATE THIS DRY BULB TEMPERATURE ON THE CONDITION LINE

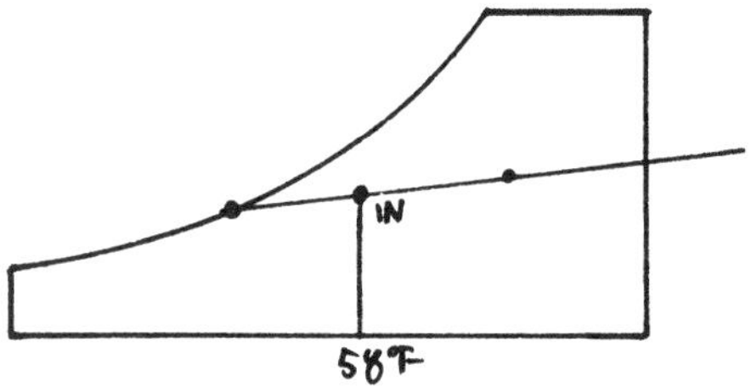


READING FROM THE CHART,

$\boxed{\omega_{IN} = 56.5\ \text{gr/POUND}}$

4 ALTHOUGH SOME HIGH-TEMPERATURE PSYCH CHARTS EXIST, IT IS NOT NECESSARY TO USE THEM.

SEE EQUATION 11.27 AT $T_i = 200°F$ AND 100% RELATIVE HUMIDITY,

$P_W = 11.526$ (FROM PAGE 6-29)

SINCE $P_t = 25$ psia,

$P_A = 25 - 11.526 = 13.474$ psia

THE SPECIFIC GAS CONSTANT OF WATER IS 85.8 {TABLE 6.4}

THE MASS OF THE WATER VAPOR IS CONSTANT

$$M_W = \frac{pV}{RT} = \frac{(11.526)(144)(1500)}{(85.8)(200+460)} = 43.96\ \text{LBM}$$

THE MASS OF THE AIR IS ALSO CONSTANT

$$M_A = \frac{(13.474)(144)(1500)}{(53.3)(200+460)} = 82.73\ \text{LBM}$$

$$\omega = \frac{43.96}{82.73} = .531$$

THE MASSES OF AIR AND WATER DO NOT CHANGE, SO THE MOLE FRACTIONS AND PARTIAL PRESSURE RATIOS ALSO DO NOT CHANGE.

{MORE}

TIMED #4, CONTINUED

$P_{w,2} = 11.526$

AT $T_2 = 400°F$, $P_{w,sat} = 247.31$ psi (PAGE 6-29)

$\phi_2 = \frac{11.526}{247.31} = \boxed{.0466}$

$\omega$ IS UNCHANGED.

$\omega_2 = \boxed{.531}$

THE ENERGY REQUIRED MUST BE OBTAINED IN TWO PARTS

AIR: FROM PAGE 6-35

$h_1 = 157.92$ BTU/LBM (AT 660°R)

$h_2 = 206.46$ BTU/LBM (AT 860 °R)

$q_1 = 206.46 - 157.92$

$= 48.54$ BTU/LBM-DRY AIR

WATER: A Mollier DIAGRAM, LOCATE POINT 1 (T = 200, P = 11.526)

$h_1 = 1146$ (APPEARS TO BE SATURATED)

NOW, FOLLOW A CONSTANT PRESSURE CURVE UP TO 400°F.

$h_2 = 1240$ BTU/LBM

$q_2 = \frac{(43.96)(1240-1146)}{82.73}$

$= 49.95$ BTU/LBM DRY AIR

$q_{total} = 48.54 + 49.95 = \boxed{98.49 \frac{\text{BTU}}{\text{LBM DRY AIR}}}$

THE DEW POINT IS THE TEMPERATURE AT WHICH THE WATER STARTS TO CONDENSE OUT IN A CONSTANT-PRESSURE PROCESS. FOLLOWING THE CONSTANT PRESSURE LINE BACK TO THE SATURATION LINE,

$\boxed{T_{dp} = 200°F}$

5 Annual savings accrue during 21-week heating season.

Assume negligible crack losses compared to volume of forced ventilation air. Also neglect humidification differences, since moisture content is not given (outside air temperature and moisture).

APPENDIX D, PAGE 10-24, AIR at 70 °F: $c_p \approx 0.240$ BTU/lbm-°F, $\rho \approx 0.075$ lbm/ft$^3$.

In one air change,

$q_{air} = m c_p dT = (801{,}000 \text{ ft}^3)(0.075 \frac{\text{lbm}}{\text{ft}^3})(0.240 \frac{\text{BTU}}{\text{lbm-°F}})(12 °F)$

$= 173{,}966$ BTU in one air change

Then on an hourly basis,

$q_{air/hr} = (173{,}966 \text{ BTU/air change})(\tfrac{1}{2} \text{ air change/hour})$

$= 86{,}983$ BTU/hr

NEXT, FOR THE ROOF, FLOOR, AND WALLS, USING EQN 11.3,

$q = UA\Delta T$

$= [(.15)(11{,}040) + (1.13)(2760) + (.05)(26{,}700) + (1.5)(690)] \times (12°F)$

$\approx 85{,}738$ BTU/hr

Unoccupied time

$t = [(14)(5) + 48] \frac{\text{hrs}}{\text{wk}} (21 \text{ weeks}) = 2478$ hrs/yr

$q_{lost/year} = (86{,}983 + 85{,}738) \frac{\text{BTU}}{\text{hr}} (2478 \frac{\text{hrs}}{\text{yr}})$

$= \frac{4.28 \text{ EE8 BTU}}{100{,}000 \text{ BTU/therm}} = 4280$ therms

$q_{req'd} = \frac{4280 \text{ therms}}{0.75} = 5707$ therms

savings = (5707 therms)(\$0.25/therm)

$= \boxed{\$1427 \text{ per year}}$

6 (a) 800 °F = 1260 °R

From AIR TABLES, APPENDIX F, PAGE 6-35

$h_{1,air,800°F} = 306.65$ BTU/lbm

$h_{2,air,350°F} = 194.25$ BTU/lbm (350°F = 810 °R)

From steam tables, APPENDIX A, PAGE 6-29, AT 80°F,

$h_{1,water} = 48.02$ BTU/lbm (or from Mollier diagram)

Using EQUATION 11.32,

$h_{2,water} = .444(350) + 1061 = 1216.4$ BTU/lbm

Using EQUATIONS 11.37 and 11.38,

$m_w = \frac{m_{air}(\Delta h_{air})}{\Delta h_w} = \frac{410 \text{ lbm/hr} (194.25 - 306.65)}{(48.02 - 1216.4)}$

$\boxed{m_w = 39 \text{ lbm/hr water}}$ USE $\Delta h = c_p \Delta T$ ONLY IF $c_p$ IS ASSUMED CONSTANT

(b) Using EQUATION 11.27, $\phi = \frac{p_w}{p_{ws}}$

From APPENDIX A, PAGE 6-29, $P_{sat,350°} = 134.63$ psia

From EQUATION 11.22, the partial pressure is proportional to the mole fraction.

#moles water $= \frac{40}{18} = 2.22$

#moles air $= \frac{410}{29} = 14.14$

$x_w = \frac{2.22}{2.22 + 14.14} = 0.136$

$p_w = (0.136)(100) = 13.6$ psia

$\phi = \frac{13.6}{134.63} = 0.101 = \boxed{10.1\ \%}$

7 cooled water flow rate $= \frac{q}{c_p \Delta T} = \frac{1 \text{ EE6 BTU/hr}}{(1 \frac{\text{BTU}}{\text{lbm-°F}})(120-110°F)}$

$= 1 \text{ EE5}$ lbm/hr

<MORE>

TIMED #7 CONTINUED

Air in, using psych diagram, $h \sim 43.0$ BTU/lbm

Air out, cannot use psych diagram (off scale), will use equations. From appendix A, page 6-29 (Sat. steam tables) at 100°F, $p_{ws} = 0.9492$ psia.

Using equation 11.27, and 11.26,

$$\omega_{air\,out} = \frac{0.622\,(0.82)(0.9492)}{14.696 - (0.82)(0.9492)} = 0.035$$

Then, for air out, using EQUATION 11.30,

$$h_2 = 0.241\,(100) + (0.035)[0.444\,(100) + 1061] = 62.55 \text{ BTU/lbm}$$

Now $\dot{m}_{air}(h_2 - h_1) = q_{removed}$, so

$$\dot{m}_{air} = \frac{1 \text{ EE6 BTU/hr}}{(62.55 - 43.0)\text{ BTU/lbm}} = \boxed{5.11 \text{ EE4 lbm/hr air}}$$

(b) Humidity ratio may be found using Mollier diagram or as follows:

Conservation of water vapor:

$$\omega_1 \dot{m}_{air,1} + \dot{m}_{make\text{-}up} = \omega_2 \dot{m}_{air,2}$$

$$\dot{m}_{make\text{-}up} = \dot{m}_{air}(\omega_2 - \omega_1)$$

from appendix A, page 6-29, at 91°F, $p_{ws} = 0.7233$ psia

Then $p_w = (0.60)(0.7233) = 0.434$ psia

Using EQUATION 11.26, $\omega_3 = \dfrac{.622\,(0.434)}{(14.696 - 0.434)} = 0.019$

make-up water:

$$\dot{m}_m = (5.11 \text{ EE4})(0.035 - 0.019) = \boxed{811 \text{ lbm/hr water}}$$

## 8

First, resistances:

walls:

| | R | Determined from: |
|---|---|---|
| 4" brick facing | 0.44 | appendix A, page 11-26 |
| 3" concrete block | 0.40 | appendix A, page 11-26 |
| 1" mineral wool | 3.33 | appendix A, page 11-25 |
| 2" furring | 0.93 | indicates 2" air space .. |

(Using appendix F, page 10-26, $e_{mineral\,wool}$ should be roughly $e_{iron\,oxide} = 0.96$. Likewise, $e_{drywall\,gypsum}$ should be about $e_{white\,paper} = 0.95$. Using EQUATION 11.2, $\frac{1}{E} = \left(\frac{1}{0.96}\right) + \left(\frac{1}{0.95}\right) - 1 = 1.09$, $E = 0.91$. Then, EXTRAPOLATING FROM TABLE 11.2 FOR A VERTICAL AIR SPACE, $C = 1.08$ AND $R = 1/C = 1/1.08 = 0.9277$.

| | R | Determined from: |
|---|---|---|
| 3/8" drywall gypsum | 0.34 | appendix A, page 11-25; $\frac{(3/8)}{(1/2)}(0.45) = 0.3375$ |
| 1/2" plaster | 0.09 | appendix A, page 11-25 |
| surface outside | 0.25 | ASHRAE, chapter 26, table 1, page 429 |
| inside | 0.61 | still air, table 10.2, $h \sim 1.65$ |

roof:

| | R | Determined from: |
|---|---|---|
| 4" concrete | 0.32 | ASHRAE, chapter 26, table 11A, page 443, sand aggregate |
| 2" insulation | 5.56 | appendix A, page 11-25 |
| felt | 0.06 | ASHRAE, chapter 26, table 10, page 442 |
| 1" air gap | 0.87 | appendix F, page 10-26, EQUATION 11.2, and table 11.2, as before |
| ceiling tile | 1.19 | assume 1/2" acoustical tile, appendix A, page 11-25 |
| surface outside | 0.25 | ASHRAE chapter 26, table 12B, page 446, for summer conditions, downward heat flow, 7.5 mph winds |
| inside | 0.92 | |

windows:

| | R | Determined from: |
|---|---|---|
| 1/4" thick, single glazing (includes film coefficient) | 0.66 | Without blinds, R = 0.88, from appendix A, page 11-26; from ASHRAE, chapter 26, GLASS AND DOOR COEFFICIENTS, page 425, with blinds, R = 0.75 (0.88) = 0.66. |

Next, heat fluxes:

walls, using EQUATION 10.7,

$$U = \frac{1}{0.25 + 0.44 + 0.40 + 3.33 + 0.93 + 0.34 + 0.09 + 0.61} = 0.16$$

from table 11.10 for 4" brick facing at 4 p.m., using EQUATION 11.71,

$$q_{walls} = 0.16[1600(17) + 1400(28) + 1500(20) + 1400(20)] = 19{,}904 \text{ BTU/hr}$$

roof,

$$U = \frac{1}{0.25 + 0.32 + 5.56 + 0.06 + 0.87 + 1.19 + 0.92} = 0.109$$

from table 11.10 for heavy - 4" concrete roof at 4 p.m.,

$$q_{roof} = (0.109)(6000)(74) = 48{,}396 \text{ BTU/hr}$$

windows, -- since all windows face east, and the calculation is for 4 p.m., assume windows are not in direct sunlight, and use EQUATION 11.72,

$$q_{windows} = \left(\frac{1}{0.66}\right)(100)(95 - 78) = 2576 \text{ BTU/hr}$$

sensible Transmission Load, then, is

$$q_{total} = 19{,}904 + 48{,}396 + 2{,}576 = \boxed{70{,}876 \text{ BTU/hr}}$$

It is indeterminate whether this is the peak cooling load; the outdoor temperature at other times must be known. See the chart below. Either 4 p.m. or 6 p.m. shows the peak cooling load.

q (BTU/hr)

| | 12 p.m. | 2 p.m. | 4 p.m. | 6 p.m. | 8 p.m. |
|---|---|---|---|---|---|
| walls | 13,984 | 18,112 | 19,904 | 24,512 | 23,328 |
| roof | 30,084 | 42,510 | 48,396 | 44,472 | 32,046 |
| windows | 1,818 (90°F) or 2,576 (95°F) | | | | |
| TOTAL (90°F outside) | 45,886 | 62,440 | 70,118 | 70,802 | 57,192 |
| TOTAL (95°F outside) | 46,644 | 63,198 | 70,876 | 71,560 | 57,950 |

# Statics

WARM-UPS

1. $T = F \cdot r = \dfrac{(3)\ \text{LBF}\ (3)\ \text{IN}}{(12)\left(\frac{\text{IN}}{\text{FT}}\right)} = .75\ \text{FT-LBF}$

2. FROM TABLE 14.1,

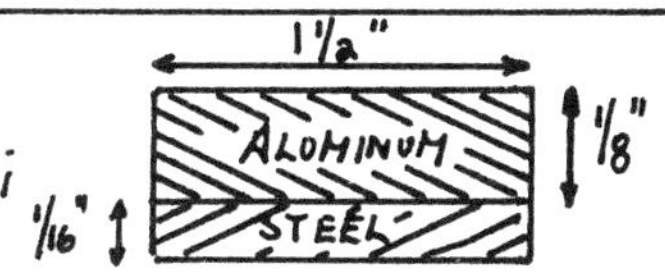


$E_{STEEL} = 30\ \text{EE6 psi}$

$E_{ALUM} = 10\ \text{EE6 psi}$

REFER TO THE PROCEDURE ON PAGE 14-7

$N = \dfrac{30\ \text{EE6}}{10\ \text{EE6}} = 3$

SO, THE EQUIVALENT, ALL ALUMINUM CROSS SECTION IS

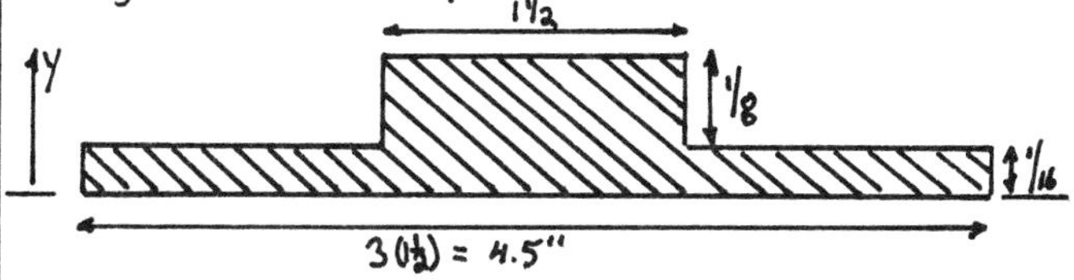


THE CENTROID OF THIS COMPOSITE STRUCTURE IS FOUND FROM EQN 12.73

$A_1 = (1/16)(4.5) = .28125$

$A_2 = (1/8)(1.5) = .1875$

$y_1 = (1/16)/2 = .03125$

$y_2 = \frac{1}{16} + (1/8)/2 = .125$

$$\bar{y} = \frac{(.28125)(.03125) + (.1875)(.125)}{.28125 + .1875} = .06875$$

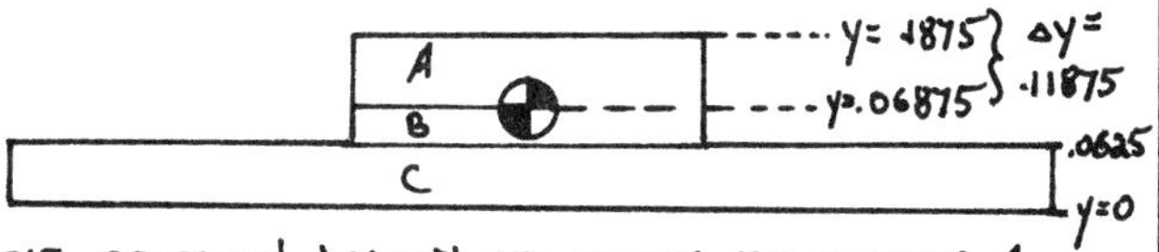


THE CENTROIDAL MOMENTS OF INERTIA FOR SECTIONS A AND B ARE

$I_A = \dfrac{bh^3}{3} = \dfrac{(1.5)(.11875)^3}{3} = 8.373\ \text{EE-4}$

$I_B = \dfrac{(1.5)(.00625)^3}{3} = 1.221\ \text{EE-7}$

THE MOMENT OF INERTIA OF SECTION C ABOUT ITS OWN CENTROID IS

$I_{ox} = \dfrac{bh^3}{12} = \dfrac{(4.5)(.0625)^3}{12} = 9.155\ \text{EE-5}$

THE DISTANCE FROM C'S CENTROID TO THE COMPOSITE CENTROID IS

$\bar{y} = \left(\dfrac{.0625}{2}\right) + .00625 = .0375$

FROM THE PARALLEL AXIS THEOREM {EQN 12.80}

$I_C = I_{ox} + A(\bar{y})^2$

$= (9.155\ \text{EE-5}) + (.28125)(.0375)^2$

$= 4.871\ \text{EE-4}$

THE TOTAL MOMENT OF INERTIA IS

$I_t = I_A + I_B + I_C = 1.324\ \text{EE-3}\ \text{IN}^4$

3. FROM TABLE 14.1

$\rho_{\text{CAST IRON}} = .256\ \dfrac{\text{LBM}}{\text{IN}^3}$

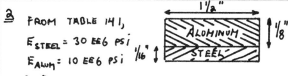

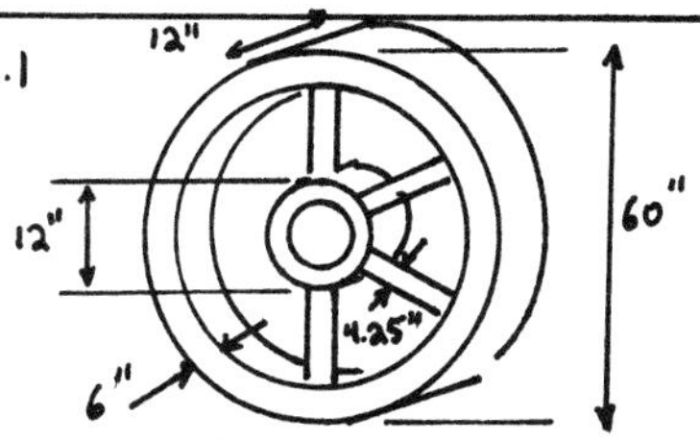


$W_{RIM} = (\text{VOL})\rho = \frac{\pi}{4}\left[(60)^2 - (48)^2\right](12)(.256)$

$= 3126.9\ \text{LBM}$

$W_{HUB} = \frac{\pi}{4}\left[(12)^2 - (6)^2\right](12)(.256)$

$= 260.6\ \text{LBM}$

THE ARMS ARE 18" LONG.

$W_{ARM} = \left(\frac{\pi}{4}\right)(4.25)^2(18)(.256)$

$= 65.4\ \text{LBM EACH}$

FOR A HOLLOW RIGHT CIRCULAR CYLINDER,

$J = \frac{1}{2}M(R_o^2 + R_i^2)$

$J_{RIM} = \left(\frac{1}{2}\right)\dfrac{3126.9}{32.2}\left[\dfrac{(60/2)^2 + (48/2)^2}{144}\right]$

$= 497.7\ \text{SLUG-FT}^2$

$J_{HUB} = \left(\frac{1}{2}\right)\dfrac{260.6}{32.2}\left[\dfrac{(12/2)^2 + (6/2)^2}{144}\right]$

$= 1.3\ \text{SLUG-FT}^2$

FROM PAGE 12-20 THE CENTROIDAL MOMENT OF INERTIA OF A CIRCULAR CYLINDER IS

$J_c = \left(\frac{1}{12}\right)m(3r^2 + L^2)$

$= \left(\frac{1}{12}\right)\left(\dfrac{65.4}{32.2}\right)\left[\dfrac{3(4.25/2)^2 + (18)^2}{144}\right]$

$= .4\ \text{EACH}$

ANALOGOUS TO EQN 12.80, THE MOMENT OF INERTIA ABOUT THE ROTATIONAL AXIS (15" AWAY FROM THE ARM'S CENTROID) IS

$J_{ARM} = J_c + Mr^2$

$= .4 + \left(\dfrac{65.4}{32.2}\right)(15/12)^2$

$= 3.57\ \text{SLUG-FT}^2\ \text{EACH}$

FOR 6 ARMS,

$J_{ARMS} = 6(3.57) = 21.4\ \text{SLUG-FT}^2$

THEN,

$J_{total} = J_{RIM} + J_{HUB} + J_{ARMS}$

$= 520.4\ \text{SLUG-FT}^2$

**CONCENTRATES**

**1**

FOR TRIANGLE EBD

LENGTH BE =

$\sqrt{(13.4)^2 - (12)^2}$

$= 5.96$

{SAY 6}

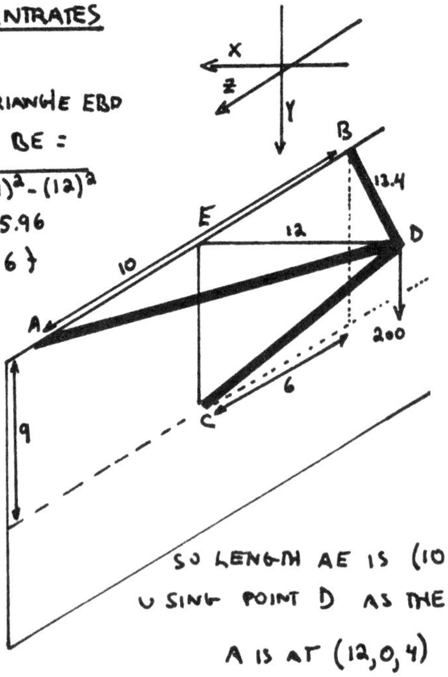


SO LENGTH AE IS $(10-6) = 4$.

USING POINT D AS THE ORIGIN,

A IS AT (12, 0, 4)

B IS AT (12, 0, −6)

C IS AT (12, 9, 0)

D IS AT (0, 0, 0)

FOLLOWING THE PROCEDURE ON PAGE 12-11, THE DIRECTION COSINES OF THE APPLIED LOAD ARE

$\cos\theta_x = 0 \quad \cos\theta_y = 1 \quad \cos\theta_z = 0$

SO $\quad F_x = 0 \quad F_y = 200 \quad F_z = 0$

THE LENGTHS OF THE LEGS ARE

$$AD = \sqrt{(12-0)^2 + (0-0)^2 + (4-0)^2} = 12.65$$

$BD = 13.4$ (GIVEN)

$$CD = \sqrt{(12)^2 + (9)^2 + (0)^2} = 15.00$$

THE LEG DIRECTION COSINES ARE

<u>FOR AD</u>

$$\cos\theta_{xA} = \frac{12}{12.65} = .949$$

$$\cos\theta_{yA} = \frac{0}{12.65} = 0$$

$$\cos\theta_{zA} = \frac{4}{12.65} = .316$$

<u>FOR BD</u>

$$\cos\theta_{xB} = \frac{12}{13.4} = .896$$

$$\cos\theta_{yB} = \frac{0}{13.4} = 0$$

$$\cos\theta_{zB} = \frac{-6}{13.4} = -.448$$

<u>FOR CD</u>

$$\cos\theta_{xC} = \frac{12}{15.00} = .8$$

$$\cos\theta_{yC} = \frac{9}{15.00} = .6$$

$$\cos\theta_{zC} = \frac{0}{15.00} = 0$$

USING EQUATIONS 12.65 – 12.67

$$\left.\begin{aligned} .949\,F_A + .896\,F_B + .8\,F_C + 0 &= 0 \\ .6\,F_C + 200 &= 0 \\ .316\,F_A - .448\,F_B + 0 &= 0 \end{aligned}\right\}$$

THE SIMULTANEOUS SOLUTION TO THESE 3 EQUATIONS IS

$F_C = -333.3$ LBF (COMPRESSION)

$F_B = 118.9$ (TENSION)

$F_A = 168.6$ (TENSION)

**2** $\quad \Sigma M_C \circlearrowleft + : \; D_y(6) - 8000(6) + 1600(16) = 0$

$D_y = 3733$

SINCE DE IS THE ONLY VERTICAL MEMBER LEAVING POINT D,

DE = 3733 {COMPRESSION}

**3** DIVIDE INTO 3 AREAS

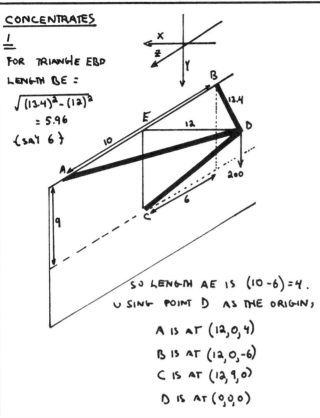

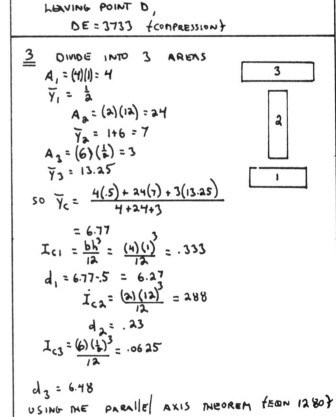

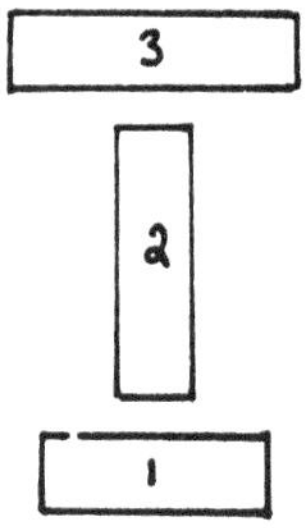


$A_1 = (4)(1) = 4$

$\bar{y}_1 = \frac{1}{2}$

$A_2 = (2)(12) = 24$

$\bar{y}_2 = 1 + 6 = 7$

$A_3 = (6)(\frac{1}{2}) = 3$

$\bar{y}_3 = 13.25$

$$\text{SO } \bar{y}_c = \frac{4(.5) + 24(7) + 3(13.25)}{4 + 24 + 3}$$

$= 6.77$

$$I_{c1} = \frac{bh^3}{12} = \frac{(4)(1)^3}{12} = .333$$

$d_1 = 6.77 - .5 = 6.27$

$$I_{c2} = \frac{(2)(12)^3}{12} = 288$$

$d_2 = .23$

$$I_{c3} = \frac{(6)(\frac{1}{2})^3}{12} = .0625$$

$d_3 = 6.48$

USING THE PARALLEL AXIS THEOREM {EQN 12.80}

$$I_{total} = .333 + 4(6.27)^2 + 288 + 24(.23)^2 + .0625 + 3(6.48)^2$$

$= 572.88$

**4** USE THE PROCEDURE ON PAGE 12-11. BUT FIRST, MOVE THE ORIGIN TO THE APEX OF THE TRIPOD SO THAT THE COORDINATES OF THE TRIPOD BASES BECOME

$A = (5, -12, 0)$

$B = (0, -8, -8)$

$C = (-4, -7, 6)$

BY INSPECTION,

$F_x = 1200 \quad F_y = 0 \quad F_z = 0$

$L_A = \sqrt{(5)^2 + (-12)^2 + (0)^2} = 13$

$\cos\theta_{xA} = \frac{5}{13} = .385$

$\cos\theta_{yA} = \frac{-12}{13} = -.923$

$\cos\theta_{zA} = \frac{0}{13} = 0$

$L_B = \sqrt{(0)^2 + (-8)^2 + (-8)^2} = 11.31$

$\cos\theta_{xB} = \frac{0}{11.31} = 0$

$\cos\theta_{yB} = \frac{-8}{11.31} = -.707$

$\cos\theta_{zB} = \frac{-8}{11.31} = -.707$

$L_C = \sqrt{(-4)^2 + (-7)^2 + (6)^2} = 10.05$

$\cos\theta_{xC} = \frac{-4}{10.05} = -.398$

$\cos\theta_{yC} = \frac{-7}{10.05} = -.697$

$\cos\theta_{zC} = \frac{6}{10.05} = .597$

FROM EQUATIONS 12.65 - 12.67

$$.385 F_A \qquad\qquad -.398 F_C = -1200$$

$$-.923 F_A - .707 F_B - .697 F_C = 0$$

$$-.707 F_B + .597 F_C = 0$$

SOLVING THESE SIMULTANEOUSLY YIELDS

$F_A = -1793$ {COMPRESSION}

$F_B = 1080$ {TENSION}

$F_C = 1279$ {TENSION}

**5** BY SYMMETRY, $A_y = L_y = 160$ KIPS

FOR DE

CUT AS SHOWN AND SUM VERTICAL FORCES.

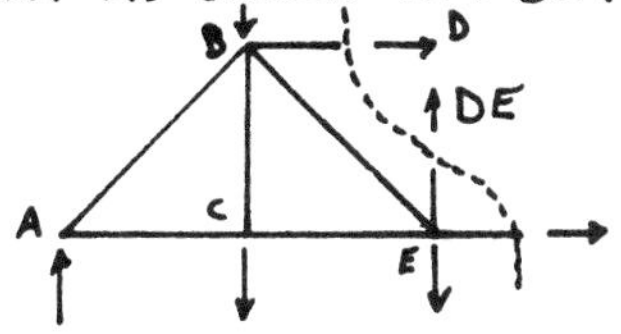


$\Sigma F_y + \uparrow: \; 160 + DE - 60 - 60 - 4 = 0$

$DE = -36$ KIPS {COMPRESSION}

FOR HJ

CUT AS SHOWN AND TAKE MOMENTS ABOUT I

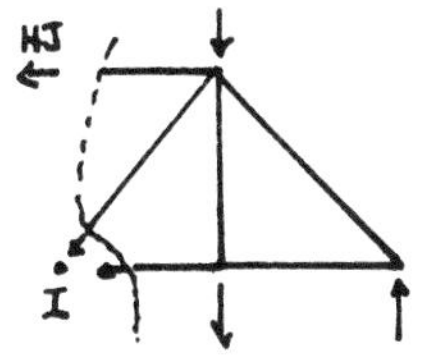


$\Sigma M_I$ ↺ : $160(60) - 60(30) - 4(30) + HJ(20) = 0$

$HJ = -384$ {COMPRESSION}

**6** IN THE X-Y PLANE

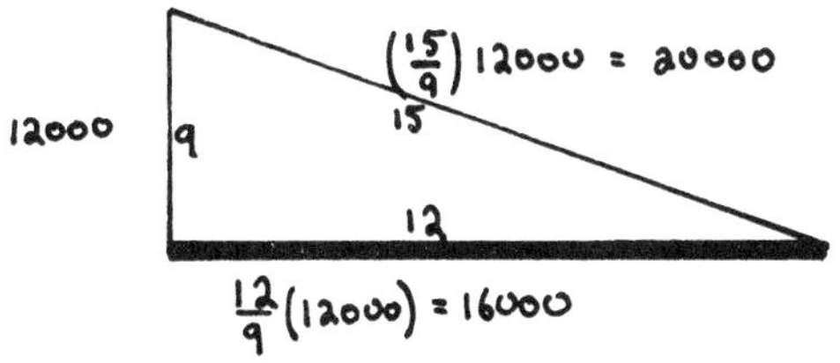


IN THE X-Z PLANE

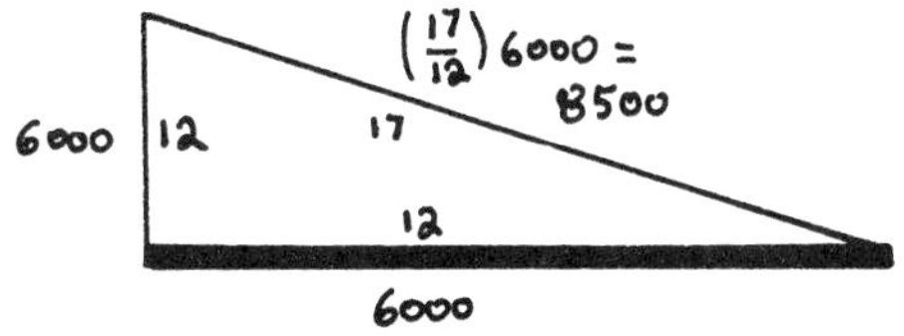


SO $A_x = 6000 + 16000 = 22000$

$A_y = 0 \qquad A_z = 0$

$B_x = 16000 \qquad B_y = 12000 \qquad B_z = 0$

$C_x = 6000 \qquad C_y = 0 \qquad C_z = 6000$

**7** REFER TO PAGE 12-10. FROM FIGURE 12-12,

$w = 2$ LBM/FT

$S = 10$

$a = 50$

SOLVING EQN 12.51 BY TRIAL AND ERROR GIVES DISTANCE C

$$S = c\left[\cosh\left(\frac{a}{c}\right) - 1\right]$$

$c = 126.6$

THE MIDPOINT {HORIZONTAL} TENSION IS GIVEN BY EQN 12.53:

$H = wC = (2)(126.6) = 253.2$ LBF

THE ENDPOINT {MAXIMUM} TENSION IS GIVEN BY EQN 12.55:

$T = wy = w(c+S)$

$= 2(126.6 + 10) = 273.2$

**8** FROM EQN 12.55,

$T = wy$ OR

$y = \frac{T}{w} = \frac{500}{2} = 250$ AT RIGHT SUPPORT

FROM EQN 12.48,

$250 = c\left(\cosh\left(\frac{50}{c}\right)\right)$

$c = 245$ BY TRIAL AND ERROR

FROM EQN 12.51,

$S = 245\left[\cosh\left(\frac{50}{245}\right) - 1\right] = 5.12$ ft

{OR, NOTICE FROM FIGURE 12.12 THAT $S = y - c = 5$}

**9** BY INSPECTION, $\bar{y} = 0$

TO FIND $\bar{x}$, DIVIDE THE OBJECT INTO 3 PARTS

$A_1 = 8(4) = 32$

$\bar{x}_1 = 2$

$A_2 = A_3 = 2(4) = 8$

$\bar{x}_2 = \bar{x}_3 = 6$

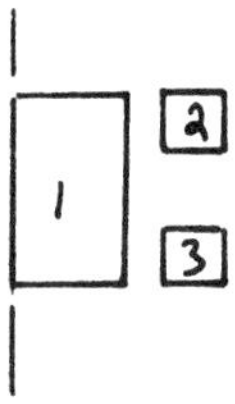


$\bar{x}_c = \frac{32(2) + 8(6) + 8(6)}{32 + 8 + 8} = 3.333$

**10** FOR THE PARABOLA

FROM PAGE 1-2,

$A = \frac{2bh}{3} = \frac{2(300)(3)}{3}$

$= 600$ LBF

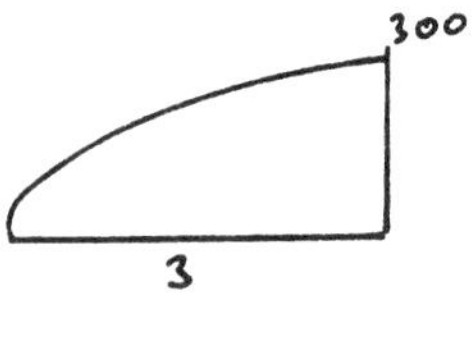


FROM PAGE 12-19, CENTROID IS

$\frac{3h}{5} = \frac{3(3)}{5} = 1.8'$ FROM TIP

DIVIDE THE REMAINING AREA INTO A TRIANGLE AND A RECTANGLE

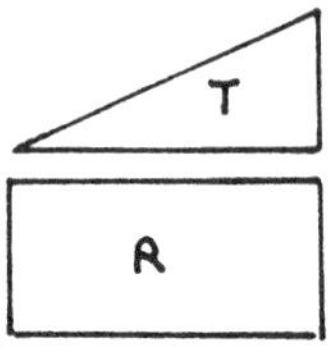


RECTANGLE

$A_R = (8)(300) = 2400$ LBF

CENTROID IS 4' FROM RIGHT END

TRIANGLE

$A_T = \frac{1}{2}(8)(700 - 300) = 1600$ LBF

CENTROID IS LOCATED

$\left(\frac{8}{3}\right) = 2.67$ FROM RIGHT END

*TIMED*

**1**

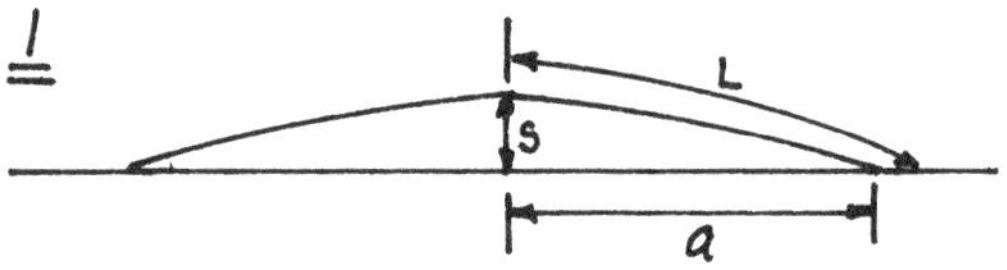


FROM TABLE 14.1, PAGE 14-3, $\alpha_{steel} = 6.5$ EE-6 1/°F

USING EQUATION 12.121,

$\delta = (6.5\text{EE-}6\ \frac{1}{°F})(5280\text{ FT.})(99.14 - 70\ °F)$

$= 1.00008$ FT $\sim 1$ FT

$L = \frac{5280 + 1}{2} = 2640.5$ FT.

FOR THERMAL EXPANSION, ASSUME THE DISTRIBUTED LOAD IS UNIFORM ALONG THE LENGTH OF THE RAIL. THIS RESEMBLES THE CASE OF A CABLE UNDER ITS OWN WEIGHT -- THE CATENARY. WHEN THE DISTANCE s IS SMALL RELATIVE TO DISTANCE a, THE PROBLEM CAN BE SIMPLIFIED BY USING THE PARABOLIC SOLUTION.

USE EQUATION 12.47 -- GIVEN L AND a (2640 FT), FIND s BY TRIAL AND ERROR.

$L \approx (2640)\left[1 + \frac{2}{3}\left(\frac{s}{2640}\right)^2 - \frac{2}{5}\left(\frac{s}{2640}\right)^4\right] = 2640.5$

| s | L (SHOULD BE 2640.5) |
|---|---|
| 1 | 2640 |
| 10 | 2640.03 |
| 100 | 2642.5 |
| 50 | 2640.6 |
| 45 | 2640.51 |
| 44.5 | 2640.50 |

**$s \approx 44.5$ FEET**

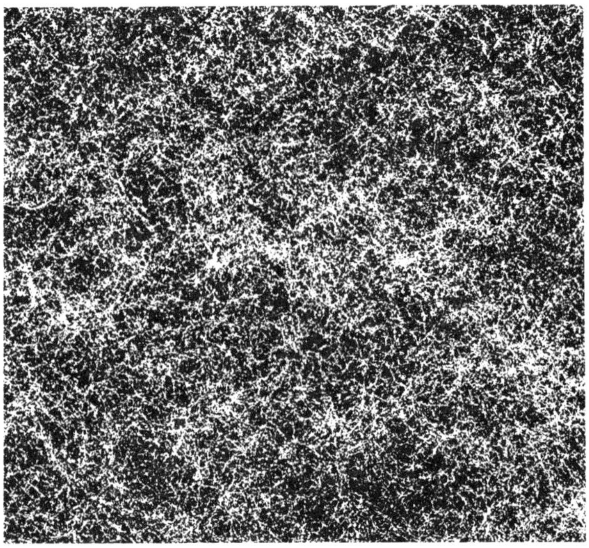

# Materials Science

1 a)

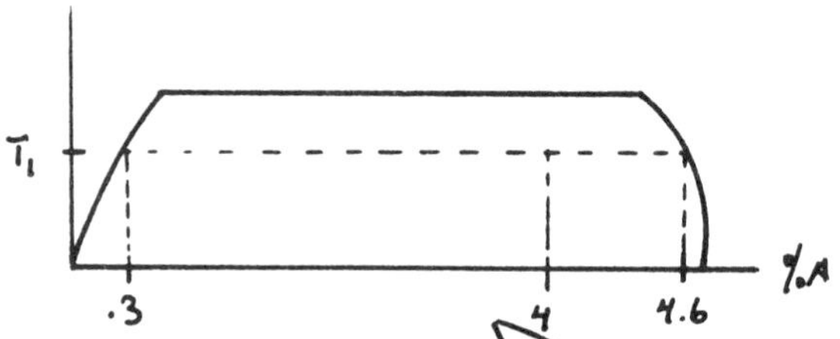


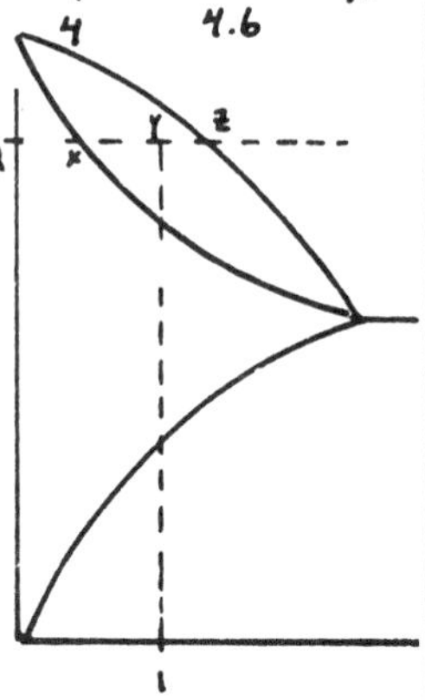


α IS .3% A, 99.7% B

β IS 4.6% A, 95.4% B

b) LINE SEGMENT XY = 9 MM

LINE SEGMENT XZ = 21 MM

$$\% \text{ LIQUID} = \frac{9}{21} = 43\%$$

$$\% \text{ SOLID} = 100 - 43 = 57\%$$

2 % CHANGE IN DIAMETER = (.3)(.020)

= .006

$$\sigma_{true} = \frac{P}{A_o(1-.006)^2} = 20242 \text{ PSI}$$

$$\epsilon_{true} = \ln\left(\frac{L}{L_o}\right) = \ln(1+.02) = .0198 \text{ IN/IN}$$

3 a)

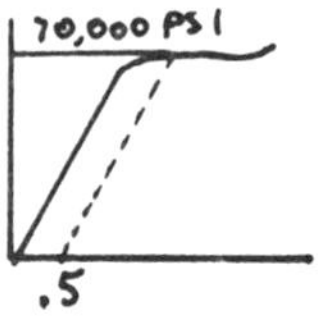


b)

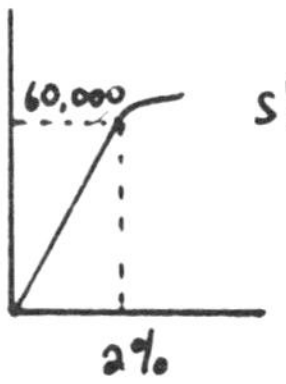


$$\text{SLOPE} = \frac{60{,}000}{.02} = 3 \text{ EE6 PSI}$$

c) THE HIGHEST STRESS REACHED IS APPROXIMATELY 80 KSI.

d) THE STRESS AT FRACTURE IS APPROXIMATELY 70 KSI.

e) STRAIN AT FRACTURE ≈ 8%

4 FROM EQN 14.77,

$$G = \frac{E}{2(1+\mu)} = \frac{3 \text{ EE6}}{2(1+.3)} = 1.15 \text{ EE6 PSI}$$

5 $$\text{DUCTILITY} = \frac{4-3.42}{4} = .145 \text{ REDUCTION IN AREA}$$

6 TOUGHNESS IS THE AREA UNDER THE CURVE: DIVIDE THE AREA INTO (20 KSI × 1%) SQUARES. THERE ARE ABOUT 25 SQUARES

$$(25)(20{,}000)\ \frac{\text{LBF}}{\text{IN}^2}\ (.01)\ \frac{\text{IN}}{\text{IN}}$$

$$= 5000\ \frac{\text{IN-LBF}}{\text{IN}^3}$$

7 PLOT THE DATA AND DRAW A STRAIGHT LINE DISREGARD THE FIRST AND LAST DATA POINTS.

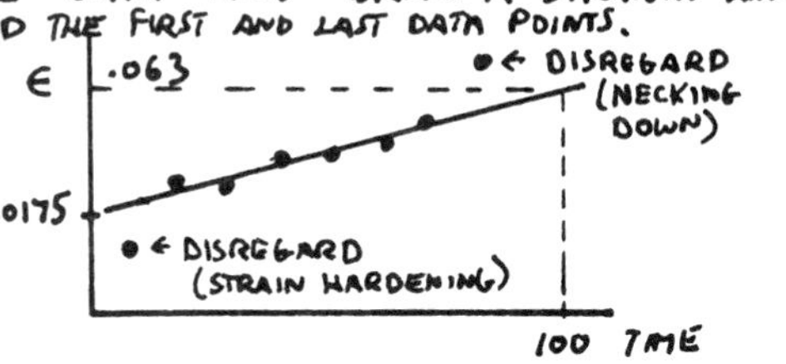


$$\text{SLOPE} = \frac{.063 - .0175}{100} = .000455\ \text{1/HR}$$

8 THE VINYL CHLORIDE MER IS

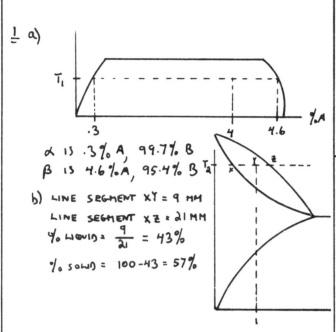

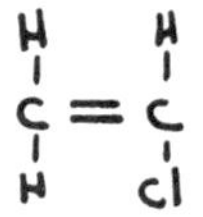

THE MOLECULAR WEIGHT IS

$$2(12) + 3(1) + 1(35.5) = 62.5$$

WITH 20% EFFICIENCY, WE WILL NEED 5 MOLECULES OF HCl PER PVC MOLECULE TO SUPPLY THE END –Cl ATOM. THIS IS THE SAME AS 5 MOLES HCl PER MOLE PVC, OR,

$$\frac{(5)(6.023 \text{ EE } 23)}{7000} = (4.3 \text{ EE } 20)\ \frac{\text{molecules HCl}}{\text{gram PVC}}$$

9 THE MOLECULAR WEIGHTS ARE

$H_2O_2$: (2)(1) + (2)(16) = 34

$C_2H_4$: (2)(12) + (4)(1) = 28

THE NUMBER OF $H_2O_2$ MOLECULES IN 10 mℓ IS

$$\frac{(10)\ \text{m}\ell\ (1)\ \text{g/m}\ell\ \left(\frac{0.2}{100}\right)\left(6.023 \text{ EE}23\right)\frac{\text{molecules}}{\text{gmole}}}{(34)\ \text{g/gmole}}$$

$$= 3.54 \text{ EE } 20$$

THE NUMBER OF ETHYLENE MOLECULES IS

$$\frac{(12)(6.023 \text{ EE } 23)}{28} = 2.58 \text{ EE } 23$$

SINCE IT TAKES 1 $H_2O_2$ MOLECULE (THAT IS, 2 $OH^-$ RADICALS) TO STABILIZE A POLYETHYLENE MOLECULE, THERE ARE 3.54 EE 20 POLYMERS. THE DEGREE OF POLYMERIZATION IS

$$\frac{2.58 \text{ EE}23}{3.54 \text{ EE}20} = 729$$

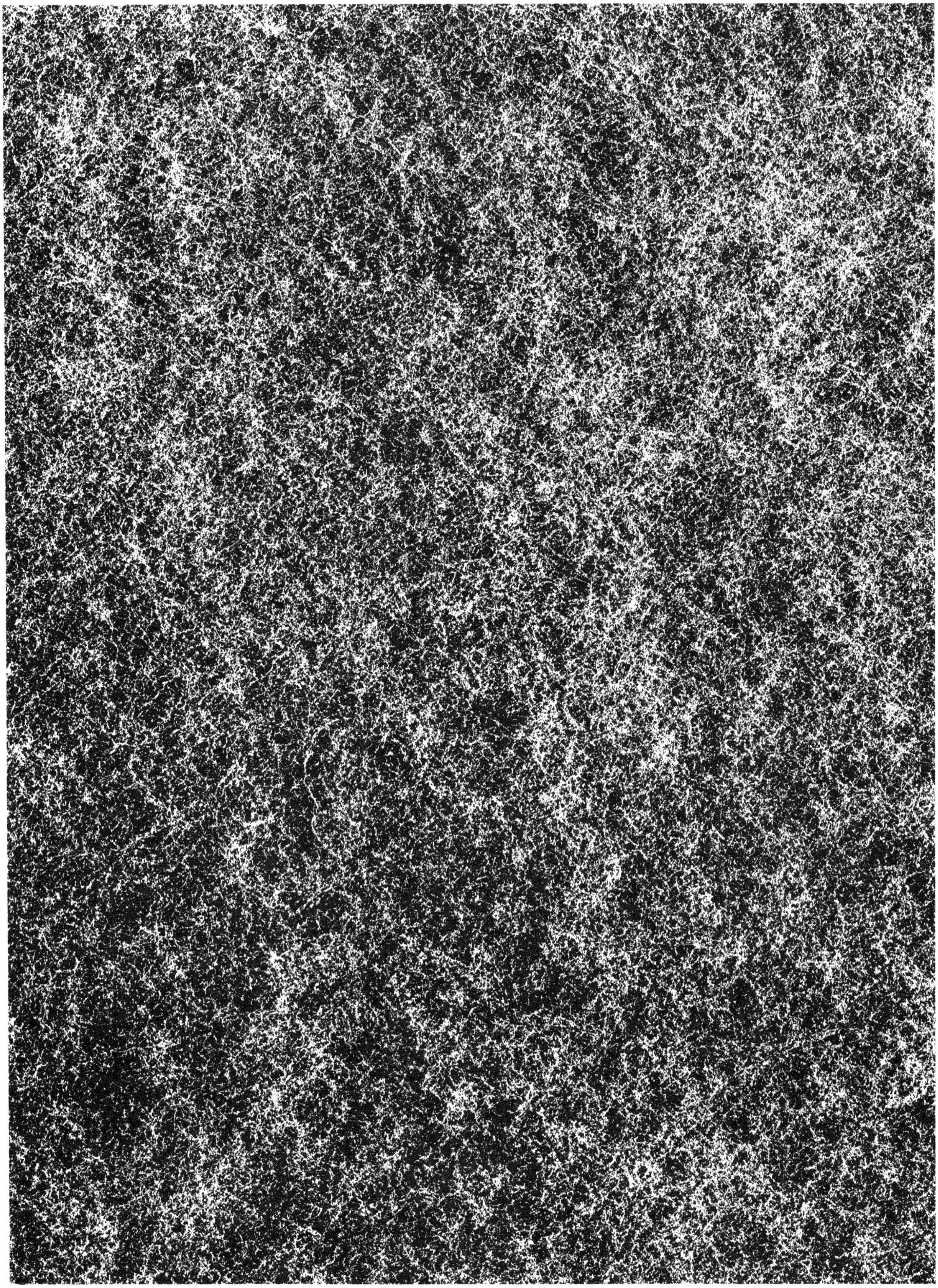

# Mechanics of Materials

**1** FIRST, FIND THE REACTIONS L AND R.

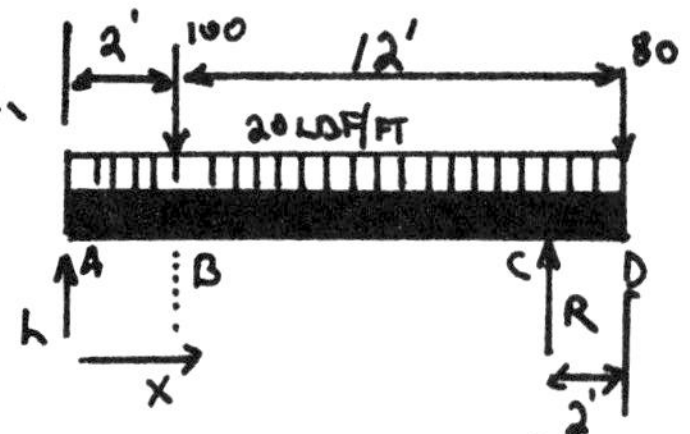


$\sum M_L$ ↻ : $100(2) + (20)(12)(6) + (80)(12) - R(10) = 0$

$R = 260$

$\sum F_y$ +↑ : $L + 260 - 100 - 80 - 20(12) = 0$

$L = 160$

SHEAR

AT $A^+$ : $V = 160$

AT $B^-$ : $V = 160 - 2(20) = 120$

AT $B^+$ : $V = 120 - 100 = 20$

AT $C^-$ : $V = 20 - 20(8) = -140$

AT $C^+$ : $V = -140 + 260 = 120$

AT $D^-$ : $V = 120 - 20(2) = 80$

AT $D^+$ : $V = 80 - 80 = 0$

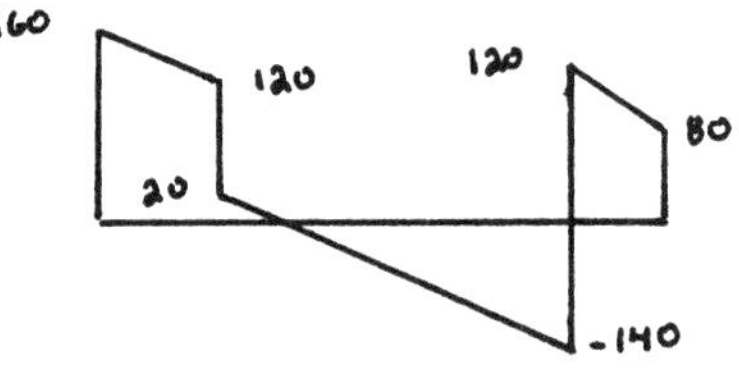


SO $V_{max} = 160$ LBF AT LEFT END

MOMENT

$M_{max}$ OCCURS WHEN $V=0$. THIS OCCURS AT

$x = 2 + \left(\frac{20}{160}\right)(8) = 3.00$

$M_{max} = 160(3) - 100(1.00) - (20)(3)\left(\frac{3}{2}\right)$

$= 290$ ft-lbf

**2**

$I = \frac{bh^3}{12}$

$= \frac{6(4)^3}{12} = 32 \text{ IN}^4$

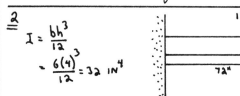

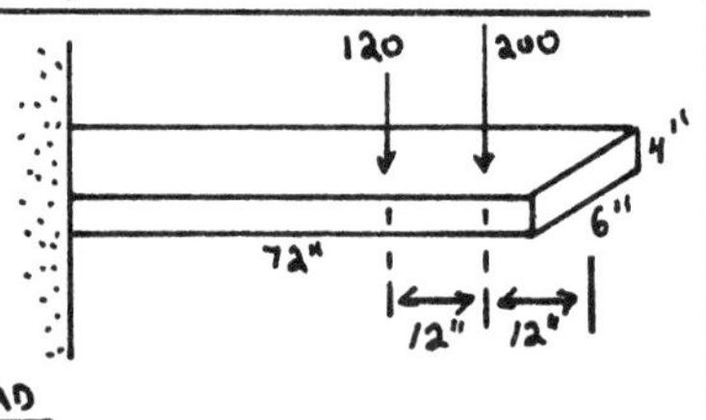


FOR THE 200 POUND LOAD

FOR CASE 1 {PAGE 14-27}, THE DEFLECTION AT THE LOAD IS

$y_1 = \frac{FL^3}{3EI} = \frac{200(60)^3}{3(1.5\text{ EE}6)(32)} = .3000''$

DIFFERENTIATING $y$ AS A FUNCTION OF $x$ GIVES THE SLOPE AT ANY POINT

$y_1' = \frac{F}{6EI}\left(0 - 3L^2 + 3x^2\right)$

AT $x=0$ AND $L=60$,

$y_1' = \frac{200}{(6)(1.5\text{ EE}6)(32)}\left(-3(60)^2\right) = -7.5\text{ EE-3 IN/IN}$

SO, THE ADDED DEFLECTION AT THE TIP IS

$y_2 = (7.5\text{ EE-3})(12) = .0900$ IN

FOR THE 120 # LOAD

$y_3 = \frac{(120)(48)^3}{(3)(1.5\text{ EE}6)(32)} = .0922$

$y_3' = \frac{120}{(6)(1.5\text{ EE}6)(32)}\left(-3(48)^2\right) = -2.88\text{ EE-3 IN/IN}$

$y_4 = (2.88\text{ EE-3})(24) = .0691$

SO

$y_{tip} = y_1 + y_2 + y_3 + y_4 = 0.5513''$

**3** FROM TABLE 14.1

$E_{steel} = 29\text{ EE}6$ PSI

$\alpha = 6.5\text{ EE-6 } 1/°\text{F}$

$\Delta L = \alpha L \Delta T$

$= (6.5\text{ EE-6})(200)(12)(70) = 1.092''$

THE CONSTRAINED LENGTH IS

$1.092 - .5 = .592''$

THE STRAIN IS

$\epsilon = \frac{\Delta L_c}{L} = \frac{.592}{\left(200 + \frac{0.5}{12}\right)(12)} = 2.466\text{ EE-4}$

THE STRESS IS

$\sigma = \epsilon E = (2.466\text{ EE-4})(2.9\text{ EE}7)$

$= 7151$ psi

**4** THE DESIGN LOAD IS

$(2.5)(75000) = 187{,}500$ LBF

FROM TABLE 14.3, $C = .5$ FOR 2 FIXED ENDS, SO FROM EQN 14.58

$L' = CL = (.5)(50)(12) = 300$ INCHES

FROM EQN 14.50, THE REQUIRED MOMENT OF INERTIA IS

$I = \frac{FL^2}{\pi^2 E} = \frac{(187{,}500)\text{ LBF }(300)^2\text{ IN}^2}{\pi^2\,(2.9\text{ EE}7)\ \frac{\text{LBF}}{\text{IN}^2}}$

$= 58.96\text{ IN}^4$

**5** ASSUME SOFT STEEL. $E = 2.9\text{ EE}7$ PSI

$\Delta L = \frac{FL}{AE}$ SO

$L_o = \frac{\Delta L A E}{F} = \frac{(.158)\text{ IN}\left(\frac{\pi}{4}\right)(1)^2\text{ IN}^2(2.9\text{ EE}7)\ \frac{\text{LBF}}{\text{IN}^2}}{15000\text{ LBF}}$

$= 239.913$ IN

$L_{total} = L_o + \Delta L$

$= 239.913 + .158 = 240.071$

**6** Assuming ASTM A36 steel, the yield strength is approximately 36000 psi

$$FS = \frac{36000}{8240} = 4.37$$

**CONCENTRATES**

**1**

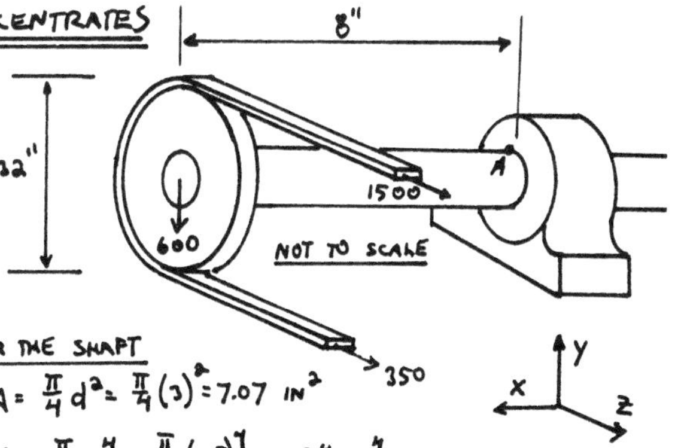


For the shaft

$$A = \frac{\pi}{4}d^2 = \frac{\pi}{4}(3)^2 = 7.07 \text{ in}^2$$

$$I = \frac{\pi}{4}r^4 = \frac{\pi}{4}(1.5)^4 = 3.98 \text{ in}^4$$

$$J = \frac{\pi}{2}r^4 = \frac{\pi}{2}(1.5)^4 = 7.98 \text{ in}^4$$

The moment at the bearing face due to the pulley weight is

$$M_y = (600) \text{ lbf } (8) \text{ in} = 4800 \text{ in-lbf}$$

The stress is

$$\sigma_y = \frac{Mc}{I} = \frac{(4800)(1.5)}{3.98} = 1809 \text{ psi}$$

The moment at the bearing face due to the belt tensions is

$$M_z = (1500 + 350)(8) = 14800 \text{ in-lbf}$$

The stress is

$$\sigma_z = \frac{(14800)(1.5)}{3.98} = 5578 \text{ psi}$$

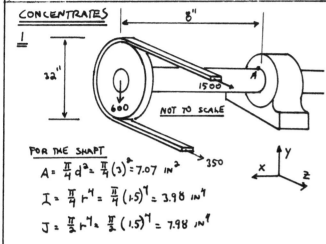

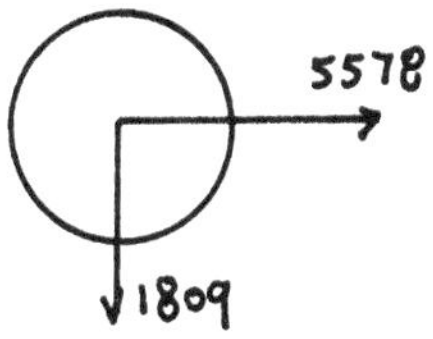


The maximum resultant bending stress is

$$\sigma_R = \pm\sqrt{(1809)^2 + (5578)^2} = \pm 5864 \text{ psi}$$

The net torque is

$$T = (1500 - 350)(16) = 18400 \text{ in-lbf}$$

The torsional shear stress is

$$\tau = \frac{Tc}{J} = \frac{(18400)(1.5)}{7.95} = 3472 \text{ psi}$$

<u>Note</u>: The direct shearing stress $\tau = \frac{V}{A}$ is zero on the shaft surface

The maximum stress is

$$\frac{1}{2}(5864) + \frac{1}{2}\sqrt{(5864)^2 + \left[(2)(3472)\right]^2}$$

$$= 7476 \text{ psi}$$

**2** The radius of gyration is

$$k = \sqrt{\frac{I}{A}} = \sqrt{\frac{350 \text{ in}^4}{25.6 \text{ in}^2}} = 3.70 \text{ in}$$

The slenderness ratio is

$$\frac{L}{k} = \frac{(25 \text{ ft})(12 \text{ in/ft})}{3.70 \text{ in}} = 81.08$$

$$\phi = \frac{1}{2}\left(\frac{L}{k}\right)\sqrt{\frac{F}{AE}}$$

$$= \left(\frac{1}{2}\right)(81.08)\sqrt{\frac{150{,}000 \text{ lbf}}{(25.6 \text{ in}^2)\left(2.9\times10^7 \frac{\text{lbf}}{\text{in}^2}\right)}}$$

$$= 0.5762 \text{ rad}$$

> 3rd and earlier printings only
>
> Calculate the eccentricity:
>
> $$\epsilon = \frac{M}{F} = \frac{(50{,}000 \text{ lbf})(10 \text{ in})}{150{,}000 \text{ lbf}} = 3.33 \text{ in}$$

(b) The largest compressive stress is

$$\sigma_{max} = \frac{F}{A}\left(1 + \frac{ec}{k^2}\sec\phi\right)$$

$$= \frac{150{,}000 \text{ lbf}}{25.6 \text{ in}^2}\left(1 + \frac{(3.33 \text{ in})(7 \text{ in})}{(3.70 \text{ in})^2}\sec(0.5762)\right)$$

$$= 17{,}757 \text{ lbf/in}^2 \text{ (psi)}$$

The stress factor of safety is

$$(FS)_{stress} = \frac{S_y}{\sigma_{max}} = \frac{36{,}000 \text{ lbf/in}^2}{17{,}757 \text{ lbf/in}^2} = \boxed{2.03}$$

(a) Find the load that would cause the maximum stress to equal the yield stress.

$$S_y = \frac{F}{A}\left(1 + \frac{ec}{k^2}\sec\left[\left(\frac{Lk}{2}\right)\sqrt{\frac{F}{AE}}\right]\right)$$

$$36{,}000 \frac{\text{lbf}}{\text{in}^2} = \frac{F_{max}}{25.6 \text{ in}^2}\left(1 + \frac{(3.33 \text{ in})(7 \text{ in})}{(3.7 \text{ in})^2} \times \sec\left[\frac{81.08}{2}\sqrt{\frac{F_{max}}{(25.6 \text{ in}^2)\left(2.9\times10^7 \frac{\text{lbf}}{\text{in}^2}\right)}}\right]\right)$$

$$36{,}000 = \frac{F_{max}}{25.6}\left(1 + 1.703\sec\left[40.54\sqrt{\frac{F_{max}}{7.424\times10^8}}\right]\right)$$

Using a calculator's equation solver, or by trial and error,

$$F_{max} = 272{,}200 \text{ lbf}$$

The load factor of safety is

$$(FS)_{load} = \frac{F_{max}}{F} = \frac{272{,}200 \text{ lbf}}{150{,}000 \text{ lbf}}$$

$$= \boxed{1.81}$$

**3** THE LARGEST FACE IS 2'×3' OR 24"×36". SINCE THE WALLS ARE FIXED, THE FLAT PLATE EQUATIONS (PAGE 14-24) MAY BE USED. FROM TABLE 14.7 FOR $(a/b) = \frac{36}{24} = 1.5$

SO $C_3 = \frac{1}{2}(.436 + .487) = .462$

$$\sigma_{MAX} = \frac{C_3 p b^2}{t^2} = \frac{(.462)(2)(24)^2}{(.25)^2} = 8515.6 \text{ PSI}$$

**4** $r_i = \frac{1.750}{2} = .875"$

IGNORE END EFFECTS. {IF THIS IS NOT TRUE, A COMBINED STRESS ANALYSIS IS REQUIRED}. FROM FIGURE 14.19, THE MAXIMUM STRESS WILL OCCUR AT THE INSIDE. USING TABLE 14.6 WITH AN ALLOWABLE STRESS OF 20,000,

$$\sigma_{c_i} = 20{,}000 = \frac{[r_o^2 + (.875)^2]\,2000}{(r_o + .875)(r_o - .875)}$$

$$20{,}000[r_o^2 - (0.875)^2] = [r_o^2 + (0.875)^2](2000)$$

$$r_o^2 = \frac{11}{9}(0.875)^2$$

$$r_o = 0.9673" \quad (d_o = 1.935")$$

**5** $r_o = \frac{1.486}{2} = .743$

$r_i = \frac{.742}{2} = .371$

FROM TABLE 14.6,

$$\sigma_{c_i} = \frac{-2(.743)^2(400)}{(.743 + .371)(.743 - .371)} = -1065.7 \text{ PSI (COMP)}$$

**6**

$A = \frac{\pi}{4}(d^2) = \frac{\pi}{4}(1)^2 = .7854$

$I = \frac{\pi}{4}(r)^4 = .04909$

$J = \frac{\pi}{2}(r)^4 = .09817$

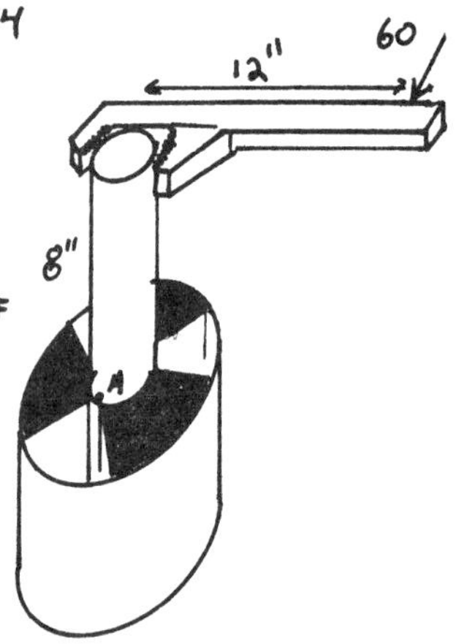


M AT CHUCK IS
(60) LBM (8) IN = 480 IN-LBF

$$\sigma = \frac{Mc}{I} = \frac{(480)(.5)}{.04909} = 4889 \text{ PSI}$$

THE APPLIED TORQUE IS

$$T = (60) \text{ LBF } (12) \text{ IN} = 720 \text{ IN-LBF}$$

SO, THE SHEAR STRESS IS

$$\tau = \frac{Tc}{J} = \frac{(720)(.5)}{.09817} = 3667 \text{ PSI}$$

FROM EQN 14.47,

$$\tau_{MAX} = \frac{1}{2}\sqrt{(4889)^2 + [(2)(3667)]^2} = 4407 \text{ PSI}$$

FROM EQN 14.46

$$\sigma_{MAX} = \frac{1}{2}(4889) + 4407 = 6852 \text{ PSI}$$

**7** $\sigma_1 = \frac{18{,}000}{1.5} = -12{,}000$ PSI

NEG. BECAUSE COMPRESSIVE.

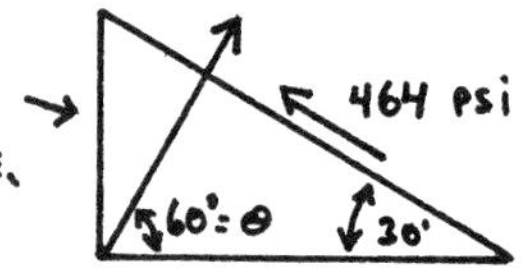


$\sigma_2 = 0$

FOR A 30° INCLINE, $\theta = 60°$. FROM EQN 14.44

$$\sigma_\theta = \frac{1}{2}(-12{,}000) + \frac{1}{2}(-12{,}000)\cos 120° + (4000)\sin 120° = 464 \text{ PSI}$$

FROM EQN 14.45,

$$\tau_\theta = -\frac{1}{2}(-12{,}000)\sin 120 + (4000)\cos 120° = 3196 \text{ PSI}$$

ALTERNATIVE INTERPRETATION

$\sigma_\theta = -6464$

$\tau_\theta = -7196$

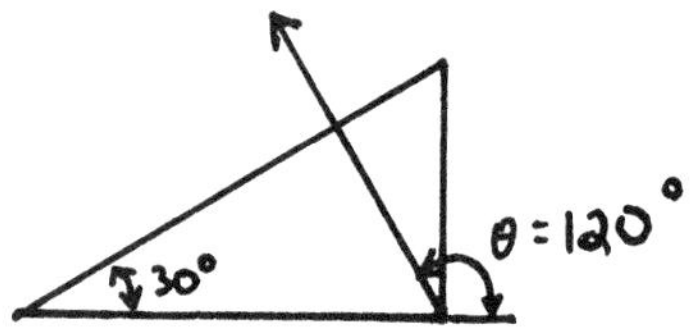


**8** THE AREA IN TENSION IS

$$\frac{\pi}{4}[(16)^2 - (15.8)^2] = 5 \text{ IN}^2$$

$$\text{SO } \sigma_N = \frac{F}{A} = \frac{40{,}000}{5} = 8000 \text{ PSI}$$

$$J = \frac{1}{32}\pi[(16)^4 - (15.8)^4] = 315.7$$

$$\text{SO } \tau = \frac{Tc}{J} = \frac{(400{,}000)(8)}{315.7} = 10{,}136 \text{ PSI}$$

FROM EQN 14.47,

$$\tau_{MAX} = \frac{1}{2}\sqrt{(8000)^2 + [(2)(10136)]^2} = 10{,}897$$

FROM EQN 14.46,

$$\sigma_N = \frac{1}{2}(8000) \pm 10897$$

OR +14,897, −6897

**TIMED**

**1**

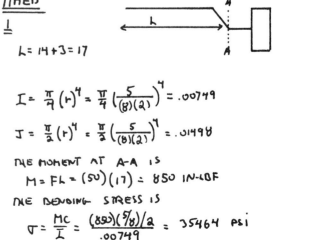

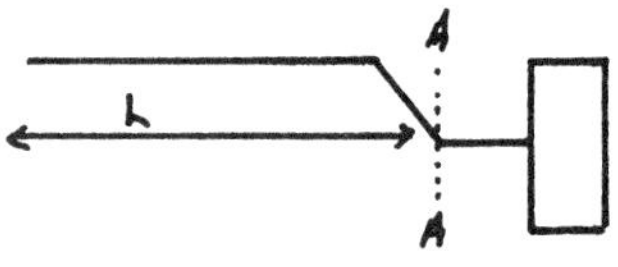


$L = 14 + 3 = 17$

$$I = \frac{\pi}{4}(r)^4 = \frac{\pi}{4}\left(\frac{5}{(8)(2)}\right)^4 = .00749$$

$$J = \frac{\pi}{2}(r)^4 = \frac{\pi}{2}\left(\frac{5}{(8)(2)}\right)^4 = .01498$$

THE MOMENT AT A-A IS

$$M = FL = (50)(17) = 850 \text{ IN-LBF}$$

THE BENDING STRESS IS

$$\sigma = \frac{Mc}{I} = \frac{(850)(5/8)/2}{.00749} = 35464 \text{ PSI}$$

THE TORQUE AT SECTION A-A DUE TO THE ECCENTRICITY IS

$$T = (50)(3) = 150 \text{ IN-LBF}$$

THE SHEAR STRESS IS

$$\tau = \frac{Tc}{J} = \frac{(150)(5/8)/2}{.01498} = 3129 \text{ psi}$$

FROM EQN 14.47, THE MAXIMUM SHEAR IS

$$\tau_{max} = \frac{1}{2}\sqrt{(35464)^2 + [(2)(3129)]^2}$$

$$= 18006 \text{ psi}$$

F/A DIRECT SHEAR IS IGNORED

FROM EQN 14.46,

$$\sigma_1, \sigma_2 = \tfrac{1}{2}(35464) \pm 18006$$

$$= 35738, -274 \text{ psi}$$

**2** FOR THE JACKET

$$r_o = \frac{12}{2} = 6'' \qquad r_i = \frac{7.75}{2} = 3.875''$$

FOR THE TUBE

$$r_o = \frac{7.75}{2} = 3.875'' \qquad r_i = \frac{4.7}{2} = 2.35$$

FOR THE JACKET, THIS IS NO DIFFERENT THAN EXPOSURE TO AN INTERNAL PRESSURE, P. FROM TABLE 14.6, PAGE 14-22, $\sigma_{ci}$ IS THE HIGHEST STRESS, SO

$$\sigma_{ci} = 18000 = \frac{[(6)^2 + (3.875)^2]\,p}{(6+3.875)(6-3.875)}$$

OR THE EQUIVALENT PRESSURE IS

$$p = 7404 \text{ psi}$$

AND $\sigma_{r_i} = -7404$

FROM EQN 14.88,

$$\Delta D = \frac{D}{E}\left[\sigma_c - \mu(\sigma_r + \sigma_L)\right]$$

$$= \frac{7.75}{(29.6\text{ EE}6)}\left[18000 - (.3)(-7404)\right]$$

$$= .00529''$$

OF COURSE, THE TUBE IS ALSO SUBJECTED TO THE PRESSURE. SO

$$\sigma_{ro} = -7404$$

$$\sigma_{co} = \frac{-[(3.875)^2 + (2.35)^2]}{[(3.875)^2 - (2.35)^2]}(7404) = -16018$$

{NOTICE THAT $(r_o + r_i)t = (r_o + r_i)(r_o - r_i) = r_o^2 - r_i^2$}

$$\Delta D = \frac{7.75}{(29.6\text{ EE}6)}\left[-16018 - (.3)(-7404)\right]$$

$$= -.00361$$

SO, THE TOTAL INTERFERENCE IS

$$.00529 + .00361 = .0089''$$

b) MAXIMUM STRESS OCCURS AT INNER JACKET FACE. $\sigma_{ci} = 18000$ psi

c) MINIMUM STRESS OCCURS AT OUTER TUBE FACE. $\sigma_{co} = -16018$ psi

**3**

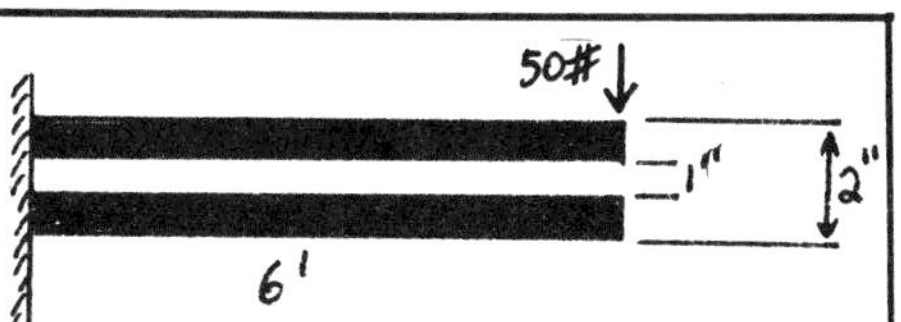


a) YES, THE SUGGESTION IS WITH MERIT. ANYTHING IN THE CENTER VOID WILL HAVE A HIGHER MODULUS OF ELASTICITY (STIFFNESS) THAN AIR.

b) CALCULATE I AS THE DIFFERENCE BETWEEN TWO CIRCLES:

$$I_{brass} = \frac{\pi}{4}\left[\left(\frac{2}{2}\right)^4 - \left(\frac{1}{2}\right)^4\right] = .736 \text{ IN}^4$$

$$E_{brass} = 15 \text{ EE}6 \text{ psi}$$

$$EI_{brass} = (1.5\text{ EE}7)(.736) = 1.104\text{ EE}7$$

$$I_{steel} = \frac{\pi}{4}\left(\frac{1}{2}\right)^4 = .0491 \text{ IN}^4$$

ASSUME SOFT STEEL, SO

$$E_{steel} = 2.9 \text{ EE}7 \text{ psi}$$

$$EI_{steel} = (2.9\text{ EE}7)(.0491) = .1424\text{ EE}7$$

FROM PAGE 14-27, CASE 1,

$$y_{tip} = \frac{FL^3}{3EI} \quad \text{\{INVERSELY PROPORTIONAL TO } EI\text{\}}$$

SO, THE % CHANGE WOULD BE

$$\% = \frac{y_{old} - y_{new}}{y_{old}} = \frac{\frac{1}{1.104} - \frac{1}{.1424 + 1.104}}{\frac{1}{1.104}}$$

$$= .114 \quad (11.4\%)$$

**4** FROM PAGE 14-30, $S_y$ FOR 6061 T4 ALUMINUM IS 19,000 psi.

THE FOLLOWING STRESSES ARE PRESENT IN THE SPOOL.

- LONG STRESS DUE TO PRESSURE ON END DISKS
- COMPRESSIVE HOOP STRESS
- BENDING STRESS AT DISK AND TUBE JUNCTION {END MOMENT}
- CIRCUMFERENTIAL (TANGENTIAL) STRESS

ASSUMPTIONS

- BENDING STRESS IS IGNORED
- CIRCUMFERENTIAL STRESS IS NEGLIGIBLE DUE TO THIN-WALL CONSTRUCTION
- FACTOR OF SAFETY CALCULATION DOES NOT DEPEND ON OBSCURE COLLAPSING PRESSURE THEORIES, BUT RATHER ON BASIC CONCEPTS.

CALCULATIONS

$$\text{END DISK AREA} = \frac{\pi}{4}\left[(2)^2 - (1)^2\right] = 2.356 \text{ IN}^2$$

$$\text{LONG FORCE} = pA = (500)(2.356) = 1178 \text{ LBF}$$

ANNULUS AREA ABSORBING THE LONG FORCE

$$= \frac{\pi}{4}\left[(1)^2 - (1 - 0.10)^2\right] = 0.149$$

$$\sigma_{LONG} = \frac{1178}{0.149} = 7906 \text{ psi (TENSILE)}$$

$$\sigma_{hoop} = \frac{pr}{t} = \frac{(500)(\frac{1}{2})}{.050} = -5000 \text{ psi (COMP)}$$

THESE ARE THE PRINCIPAL STRESSES. FROM DISTORTION ENERGY THEORY, THE VON MISES STRESS IS

$$\sigma' = \sqrt{\sigma_1^2 - \sigma_1\sigma_2 + \sigma_2^2}$$

$$= \sqrt{(7906)^2 - (7906)(-5000) + (-5000)^2} = 11{,}181$$

$$FS = \frac{19{,}000}{11{,}181} = 1.70$$

ANSWER WILL DEPEND ON THE VALUE OF $S_y$ USED.

**5** PUT ALL LENGTHS IN TERMS OF RADIUS IN METERS

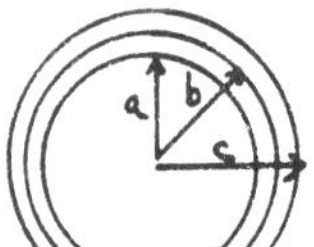


$$a = \left[\frac{120-2-2-3-3}{2}\right]10^{-3}$$

$= 5.5$ EE-2 M

$$b = \left[\frac{120-2-2}{2}\right] \text{EE-3}$$

$= 5.8$ EE-2 M

$$c = \left[\frac{120}{2}\right] \text{EE-3} = 6.0 \text{ EE-2 M}$$

$I = .3$ EE-3 $= 3$ EE-4 M

IN GENERAL, THE INTERFERENCE ($I$) IS

$$I = |\Delta D_{INNER}| + |\Delta D_{OUTER}|$$

$\Delta D$ IS CALCULATED THE SAME FOR BOTH. THERE IS NO LONG STRESS, SO

$$\Delta D = \frac{D}{E}\left[\sigma_c - \mu\sigma_r\right]$$

INNER CYLINDER

- EXTERNAL PRESSURE
- $r_i = a = 5.5$ EE-2
- $r_o = b = 5.8$ EE-2
- $t = b - a = .3$ EE-2
- $D = 2b = .116$ m

$$\sigma_{c,o} = \frac{-(r_o^2 + r_i^2)P}{(r_o + r_i)t} = \frac{-[(5.8)^2 + (5.5)^2](\text{EE-2})^2 P}{(5.8+5.5)(.3)(\text{EE-2})^2}$$

$= -18.85P$

$\sigma_{r,o} = -P$

$$\Delta D = \frac{.116}{207 \text{ EE9}}\left[-18.85P - (.3)(-P)\right]$$

$= -(1.04 \text{ EE-11})P$

OUTER CYLINDER

- INTERNAL PRESSURE
- $r_i = b = 5.8$ EE-2 M
- $r_o = c = 6.0$ EE-2 M
- $t = .2$ EE-2 M
- $D = 2c = 0.120$ M

$\sigma_{r,i} = -P$

$$\sigma_{c,i} = \frac{(r_o^2 + r_i^2)P}{(r_o + r_i)t} = \frac{[(6)^2 + (5.8)^2]P}{(6.0+5.8)(.2)}$$

$= 29.51P$

$$\Delta D = \frac{.116}{207 \text{ EE9}}\left[29.51P - (.3)(-P)\right]$$

$= +(1.64 \text{ EE-11})P$

SINCE $I$ IS KNOWN TO BE 3 EE-4 M,

$$3 \text{ EE-4} = P(1.04 \text{ EE-11} + 1.64 \text{ EE-11})$$

$$P = 1.12 \text{ EE7 N/M}^2 \text{ (PASCALS)}$$

NOW THAT P IS KNOWN, THE CIRCUMFERENTIAL STRESSES CAN BE FOUND:

INNER CYLINDER

$$\sigma_{c,o} = 18.85P = 2.11 \text{ EE8 N/M}^2$$

OUTER CYLINDER

$$\sigma_{c,i} = 29.51P = 3.31 \text{ EE8 N/M}^2$$

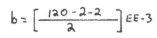

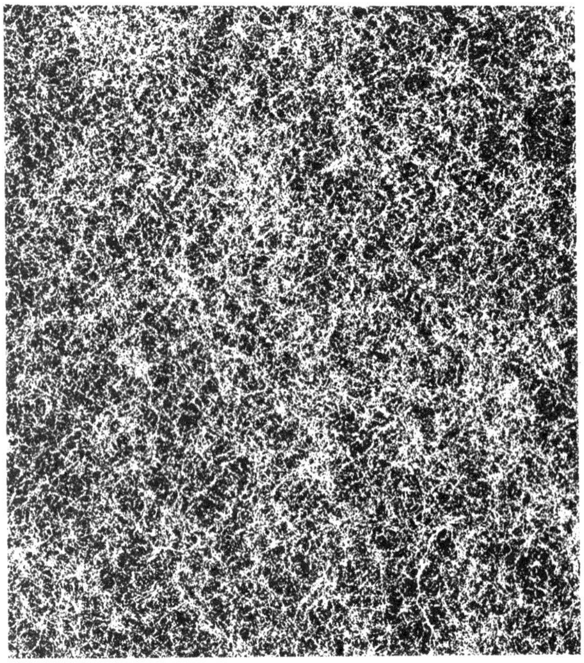

# Machine Design

WARM-UPS

1

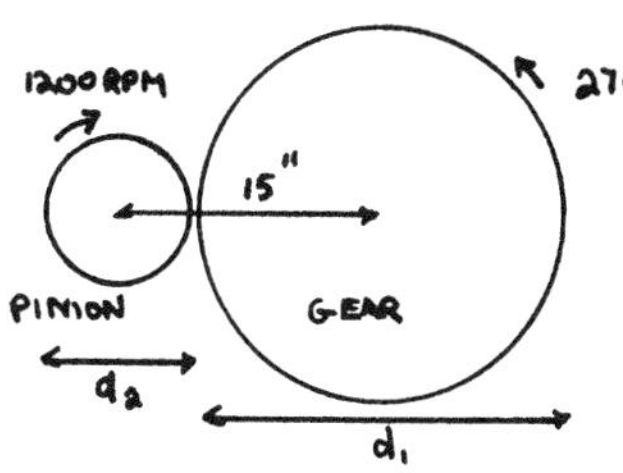


FIRST, FIND THE PITCH DIAMETERS. LET GEAR 2 BE THE PINION, AND GEAR 1 BE THE GEAR. THEN

$$\frac{d_1}{2} + \frac{d_2}{2} = 15$$

$$\frac{d_1}{d_2} = \frac{1200}{270}$$

SOLVING THESE SIMULTANEOUSLY,

$d_1 = 24.49$ AND $d_2 = 5.51$

FROM EQUATION 15.75, THE PITCH LINE VELOCITY IS

$$v = \frac{\pi (\text{rpm}) d}{12} = \frac{\pi (270)(24.49)}{12} = 1731.1 \text{ FPM}$$

THIS VELOCITY IS THE SAME FOR BOTH GEARS.

TO GET STARTED, ASSUME FOR THE PINION:
$W = 6''$ AND $Y = 0.2$

THE ENDURANCE LIMIT IS CHOSEN AS THE MAXIMUM STRESS SINCE NO HARDNESS DATA WAS GIVEN. ASSUME FOR THE 1045 STEEL

$$S_{\text{endurance}} = 90 \text{ ksi}$$

USE A FACTOR OF SAFETY OF 3. SO, THE allowable STRESS IS

$$\sigma_{\text{allowable}} = \frac{90}{3} = 30 \text{ ksi}$$

THE BARTH SPEED FACTOR (EQUATION 15.103) IS

$$K_d = \frac{600}{600 + 1731.1} = .257$$

NEGLECTING THE small AMOUNT OF RADIAL STRESS, FROM EQUATIONS 15.98 AND 15.104

$$\sigma = \frac{(\text{HP})(33{,}000)\,P}{v K_d w Y}$$

OR

$$P = \frac{(.257)(30{,}000)(6)(.2)(1731.1)}{(550)(33{,}000)} = 0.88$$

SO, CHOOSE THE NUMBER OF TEETH PER INCH AS 1. THEN, THE NUMBER OF TEETH ON THE PINION AND GEAR ARE

$$N_{\text{pinion}} = P d_p = (1)(5.51) = 5.51 \text{ (SAY 6)}$$

$$N_{\text{gear}} = (1)(24.49) = 24.49 \text{ (SAY 25)}$$

NOW THAT THE APPROXIMATE NUMBERS OF TEETH ARE KNOWN, TABLE 15.6 GIVES THE ACTUAL FORM FACTOR.

$Y_{\text{pinion}} \approx .15$ $Y_{\text{gear}} = .303$

EQUATION 15.98 (AND 15.104) MAY BE USED TO FIND W BASED ON THE NEW INFORMATION.

$$W = \frac{(\text{HP})(33000)(P)}{v \sigma Y K_d}$$

$$W_{\text{pinion}} = \frac{(550)(33000)(1)}{(1731.1)(30{,}000)(.15)(.257)} = 9.07''$$

ASSUME $\sigma_{max} = 50{,}000$ PSI FOR THE CAST STEEL GEAR

$$W_{\text{gear}} = \frac{(550)(33000)(1)}{(1731.1)\left(\frac{50{,}000}{3}\right)(.303)(.257)} = 8.08''$$

THIS IGNORES STRESS CONCENTRATION FACTORS. BOTH GEARS would BE 9" OR LARGER.

2 THE MAXIMUM SPEED RATIO IS $\frac{96}{12} = 8$.

WITH 3 STAGES: $(8)^3 = 512$ (TOO SMALL)

WITH 4 STAGES $(8)^4 = 4096$

SO 4 SETS OF GEARS ARE NEEDED. TRY TO KEEP all GEAR SETS AS close AS possible

$$\sqrt[4]{600} \approx 5$$

$$\left(\frac{60}{12}\right)\left(\frac{60}{12}\right)\left(\frac{60}{12}\right)\left(\frac{72}{15}\right) = 600$$

OTHER COMBINATIONS ARE POSSIBLE

3 THIS SUBJECT IS NOT COVERED IN THE MECHANICAL ENGINEERING REFERENCE MANUAL.

(a) FOR A B V-BELT, THE CORRECTION TO THE INSIDE CIRCUMFERENCE IS 1.8 INCHES. THE PITCH LENGTH IS

$$L_p = L_{\text{inside}} + \text{CORRECTION} = 90 \text{ IN} + 1.8 \text{ IN}$$

$$= 91.8 \text{ IN}$$

(b) THE RATIO OF SPEEDS DETERMINES THE SHEAVE SIZES

$$d_{\text{equipment}} = d_{\text{motor}}\left(\frac{n_{\text{motor}}}{n_{\text{equipment}}}\right)$$

$$= (10 \text{ IN})\left(\frac{1750 \text{ RPM}}{800 \text{ RPM}}\right) = 21.875 \text{ IN}$$

THE CLOSEST STANDARD SHEAVE SIZE IS 21.8 IN.

THIS IS MORE THAN THE MINIMUM DIAMETER (5.4 IN) FOR A B V-BELT

4

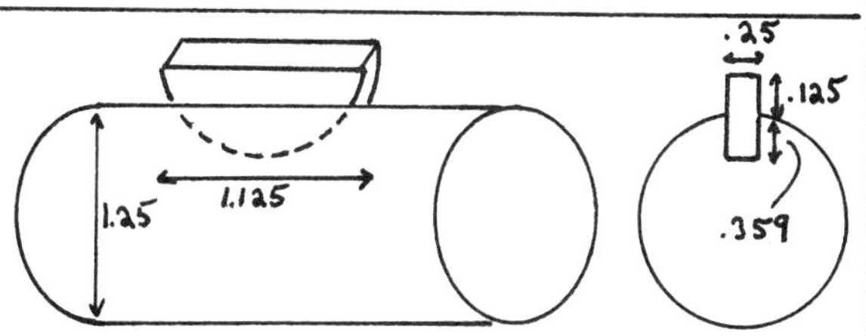


THE ABOVE DIMENSIONS ARE TAKEN FROM MACHINERY'S HANDBOOK

FROM PAGE 14-29, FOR 1030 STEEL

$$S_{yt} = 73{,}900 \text{ PSI}$$

SO $S_{ys} = .577(73{,}900) = 42{,}640$ PSI

$$S_{yc} \approx S_{yt} = 73{,}900 \text{ PSI}$$

THERE ARE 2 POSSIBLE APPROACHES TO THIS PROBLEM. THE FIRST METHOD WOULD HAVE YOU calculate THE STRESSES BASED ON THE ACTUAL AREAS IN SHEAR AND BEARING. THIS IS ALMOST NEVER DONE. RATHER, APPROXIMATIONS ARE USED. IF THE FACTOR OF SAFETY IS TOO small, USE MORE THAN 1 KEY.

(MORE)

WARM-UP #4 CONTINUED

THE USUAL APPROACH IS TO ASSUME THE ENTIRE LENGTH OF THE KEY CARRIES THE SHEAR LOAD. THE AREA IN SHEAR IS

$$A = (L)(W) = (1.125)(.25) = .2813 \text{ IN}^2$$

ASSUMING THE TORQUE IS CARRIED AT THE SHAFT SURFACE, THE MAXIMUM TORQUE WOULD BE

$$\tau = F\cdot r = (S_{ys})A(r)$$
$$= (42{,}640)(.2813)\left(\frac{1.25}{2}\right)$$
$$= 7497 \text{ IN-LBF}$$

SO, THE FACTOR OF SAFETY IN SHEAR IS

$$FS_s = \frac{7497}{4200} = 1.78$$

THE SMALLEST AREA IN BEARING OCCURS IN THE HUB AND IS

$$A = (L)(d) = (1.125)(.125) = .1406 \text{ IN}^2$$

THE ACTUAL APPLIED FORCE IS

$$F = \frac{\tau}{r} = \frac{4200 \text{ IN-LBF}}{\left(\frac{1.25}{2}\right)\text{ IN}} = 6720 \text{ LBF}$$

THE BEARING STRESS IS

$$\sigma = \frac{F}{A} = \frac{6720}{.1406} = 47795 \text{ PSI}$$

USING THE COMPRESSIVE YIELD STRENGTH AS THE MAXIMUM BEARING STRESS, THE FACTOR OF SAFETY IN BEARING IS

$$FS_b = \frac{73{,}900}{47{,}795} = \boxed{1.55}$$

5

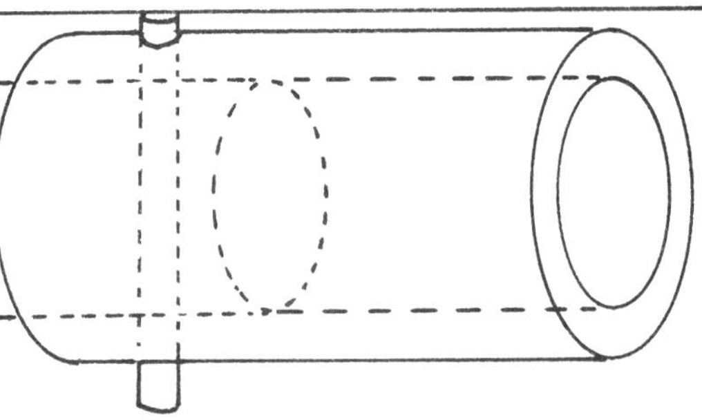

THE AREA OF THE PIN IN SHEAR IS

$$(2)\text{ SURFACES}\left(\frac{\pi}{4}\right)d^2 = (1.571\, d^2) \text{ IN}^2$$

THE TOTAL SHEAR IS

$$V = \frac{\tau}{r} = \frac{(400)\text{ IN-LBF}}{\left(\frac{1.125}{2}\right)\text{ IN}} = 711.1 \text{ LBF}$$

FROM PAGE 14-29 FOR 1030 STEEL

$$S_{yt} = 73{,}900 \text{ psi}$$

SO $S_{ys} = .577(73{,}900) = 42{,}640$ psi

USE A FACTOR OF SAFETY OF 2.5

$$\tau_{MAX} = \frac{42{,}640}{2.5} = 17{,}060 \text{ psi}$$

THE MAXIMUM SHEAR STRESS EXPERIENCED BY A ROUND BAR IN SHEAR IS

$$\frac{4V}{3A} = \frac{(4)(711.1 \text{ LBF})}{(3)(2)\left(\frac{\pi}{4}\right)d^2} = 17{,}060 \text{ LBF/IN}^2$$

SOLVING, $d = 0.188$ IN

6

ANNEALING: HEAT TO ABOVE THE RECRYSTALLIZATION TEMPERATURE {600°F} AND AIR COOL TO ROOM TEMPERATURE. THE ALCOA ALUMINUM HANDBOOK RECOMMENDS 775°F HELD FOR 2-3 HOURS, COOLING AT 50°F/HR DOWN TO 500°F WITH SUBSEQUENT COOLING RATE NOT IMPORTANT.

PRECIPITATION HARDENING {BASED ON THE ALCOA ALUMINUM HANDBOOK}

1. SOLUTION TREATMENT AT 950°F FOR ABOUT 4 HOURS FOLLOWED BY RAPID QUENCH IN COLD WATER
2. REHEAT TO 320°F FOR 12-16 HOURS, FOLLOWED BY COOLING AT A RATE WHICH IS "NOT UNDULY SLOW"

7

ASSUME $E = 2.9$ EE7 PSI.

NEGLECT PLATE COMPRESSION
NEGLECT NUT AND HEAD DEFORMATION

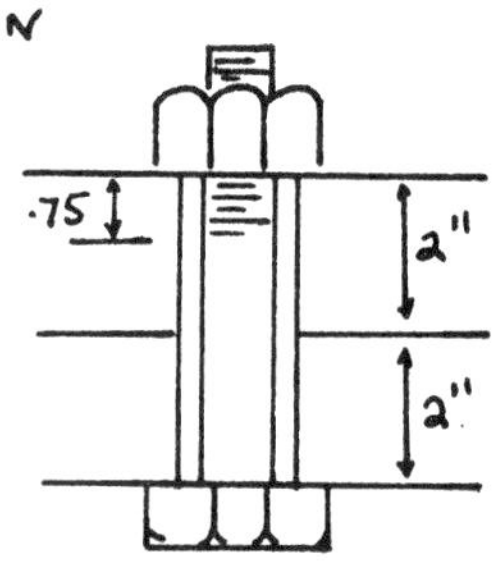


THE BODY OF THE BOLT HAS A DIAMETER AND AREA OF

$$D = .75''$$
$$A = \frac{\pi}{4}(.75)^2 = .4418 \text{ IN}^2$$

THE STRESS AREA IS FOUND FROM PAGE 14-34 TO BE

$$A = .3724 \text{ IN}^2$$

THE ELONGATION IN THE UNTHREADED PART OF THE BOLT IS

$$\delta = \frac{FL}{AE} = \sigma\frac{L}{E} = \frac{(40000)\text{ PSI}(3.25)\text{ IN}}{2.9\text{EE7 PSI}} = .00448''$$

THE STRESS IN THE THREADED PART OF THE BOLT IS

$$40{,}000\left(\frac{.4418}{.3724}\right) = 47454 \text{ psi}$$

THE ELONGATION IN THE THREADED PART, INCLUDING 3 THREADS IN THE NUT, IS

$$\delta = \sigma\frac{L}{E} = \frac{(47454)\left[.75 + 3\left(\frac{1}{16}\right)\right]}{2.9\text{EE}7} = .00153$$

THE TOTAL ELONGATION IS

$$\delta_t = .00448 + .00153 = .00601$$

8

$$F_{AVE} = \tfrac{1}{2}(1000 + 8000) = 4500 \text{ LBF}$$
$$F_{ALT} = \tfrac{1}{2}(8000 - 1000) = 3500$$

THE THREADED SECTION HAS A

$$A_{MIN} = .2256 \text{ IN}^2 \text{ \{PAGE 14-34\}}$$

THE STRESS CONCENTRATION FACTOR FOR ROLLED THREADS IS 2.2 {SHIGLEY, MECHANICAL ENGINEERING DESIGN, 3RD EDITION, PAGE 252}

SO $\sigma_{AVE} = \dfrac{4500}{.2256} = 19947$ PSI

$$\sigma_{ALT} = \frac{3500(2.2)}{.2256} = 34131 \text{ PSI}$$

SINCE $S_{ut}$ IS NOT KNOWN, THE SODERBERG DIAGRAM MUST BE USED.

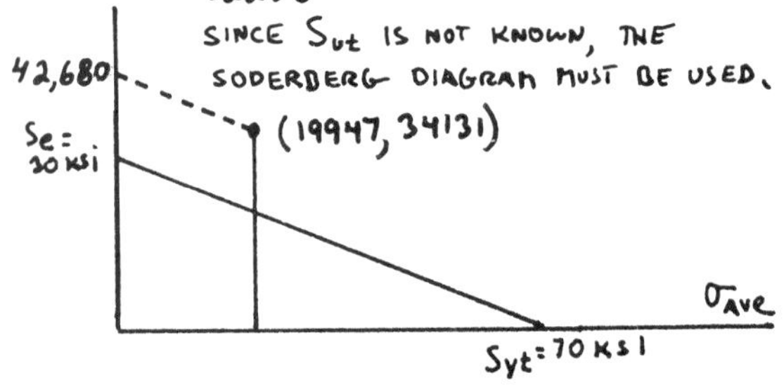


FAILURE WILL OCCUR BECAUSE THE POINT IS ABOVE THE LINE. THE FACTOR OF SAFETY IS

$$FS \approx \frac{30{,}000}{42{,}680} = 0.703$$

**9** FROM PAGE 14-27 {CASE 1}

$$y = \frac{FL^3}{3EI}$$

$$I = \frac{bh^3}{12} = \frac{6(h)^3}{12} = .5h^3$$

$$L = 2(12) = 24 \text{ IN}$$

$$2'' = \frac{800(24)^3}{(3)(2.9EE7)(.5)h^3}$$

OR $h = .503''$

A COMPLETE SOLUTION WOULD ALSO CHECK BENDING STRESS.

**10** THE CHANGE IN KINETIC ENERGY IS

$$\Delta K.E. = \tfrac{1}{2}J[\omega_1^2 - \omega_2^2] = 2J\left(\frac{\pi}{60}\right)^2[n_1^2 - n_2^2]$$

n IS IN rpm

SO,

$$J = \frac{\Delta K.E.}{2\left(\frac{\pi}{60}\right)^2(n_1^2 - n_2^2)} = \frac{1500 \text{ FT-LBF}}{2\left(\frac{\pi}{60}\right)^2[(200)^2 - (175)^2]}$$

$$= 29.18 \text{ FT-LBF-SEC}^2 \quad (\text{SLUG-FT}^2)$$

ASSUMING ALL THE MASS IS CONCENTRATED AT THE MEAN RADIUS $r_m = 15''$,

$$J = mr_m^2 = \frac{wtL_m \rho r_m^2}{g_c} \quad \text{WHERE } L_m = 2\pi r_m$$

SO, THE THICKNESS IS

$$t = \frac{g_c J}{2\pi w \rho r_m^3}$$

INCLUDING THE 10% FACTOR,

$$t = \frac{(32.2)(29.18)}{(1.10)\left(\frac{12}{12}\right)(2\pi)\left(\frac{15}{12}\right)^3[(0.26)(12)^3]}$$

$$= 0.155 \text{ FT} = \boxed{1.86''}$$

**11** ASSUME A COEFFICIENT OF FRICTION OF $\mu = .12$

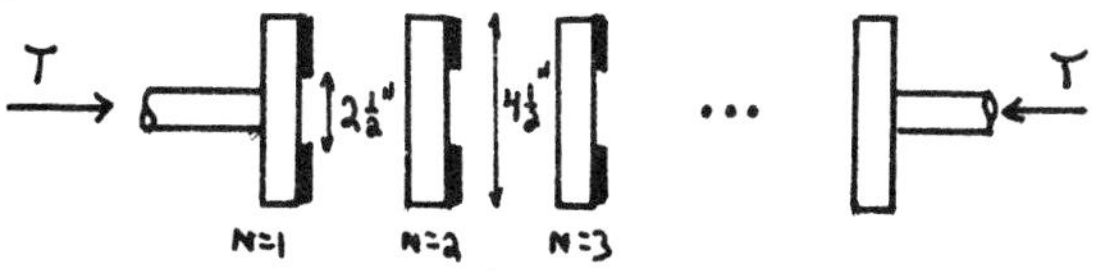


WITH N FRICTION PLANES, THE CONTACT SURFACE AREA IS

$$A = N\frac{\pi}{4}\left((4.5)^2 - (2.5)^2\right) \text{ IN}^2 = 11N \text{ IN}^2$$

WITH A CONTACT PRESSURE OF P = 100 PSI, THE NORMAL FORCE IS

$$F_N = AP = (11N) \text{ IN}^2 (100) \frac{\text{LBF}}{\text{IN}^2} = 1100N \text{ LBF}$$

THE FRICTIONAL FORCE IS

$$F_f = \mu F_N = (.12)(1100N) \text{ LBF} = 132N \text{ LBF}$$

ASSUME THE FRICTIONAL FORCE IS APPLIED AT THE MEAN RADIUS

$$r_m = \tfrac{1}{2}\left(\frac{2.5 + 4.5}{2}\right) = 1.75 \text{ IN}$$

THE RESISTING TORQUE IS

$$T_R = r_m F_f = (1.75) \text{ IN} (132N) \text{ LBF} = 231N$$

SETTING THE SLIPPING TORQUE EQUAL TO THE RESISTING TORQUE,

$$3(300) \text{ IN-LBF} = (231N) \text{ IN-LBF}$$

$$N = 3.9$$

USE 4 CONTACT SURFACES - 3 PLATES AND 2 DISCS

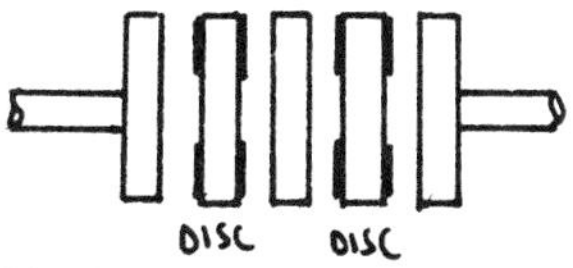


## CONCENTRATES

**1** DESIGN FOR STATIC LOADING. CHOOSE A SPRING INDEX OF 9 {FROM PAGE 15-11}. FROM EQUATION 15.58

$$d = \frac{8FC^3N_a}{G\delta} = \frac{(8)(50)(9)^3(12)}{(1.2EE7)(0.5)} = 0.583''$$

SO $D = 9d = 9(0.583) = 5.249''$

A NON-STANDARD WIRE DIAMETER

**2** ASSUME ASTM A230 WIRE. TO GET STARTED, ASSUME $d = .15''$ WIRE.

FROM TABLE 15.4,

$$S_{ut} = 205{,}000 \text{ PSI}$$

BASED ON EQUATION 15.74 WITH A SAFETY FACTOR 1.5, THE ALLOWABLE WORKING STRESS IS

$$\tau = \frac{(.3)(S_{ut})}{FS} = \frac{(.3)(205{,}000)}{1.5} = 41{,}000$$

FROM EQUATION 15.62,

$$W = \frac{4(10) - 1}{4(10) - 4} + \frac{.615}{10} = 1.145$$

FROM EQUATION 15.61,

$$d = \sqrt{\frac{8CFW}{\pi\tau}} = \sqrt{\frac{8(10)(30)(1.145)}{\pi(41{,}000)}} = 0.146$$

FROM TABLE 15.2, USE W+M WIRE #9 WITH $d = 0.1483$.

$$D = 10d = (10)(0.1483) = 1.483$$

FROM EQUATION 15.60, THE SPRING CONSTANT IS

$$K = \frac{\Delta F}{\Delta\delta} = \frac{30 - 20}{.3} = 33.33 \text{ LBF/IN}$$

FROM EQN 15.60,

$$N_a = \frac{Gd}{8KC^3} = \frac{(11.5EE6)(0.1483)}{(8)(33.33)(10)^3} = 6.4$$

SPECIFY A SQUARED AND GROUND SPRING. USE 8.4 COILS WITH 6.4 ACTIVE.

THE SOLID HEIGHT IS

$$(6.4)(0.1483) = 0.949''$$

AT SOLID, THE MAXIMUM STRESS SHOULD NOT EXCEED $S_{us}$. FROM TABLE 15.4,

$$S_{UT} = 205{,}000 \text{ psi}$$

FROM TABLE 15.3 FOR UNPEENED WIRE,

$$\tau_{max,solid} \approx 0.3S_{UT} = (0.30)(205{,}000 \text{ psi}) = 61{,}500 \text{ psi}$$

FROM EQN 15.61, THE FORCE AT SOLID HEIGHT IS

$$F_{solid} = \frac{\tau_{solid}\pi d^3}{8DW} = \frac{(61{,}500)(\pi)(0.1483)^3}{(8)(1.483)(1.145)} = 46.39 \text{ LBF}$$

SO, THE DEFLECTION AT SOLID IS

$$\delta_{solid} = \frac{F_{solid}}{k} = \frac{46.39 \text{ LBF}}{33.33 \frac{\text{LBF}}{\text{IN}}} = 1.39 \text{ IN}$$

THE MINIMUM FREE HEIGHT IS

$$0.949 \text{ IN} + 1.39 \text{ IN} = \boxed{2.34 \text{ IN}}$$

3 THE POTENTIAL ENERGY ABSORBED IS

$$E_p = mg\Delta h = \frac{(700)(32.2)(10+46)}{(32.2)} = 39200 \text{ IN-LBF}$$

THE WORK DONE BY THE SPRING IS

$$W = \tfrac{1}{2}kx^2$$

$$39200 = \tfrac{1}{2}(k)(10)^2$$

$$k = 784 \text{ LBF/IN}$$

THE EQUIVALENT FORCE IS

$$F = kx = (784)(10) = 7840 \text{ LBF}$$

FROM EQUATION 15.62

$$W = \frac{(4)(7)-1}{(4)(7)-4} + \frac{.615}{7} = 1.213$$

FROM EQUATION 15.61

$$d = \sqrt{\frac{8CFW}{\pi\tau}} = \sqrt{\frac{8(7)(7840)(1.213)}{(\pi)(50{,}000)}} = 1.841$$

$$D = 7d = 12.89$$

FROM EQUATION 15.60

$$N_a = \frac{Gd}{8kC^3} = \frac{(1.2\text{EE}7)(1.841)}{(8)(784)(7)^3} = 10.27 \quad \{\text{SAY } 10.3\}$$

1 FOR THE STEEL

$$E_s = 2.9 \text{ EE7 psi}$$

$$\mu_s = .3$$

FOR THE CAST IRON

$$E_c = 1.45 \text{ EE7 psi} \quad (\text{P.}23\text{-}3)$$

$$\mu_c = .27$$

$$S_{UT} = 30{,}000 \text{ psi}$$

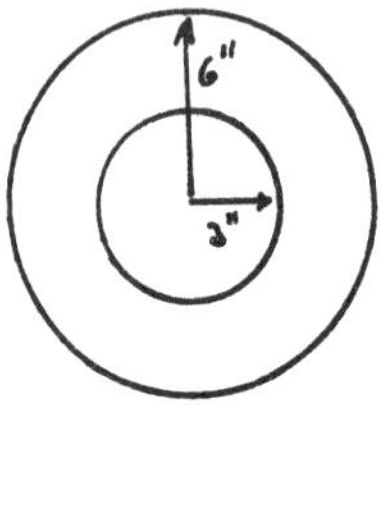


ASSUME A SAFETY FACTOR OF 3. THE TENSILE STRENGTH OF THE CAST IRON GOVERNS, SO KEEP

$$\sigma < \frac{30{,}000}{3} = 10{,}000 \text{ psi}$$

FROM PAGE 14-22 FOR THE CAST IRON HUB UNDER INTERNAL PRESSURE,

$$\sigma_{MAX} = \sigma_{ci} = 10{,}000 \text{ psi}$$

NOW,

$$\sigma_{ci} = 10{,}000 = \frac{[(6)^2+(3)^2]\,p}{[(6)^2-(3)^2]}$$

OR $p = 6000$ psi

$$\sigma_{ri} = -p = -6000 \text{ AT INNER SURFACE}$$

FROM EQUATION 14.88, THE RADIAL INTERFERENCE IN THE CAST IRON IS

$$\Delta r = \frac{r}{E}\left[\sigma_c - \mu(\sigma_r + \sigma_L)\right]$$

$$= \frac{3}{1.45\text{EE}7}\left[10{,}000 - .27(-6000)\right] = 2.40 \text{ EE-3 IN}$$

FROM TABLE 14.6 FOR A CYLINDER UNDER EXTERNAL PRESSURE,

$\sigma_{ro}$ IS MAX

$$\sigma_{co} = 6000\frac{-[(3)^2+(0)^2]}{[(3)^2-(0)^2]} = -6000 \text{ PSI}$$

$$\sigma_{ro} = -p = -6000$$

FROM EQUATION 14.88 THE RADIAL INTERFERENCE IN THE STEEL IS

$$\Delta r = \frac{3}{2.9\text{EE}7}\left[-6000 - (.3)(-6000)\right] = -4.34 \text{ EE-4}$$

THE TOTAL RADIAL INTERFERENCE IS

$$2.40 \text{ EE-3} + 4.34 \text{ EE-4} = 2.83 \text{ EE-3}$$

5 PROCEED AS IN WARM-UP #1. FIRST FIND THE PITCH DIAMETERS

$$\frac{d_1}{2} + \frac{d_2}{2} = 15$$

$$\frac{d_1}{d_2} = \frac{250}{83.33}$$

SOLVING THESE SIMULTANEOUSLY,

$$d_1 = 22.5 \qquad d_2 = 7.5$$

FROM EQUATION 15.75, THE PITCH LINE VELOCITY IS

$$V = \frac{\pi(\text{RPM})d}{12} = \frac{\pi(7.5)(250)}{12} = 490.9 \text{ FPM}$$

FOR THE PINION, ASSUME $Y = .3$, AND $w = 1"$. USE $\sigma_{max} = 30{,}000$ psi.

THE BARTH SPEED FACTOR IS

$$K_d = \frac{600}{600 + 490.9} = .55$$

FROM EQUATIONS 15.98 AND 15.104

$$P = \frac{k_d \sigma w Y v}{(HP)(33{,}000)} = \frac{(.55)(30000)(1)(.3)(490.9)}{(40)(33000)} = 1.84$$

SO CHOOSE $P = 2$. THE NUMBERS OF TEETH ARE

$$N_{PINION} = Pd = 2(7.5) = 15$$

$$N_{GEAR} = 2(22.5) = 45$$

FROM TABLE 15.6, ACTUAL VALUES OF $Y$ ARE

$$Y_{PINION} = 0.290$$

$$Y_{GEAR} \approx 0.401$$

FROM EQUATIONS 15.98 AND 15.104, THE WIDTHS ARE

$$w = \frac{(33000)P(HP)}{v(\sigma)Y k_d}$$

$$w_{PINION} = \frac{(33000)(2)(40)}{(490.9)(30{,}000)(.290)(.55)} = 1.12"$$

USE $\sigma_{MAXIMUM} = 50{,}000$ FOR THE CAST STEEL GEAR:

$$w_{GEAR} = \frac{(33000)(2)(40)}{(490.9)\left(\frac{50{,}000}{2}\right)(.401)(.55)} = 1.46"$$

6 THE CENTROID OF THE RIVET GROUP IS

$\left(\frac{2}{3}\right)(9) = 6"$ FROM EACH RIVET

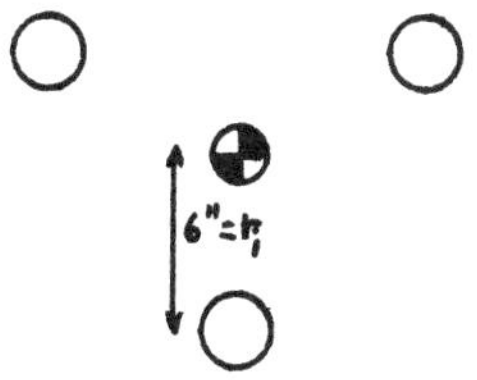


THE VERTICAL SHEAR LOAD CARRIED BY EACH RIVET IS $F_v = P/3$

FOR EACH RIVET, THE RESISTING MOMENT OF INERTIA IS

$$J = A r_i^2 = (\pi r^2) r_i^2 = \pi\left(\frac{.75}{2}\right)^2 (6)^2 = 15.9 \text{ in}^4$$

THE APPLIED MOMENT IS $20P$

THE AREA OF EACH RIVET IS

$$A = \frac{\pi}{4} d^2 = \frac{\pi}{4}(.75)^2 = .442$$

SO, THE VERTICAL SHEAR STRESS IS

$$\tau_v = \frac{F_v}{A} = \frac{P}{3(.442)} = .754\,P$$

THE TWISTING SHEARING STRESS IN THE RIGHT-MOST RIVET (WHICH BY INSPECTION IS THE HIGHEST STRESSED) IS

$$\tau = \frac{Mr}{J} = \frac{\left(\frac{20P}{3}\right)6}{15.9} = 2.516\,P$$

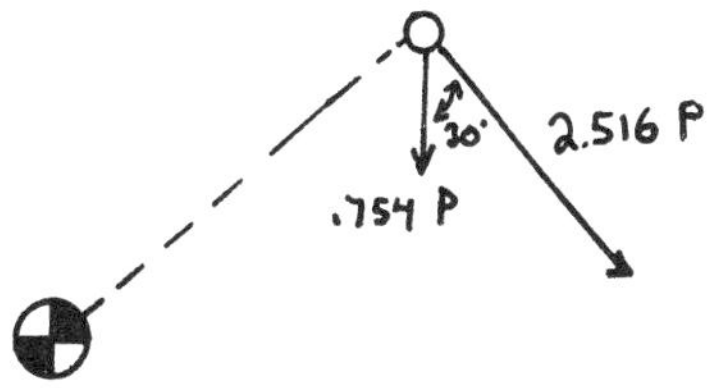


COMBINING THE TWO STRESSES,

$$\tau_{max} = P\sqrt{[(2.516)(\sin 30°)]^2 + [0.754 + (2.516)(\cos 30°)]^2} = 3.191\,P$$

KEEPING $\tau_{max} < 15{,}000$ psi,

$$P = \frac{15000}{3.191} = 4700 \text{ LBF}$$

7 ASSUME THE WELD THROAT THICKNESS IS $t$.

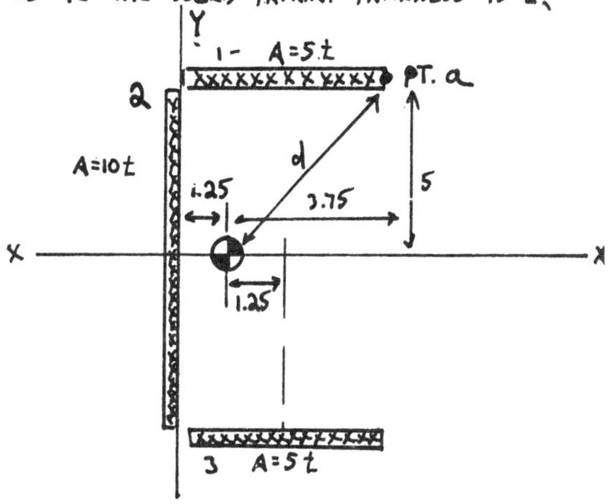


BY INSPECTION, $\bar{y} = 0$

| | | |
|---|---|---|
| $A_1 = 5t$ | $A_2 = 10t$ | $A_3 = 5t$ |
| $\bar{x}_1 = 2.5$ | $\bar{x}_2 = 0$ | $\bar{x}_3 = 2.5$ |

$$\bar{x}_c = \frac{5t(2.5) + 10t(0) + 5t(2.5)}{5t + 10t + 5t} = 1.25$$

THE CENTROIDAL MOMENT OF INERTIA IN THE X DIRECTION IS

$$I_x = \frac{t(10)^3}{12} + 2\left[\frac{5(t)^3}{12} + (5t)(5)^2\right] = 333.33\,t + .833\,t^3$$

SINCE $t$ WILL BE SMALL (PROBABLY LESS THAN .5") THE $t^3$ TERM MAY BE OMITTED. SO,

$$I_x = 333.33t$$

SIMILARLY THE CENTROIDAL MOMENT OF INERTIA IN THE Y DIRECTION IS

$$I_y = \frac{(10)t^3}{12} + (10t)(1.25)^2 + 2\left[\frac{t(5)^3}{12} + (5t)(1.25)^2\right]$$

$$= .833t^3 + 15.625t + 20.833t + 15.625t$$

$$= .833t^3 + 52.08t$$

$$\approx 52.08t$$

THE POLAR MOMENT OF INERTIA IS

$$J = I_x + I_y = 385.4t$$

THE MAXIMUM SHEAR WILL OCCUR AT POINT a SINCE IT IS FARTHEST AWAY.

$$d = \sqrt{(3.75)^2 + (5)^2} = 6.25$$

THE APPLIED MOMENT IS $M = Fx$:

$$M = (10000)(12 + 3.75) = 157{,}500 \text{ IN-LBF}$$

SO THE TORSIONAL SHEAR STRESS IS

$$\tau = \frac{Mc}{J} = \frac{(157{,}500)(6.25)}{385.4t} = \frac{2554.2}{t} \text{ psi}$$

THIS SHEAR STRESS IS AT RIGHT ANGLES TO THE LINE d. THE STRESS CAN BE DIVIDED INTO X AND Y COMPONENTS:

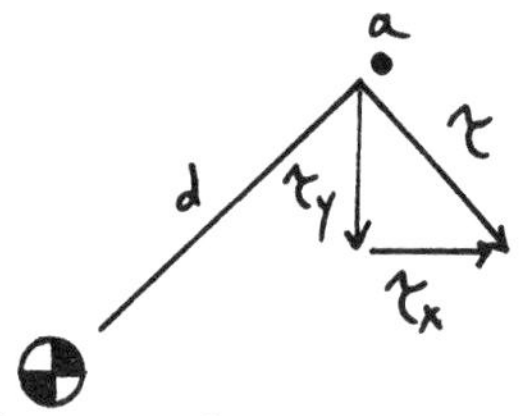


$$\tau_y = \frac{3.75}{6.25}\left(\frac{2554.2}{t}\right) = \frac{1532.5}{t}$$

$$\tau_x = \frac{5}{6.25}(\tau) = \frac{2043.4}{t}$$

IN ADDITION, THE VERTICAL LOAD OF 10,000 LBF MUST BE SUPPORTED. THE VERTICAL SHEAR STRESS IS

$$A = 10t + 5t + 5t = 20t$$

$$\tau_y = \frac{10{,}000}{20t} = \frac{500}{t}$$

THE RESULTANT SHEAR STRESS IS

$$\tau = \sqrt{\left(\frac{2043.4}{t}\right)^2 + \left(\frac{1532.5 + 500}{t}\right)^2} = \frac{2882.1}{t}$$

ASSUMING AN ALLOWABLE LOAD OF 8000 PSI FOR AN UNSHIELDED WELD IN SHEAR, THE REQUIRED THROAT THICKNESS IS

$$\frac{2882.1}{t} = 8000 \quad \text{OR} \quad t = 0.360"$$

$$\text{OR } w = \frac{0.360}{0.707} = 0.51" \text{ (SAY, } 1/2")$$

8 THE ALUMINUM STRENGTH PER INCH OF BOND IS

$$S = (15000)(1)(.020) = 300 \text{ LBF}$$

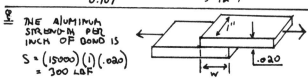

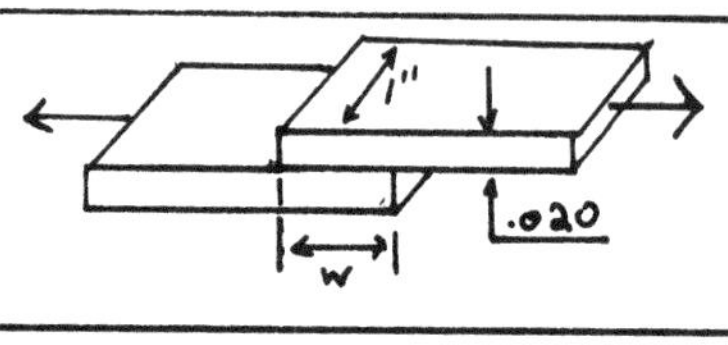


THE WIDTH OF BOND REQUIRED IS

$$300 = \left(\frac{1500}{2}\right)(1)(w)$$

$$w = .4"$$

9 THE CENTROIDAL MOMENT OF INERTIA IS

$$\frac{\pi}{4}r^4 = \frac{\pi}{4}(1)^4 = .7854$$

AT THE STEP, THE MOMENT IS

$$M = (2500)\left(4 - \frac{5}{16}\right) = 9218.8 \text{ IN-LBF}$$

$$\frac{D}{d} = \frac{3}{2} = 1.5$$

$$\frac{r}{d} = \frac{\frac{5}{16}}{2} = .156$$

> ASSUME DIRECT SHEAR IS LOW BY COMPARISON

FROM PAGE 14-33 $K \approx 1.5$

THE BENDING STRESS IS

$$\sigma = K\frac{Mc}{I} = 1.5\left(\frac{(9218.8)(1)}{.7854}\right) = 17{,}607 \text{ psi}$$

(IF DIRECT SHEAR STRESS IS INCLUDED, $\sigma$ = 17,643 PSI.)

10 NEGLECT THE EXPANSION OF THE SHAFT DIAMETER, WHICH IS SMALL ANYWAY. (THIS IS A CONSERVATIVE ASSUMPTION.)

REFER TO FORMULAS FOR STRESS AND STRAIN (ROARK AND YOUNG). THE CHANGE IN INNER RADIUS IS

$$\Delta R_o = \frac{1}{4}\left[\frac{\rho\omega^2}{386.4}\right]\left[\frac{R_o}{E}\right]\left[(3+\mu)R^2 + (1-\mu)R_o^2\right] \quad \text{IN INCHES}$$

$\mu$ = Poisson's ratio = 0.3

$\rho$ = steel density = 0.283 lbm/in³

$$\omega = \frac{(2\pi)(3500)}{60} = 366.5 \text{ rad/sec}$$

E = 2.9 EE7 psi

R = OUTER RADIUS = 8"

$R_o$ = INNER RADIUS = 1"

$$\Delta R_o = \frac{1}{4}\left[\frac{(0.283)(366.5)^2}{386.4}\right]\left[\frac{1}{2.9\text{EE}7}\right]\left[(3+0.3)(8)^2 + (1-0.3)(1)^2\right]$$

$$= 1.80 \text{ EE-4}$$

SINCE BOTH THE SHAFT AND THE DISK ARE STEEL, EQUATION 14.90 CAN BE USED TO CALCULATE THE INTERFERENCE REQUIRED TO OBTAIN A RESIDUAL PRESSURE OF 1250 PSI.

$$I_{\text{DIAMETRAL}} = \Delta D_o = \frac{(4)(1)(1250)}{2.9 \text{ EE7}}\left[\frac{1}{1 - \left(\frac{1}{8}\right)^2}\right]$$

$$= 1.75 \text{ EE-4}$$

$$I_{\text{RADIAL}} = \left(\frac{1}{2}\right) I_{\text{DIAMETRAL}} = 8.76 \text{ EE-5}$$

THE REQUIRED INITIAL INTERFERENCE IS

$$I_{\text{RADIAL}} = 1.80 \text{ EE-4} + 8.76 \text{ EE-5}$$

$$= \boxed{2.68 \text{ EE-4 IN}}$$

**11**

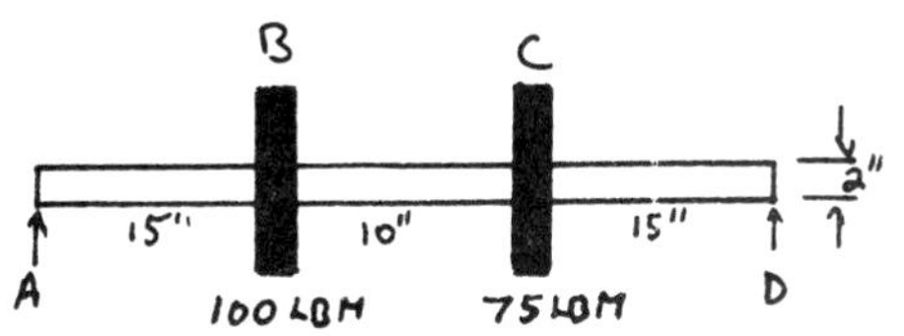


$I_{SHAFT} = \frac{1}{4}\pi r^4 = (.25)(\pi)(1)^4\ \text{IN}^4 = .7854\ \text{IN}^4$

CALCULATE DEFLECTIONS DUE TO 100# LOAD

$$\delta_{B,100} = \frac{Fa^2b^2}{3EIL} = \frac{(100)(15)^2(25)^2}{(3)(3\ EE7)(.7854)(40)}$$

$$= 4.97\ EE\text{-}3\ \text{IN}$$

$$\delta_{C,100} = \left(\frac{Fbx}{6EIL}\right)(L^2-b^2-x^2)$$

$$= \frac{(100)(15)(15)}{(6)(3\ EE7)(.7854)(40)}\left((40)^2-(15)^2-(15)^2\right)$$

$$= 4.58\ EE\text{-}3\ \text{IN}$$

CALCULATE DEFLECTIONS DUE TO 75 # LOAD

$$\delta_{C,75} = \frac{(75)(25)^2(15)^2}{(3)(3\ EE7)(.7854)(40)} = 3.73\ EE\text{-}3\ \text{IN}$$

$$\delta_{B,75} = \frac{(75)(15)(15)}{(6)(3\ EE7)(.7854)(40)}\left((40)^2-(15)^2-(15)^2\right)$$

$$= 3.43\ EE\text{-}3\ \text{IN}$$

THE TOTAL DEFLECTIONS ARE

$$\delta_B = 4.97\ EE\text{-}3 + 3.43\ EE\text{-}3\ \text{IN} = 8.4\ EE\text{-}3\ \text{IN}$$

$$\delta_C = 8.31\ EE\text{-}3\ \text{IN}$$

THE CRITICAL SHAFT SPEED IS

$$f = \frac{1}{2\pi}\sqrt{\frac{g\sum w_i\delta_i}{\sum w_i\delta_i^2}}$$

$$= \frac{1}{2\pi}\sqrt{\frac{386\left[(100)(8.4\ EE\text{-}3)+(75)(8.3\ EE\text{-}3)\right]}{(100)(8.4\ EE\text{-}3)^2+(75)(8.3\ EE\text{-}3)^2}}$$

$$= \boxed{34.2\ \text{HZ} = 2052\ \text{RPM}}$$

**12** $N_S = 24$ $\quad \omega_S = +50$ RPM

$N_P = 40$

$N_R = 104$ $\quad \omega_R = 0$

$$TV = -\frac{N_R}{N_S} = -\frac{104}{24} = -4.333$$

$$\omega_S = (TV)(\omega_R) + \omega_C(1-TV)$$

$$\omega_C = \frac{+50}{(1-(-4.333))} = \boxed{+9.376\ \text{RPM CLOCKWISE}}$$

**13**

$$\omega_R = \omega_A\left(\frac{-N_A}{N_R}\right) = (-100)\left(\frac{-30}{100}\right) = 30\ \text{RPM (CLOCKWISE)}$$

$$TV = -\frac{N_R}{N_S} = -\frac{80}{40} = -2$$

$$\omega_S = (TV)(\omega_R) + \omega_C(1-TV)$$

$$\omega_S = (-2)(+30) + 60(1-(-2)) = \boxed{+120\ \text{RPM (CLOCKWISE)}}$$

**14**

$N_S = 10(5) = 50$ $\quad \omega_S = 0$

$N_P = 10(2.5) = 25$

$N_R = 10(10) = 100$ $\quad \omega_R = +1500$ RPM

$$TV = -\frac{N_R}{N_S} = \frac{-100}{50} = -2$$

$$\omega_S = (TV)(\omega_R) + \omega_C(1-TV)$$

$$0 = (-2)(1500) + \omega_C(1-(-2))$$

$$\omega_C = +1000\ \text{RPM}$$

$$T_{INPUT} = \frac{(33000)(HP)}{(RPM)(2\pi)} = \frac{(33000)(15)}{(1500)(2\pi)} = \boxed{52.52\ \text{FT-LBF}}$$

$$T_{OUTPUT} = \frac{(33000)(15)}{(1000)(2\pi)} = \boxed{78.78\ \text{FT-LBF}}$$

**TIMED**

**1**

$\mu = 1.184\ EE\text{-}6$ REYNS

$N = \frac{1200\ \text{RPM}}{60} = 20$ RPS

$W = 880$ LBF

$\frac{c}{r} = \frac{\text{RADIAL CLEARANCE}}{\text{JOURNAL RADIUS}} = \frac{1}{1000}$ *

$r = 1.5$

$\ell = 3.5$

* NOTICE THAT $\frac{C_d}{d} = \frac{2C_r}{2r} = \frac{C_r}{r}$

THE UNIT LOAD IS

$$p = \frac{W}{2(r)l} = \frac{880}{(2)(1.5)(3.5)} = 83.81 \text{ psi}$$

THE BEARING CHARACTERISTIC NUMBER IS

$$S = \left(\frac{r}{c}\right)^2 \frac{\mu N}{p} = (1000)^2 \frac{(1.184\text{ EE-}6)(20)}{83.81} = .283$$

$$\frac{l}{d} = \frac{l}{2r} = \frac{3.5}{3} = 1.17$$

FROM FIGURE 15.45 THE FRICTION VARIABLE IS 6, SO THE COEFFICIENT OF FRICTION IS

$$f = 6\left(\frac{c}{r}\right) = \frac{6}{1000} = .006$$

THE FRICTION TORQUE IS

$$T = fWr = (.006)(880)(1.5) = 7.92 \text{ IN-LBF}$$

THE FRICTION HORSEPOWER IS

$$HP = \frac{(7.92)\text{ IN-LBF }(20)\frac{1}{\text{sec}}(2\pi)}{(12)\frac{\text{IN}}{\text{FT}}(550)\frac{\text{FT-LBF}}{\text{HP-sec}}} = .151 \text{ HP}$$

FROM FIGURE 15.46 THE FILM THICKNESS VARIABLE IS .64. THE FILM THICKNESS IS GIVEN BY EQN 15.177.

$$h_o = 1.5\left(\frac{1}{1000}\right)(.64) = .00096''$$

BECAUSE THIS IS OUT OF THE RECOMMENDED REGION THE LOAD IS PROBABLY TOO MUCH FOR THE BEARING

2 BECAUSE OF THEIR HARDNESS, ASSUME

E = 3.0 EE7 FOR THE BOLTS

E = 2.9 EE7 FOR THE VESSEL

THE BOLT STIFFNESS IS

$$K_b = \frac{AE}{L} = \frac{6\left(\frac{\pi}{4}\right)\left(\frac{3}{8}\right)^2 3\text{ EE7}}{\frac{3}{8}+\frac{3}{4}} = 1.767\text{ EE7 } \frac{\text{LBF}}{\text{IN}}$$

THE PLATE/VESSEL CONTACT AREA IS AN ANNULUS WITH INSIDE AND OUTSIDE DIAMETERS OF 8.5" AND 11.5" RESPECTIVELY

$$\text{AREA} = \frac{\pi}{4}\left[(11.5)^2 - (8.5)^2\right] = 47.12 \text{ IN}^2$$

THE VESSEL/PLATE STIFFNESS IS

$$K_m = \frac{AE}{L} = \frac{(47.12)(2.9\text{ EE7})}{\frac{3}{8}+\frac{3}{4}} = 1.215\text{ EE9}$$

LET X BE THE DECIMAL PORTION OF THE PRESSURE LOAD (P) TAKEN BY THE BOLTS. THE INCREASE IN THE BOLT LENGTH IS

$$\delta_b = \frac{xP}{K_b}$$

THE PLATE/VESSEL DEFORMATION IS

$$\delta_m = \frac{(1-x)P}{K_m}$$

BUT $\delta_b = \delta_m$

$$\text{SO } \frac{x}{1.767\text{ EE7}} = \frac{1-x}{1.215\text{ EE9}} \quad \text{OR } x = .0143$$

THE END PLATE AREA EXPOSED TO PRESSURE IS

$$\frac{\pi}{4}(8.5)^2 = 56.75 \text{ IN}^2$$

THE FORCES ON THE END PLATE ARE

$$F_{max} = pA = 350(56.75) = 19860 \text{ LBF}$$

$$F_{min} = 50(56.75) = 2838 \text{ LBF}$$

THE STRESS AREA IN THE BOLT (PAGE 14-34) IS .0876 IN$^2$

THE STRESSES IN EACH BOLT ARE

$$\sigma_{max} = \frac{3700 + \frac{(.0143)(19860)}{6}}{.0876} = 42778 \text{ psi}$$

$$\sigma_{min} = \frac{3700 + \frac{(.0143)(2838)}{6}}{.0876} = 42,315 \text{ psi}$$

THE MEAN STRESS IS

$$\sigma_{mean} = \tfrac{1}{2}(42778 + 42,315) = 42,547 \text{ psi}$$

THE ALTERNATING STRESS IS (USING THE THREAD CONCENTRATION FACTOR OF 2)

$$\sigma_{alt} = \left(\tfrac{1}{2}\right)(2)(42778 - 42,315) = 463 \text{ psi}$$

FOR THE BOLT MATERIAL, USE EQN 15.20

$$S'_{end} = .5\, S_{ut} = .5(110,000) = 55,000$$

DERATING FACTORS FROM OTHER SOURCES ARE

$K_a = .72$ (SURFACE FINISH)

$K_b = .85$ (SIZE)

$K_c = .90$ (RELIABILITY)

$$S_e = (.72)(.85)(.90)(55000) = 30,294 \text{ psi}$$

PLOTTING THE POINT ON THE MODIFIED GOODMAN DIAGRAM,

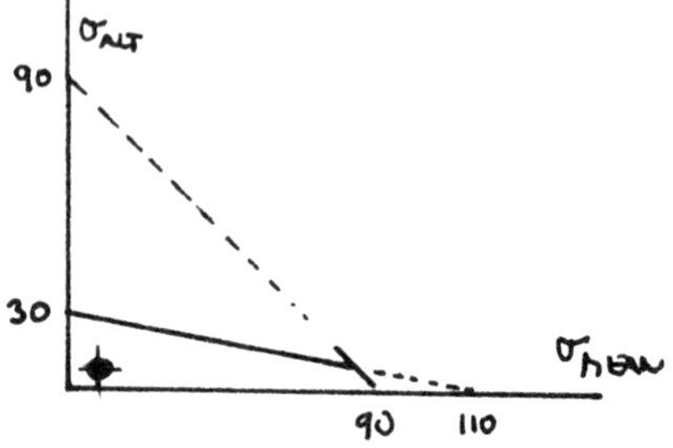


IT IS APPARENT THAT THE FACTOR OF SAFETY IS QUITE HIGH AND THE BOLTS ARE IN NO DANGER OF FAILING. HOWEVER, THE PLATE SHOULD ALSO BE CHECKED.

THE STRESS IN THE FLANGE/VESSEL IS

$$\sigma_{min} = \frac{(6)(3700) - (1-.0143)(19860)}{47.12} = 55.69 \text{ psi}$$

$$\sigma_{max} = \frac{(6)(3700) - (1-.0143)(2838)}{47.12} = 411.8 \text{ psi}$$

THESE STRESSES ARE VERY LOW.

3 ASSUME THE DIAMETERS ARE MEAN DIAMETERS.

THE SPRING INDEXES ARE

$$C_{INNER} = \frac{D}{d} = \frac{1.5}{.177} = 8.475$$

$$C_{outer} = \frac{2.0}{.2253} = 8.877$$

FOR OIL-HARDENED STEEL, $G = 11.5$ EE6 PSI.

ASSUME THE NUMBER OF COILS IS THE NUMBER OF TOTAL COILS. THE NUMBER OF ACTIVE COILS IS 2 LESS THAN THE TOTAL SINCE THE SPRINGS HAVE SQUARED AND GROUND ENDS.

$$N_{a,inner} = 12.75 - 2 = 10.75$$

$$N_{a,outer} = 10.25 - 2 = 8.25$$

FROM EQUATION 15.60, THE SPRING RATES (STIFFNESSES) ARE

$$k_{INNER} = \frac{(11.5 \text{ EE6})(.177)}{(8)(8.475)^3(10.75)} = 38.88 \text{ LB/IN}$$

$$k_{outer} = \frac{(11.5 \text{ EE6})(.2253)}{(8)(8.877)^3(8.25)} = 56.12 \text{ LB/IN}$$

SINCE THE INNER SPRING IS LONGER BY $(4.5 - 3.75) = .75''$ THE INNER SPRING WILL ABSORB THE FIRST

$$(38.88)(.75) = 29.16 \text{ LB}$$

THE REMAINING $(150 - 29.16) = 120.84$ WILL BE SHARED BY BOTH SPRINGS.

THE COMPOSITE SPRING CONSTANT IS

$$(38.88) + 56.12 = 95.0 \text{ LB/IN}$$

THE DEFLECTION OF THE OUTER SPRING IS

$$\delta_{outer} = \frac{120.84}{95} = 1.272 \text{ IN}$$

THE TOTAL DEFLECTION OF THE INNER SPRING IS

$$\delta_{INNER} = .75 + 1.272 = \boxed{2.022''}$$

THE TOTAL FORCE EXERTED BY THE INNER SPRING IS

$$F = k\delta = (38.88)(2.022) = 78.62 \text{ LB}$$

THE WAHL FACTOR IS

$$W = \frac{(4)(8.475) - 1}{(4)(8.475) - 4} + \frac{.615}{8.475} = 1.173$$

FROM EQUATION 15.61, THE SHEAR STRESS IS

$$\tau_{INNER} = \frac{(8)(8.475)(78.62)(1.173)}{\pi(.177)^2} = 63{,}528 \text{ psi}$$

ACCORDING TO THE MAXIMUM SHEAR STRESS THEORY,

$$\tau_{max} = .5 S_{yt}$$

UNFORTUNATELY, THE YIELD STRENGTH IS NOT GIVEN. FROM TABLE 15.4 FOR OIL TEMPERED STEEL,

$$S_{UT} = 215 - \left(\frac{.177 - .15}{.20 - .15}\right)(215 - 195)$$

$$= 204 \text{ ksi}$$

FROM FOOTNOTE 17 ON PAGE 15-14,

$$S_{yt} = .75 S_{UT} = .75(204) = 153 \text{ ksi}$$

THEN, $\tau_{max} = .5(153) = 76.5$ KSI

THE .577 MULTIPLIER FACTOR IS NOT USED BECAUSE IT WAS DERIVED FROM DISTORTION ENERGY THEORY, NOT FROM THE MAXIMUM SHEAR STRESS THEORY.

THE FACTOR OF SAFETY IS

$$FS = \frac{76.5}{63.5} = \boxed{1.2}$$

ANSWER WILL VARY DEPENDING ON YOUR METHOD OF CALCULATING $S_{yt}$

THE INNER AND OUTER SPRINGS SHOULD BE WOUND WITH OPPOSITE DIRECTION HELIXES. THIS WILL MINIMIZE RESONANCE AND PREVENT COILS FROM ONE SPRING ENTERING THE OTHER SPRING'S GAPS.

4 THIS IS CLEARLY AN AUTOMOBILE DIFFERENTIAL, SO BOTH OUTPUT SHAFTS MUST TURN IN THE SAME DIRECTION

THIS IS IDENTICAL TO THE FOLLOWING GEAR SET

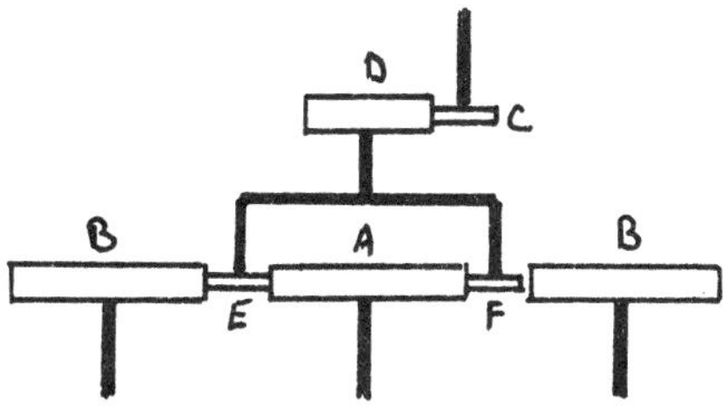


THE DIFFERENCE BETWEEN THIS AND A CONVENTIONAL EPICYCLIC GEAR TRAIN IS THAT THE RING GEAR IS REPLACED WITH TWO EXTERNAL GEARS, B.

$$\omega_C = +600 \text{ RPM}$$

$$\omega_D = \frac{-18}{54}(600) = -200 \text{ RPM}$$

$$\omega_A = -50 \text{ RPM}$$

FROM EQN 15.91,

$$TV = \frac{N_B}{N_A} = \frac{30}{30} = -1$$

TV IS CLEARLY NEGATIVE. LOOK AT THE ACTUAL GEAR SET (NOT THE EQUIVALENT GEAR SET ABOVE) HOLD GEAR D STATIONARY, TURN GEAR A, THEN GEAR B MOVES IN THE OPPOSITE DIRECTION OF GEAR A.

FROM EQUATION 15.92

$$\omega_A = (TV)\omega_B + \omega_D(1-TV)$$

$$-50 = (-1)\omega_B - 200(1-(-1))$$

$$\omega_B = -350$$

**5** $\omega_r = 0$

$\omega_s = 1000$ RPM

$\omega_c = \frac{\omega_s}{3} = 333.3$ RPM

FROM EQUATION 15.92

$$1000 = (TV)\,0 + 333.3(1-TV)$$

$$TV = -2$$

FROM EQUATION 15.91

$$2 = \frac{N_{RING}}{N_{SUN}}$$

ASSUME $N_s = 40$ AND $N_R = 80$. SINCE $P = 10$, THEN $D_s = 4$, $D_R = 8$, $D_p = 2$

CHECK: $8 = 4 + (2\times 2)$ ✓

FROM EQUATION 15.93

$$\frac{20}{40} = -\frac{1000 - 333.3}{\omega_p - 333.3}$$

$$\boxed{\omega_p = -1000 \text{ RPM}}$$

**6** FIRST, SIMPLIFY THE PROBLEM. INPUT #2 CAUSES GEAR D-E TO TURN AT

$$(-75)\left(\frac{-28}{32}\right) = 65.625 \text{ RPM (CW)}$$

SO, $\omega_D = 65.625$

NOW, FOLLOW THE PROCEDURE ON PAGE 15-17

STEP 1: GEARS A AND D HAVE THE SAME CENTER.

| A | D |
|---|---|

STEP 2:

| | A | D |
|---|---|---|
| 1 | $\omega_{ARM}$ | $\omega_{ARM}$ |
| 2 | | |
| 3 | | |

STEP 3: SINCE ALL SPEEDS ARE KNOWN CHOOSE GEAR D ARBITRARILY AS THE "UNKNOWN".

| | A | D |
|---|---|---|
| 1 | $\omega_{ARM}$ | $\omega_{ARM}$ |
| 2 | | |
| 3 | | $\omega_D$ |

STEP 4:

| | A | D |
|---|---|---|
| 1 | $\omega_{ARM}$ | $\omega_{ARM}$ |
| 2 | | $\omega_D - \omega_{ARM}$ |
| 3 | | $\omega_D$ |

STEP 5: WE WANT THE RATIO $\left(\frac{\omega_A}{\omega_D}\right)$.

THE TRANSMISSION PATH D→A IS A COMPOUND MESH.

$$\omega_A = \omega_D\left(\frac{N_B N_D}{N_A N_C}\right) = \omega_D\left(\frac{(63)(56)}{(68)N_C}\right)$$

$$= \frac{51.8823\,\omega_D}{N_C}$$

SO, PUT $\frac{51.8823}{N_C}(\omega_D - \omega_{ARM})$ INTO COLUMN A

| | A | D |
|---|---|---|
| 1 | $\omega_{ARM}$ | $\omega_{ARM}$ |
| 2 | $\frac{51.8823}{N_C}(\omega_D - \omega_{ARM})$ | $\omega_D - \omega_{ARM}$ |
| 3 | | $\omega_D$ |

STEP 6: INSERT THE KNOWN VALUES INTO $\omega$

| | A | D |
|---|---|---|
| 1 | 150 | 150 |
| 2 | $\frac{51.8823}{N_C}(65.625-150)$ | 65.625 − 150 |
| 3 | −250 | 65.625 |

STEP 7:

ROW 1 + ROW 2 = ROW 3

$$150 + \frac{51.8823}{N_C}(65.625 - 150) = -250$$

SOLVING FOR $N_C$ GIVES

$$N_C = 10.94$$

SAY, $\boxed{N_C = 11}$

**7** (a) USING EQUATION 15.76,

$$\text{OUTPUT SPEED} = 1800\left(\frac{50}{25}\right)\left(\frac{60}{20}\right) = \boxed{10{,}800 \text{ rpm}}$$

(b) USING EQUATION 15.116,

$$T_4 = \frac{63{,}025(50)(0.98)^2}{10{,}800} = \boxed{280 \text{ in-lbf}}$$

(c) SHAFT TORQUE $= T_{2\text{-}3} = \frac{63{,}025(50)(.98)}{3600} = 857.84$ in-lbf

FROM M.F. SPOTTS, *Design of Machine Elements*, TABLE 14-4, PAGE 459, FOR COLD-DRAWN 1045 STEEL, $\sigma_{ys} = 69{,}000$ psi, $\tau_{max} = \frac{1}{2}(69{,}000) = 34{,}500$ psi

FOR DYNAMIC LOADING, $P = 5 = $ no. teeth$/d_p$

FOR GEAR 3, $d_p = \frac{60}{5} = 12$ in. DIAMETER

PITCH CIRCLE VELOCITY AT 60 TOOTH GEAR SET,

USING EQUATION 15.75,

$$V_r = \frac{3600\,\pi\,(12)}{12} = 11{,}310 \text{ fpm}$$

USING EQUATION 15.107,

$$F_t = \frac{33{,}000(0.98)(50)}{11{,}310} = 143 \text{ lbf}$$

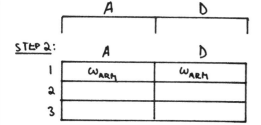

TIMED #7 CONTINUED

USING EQUATION 15.106, $\phi_t = \arctan\left(\frac{\tan 20°}{\cos 25°}\right) = 21.9°$

USING EQUATION 15.108, $F_r = 143 \tan 21.9° = 57.4$ lbf

AND EQUATION 15.109, $F_a = 143 \tan 25° = 66.7$ lbf

THE RESULTING BENDING MOMENT ON THE SHAFT IS

$M = (57.4 \text{ lbf})(4 \text{ in}) = 229.6 \text{ in-lbf}.$

FROM TABLE 15.9, FOR A STEADY LOAD, $K_m = 1.5$, $K_t = 1.0$, AND USING EQUATION 15.120 (WITH A FACTOR OF SAFETY OF 3) REARRANGED,

$$d^3 = \frac{(16)(3)}{\pi \cdot \frac{1}{2}(69{,}000)}\sqrt{\left[(1.5)(229.6)\right]^2 + \left[(1)(857.84)\right]^2}$$

$$= 0.40938 \text{ in}^3$$

$d = 0.74$ in. (VALUE WILL DEPEND ON CHOICE OF $\tau_{max}$)

CHECK AXIAL STRESSES

$$\sigma_{axial} = \frac{F_a}{A} = \frac{66.7 \text{ lbf}}{\frac{\pi}{4}(0.74 \text{ in})^2} = 154 \text{ psi}$$

THIS STRESS IS MUCH LESS THAN THE YIELD STRENGTH (69,000 psi), THUS IS INSIGNIFICANT IN THIS PROBLEM. THE MINIMUM DIAMETER, THEN, SHOULD BE 0.74 in. CONSULTING TABLE 15.10 FOR TRANSMISSION SHAFTING, THE STANDARD DIAMETER OF $\boxed{\frac{15}{16}\text{"}}$ (= 0.9375 in.) WOULD BE ADEQUATE.

**8** FROM APPENDIX B, PAGE 14-31, $S_{ut} = 30{,}000$ psi

CHOOSE A SAFETY FACTOR OF 10.

FROM TABLE 14.1, PAGE 14-3, $\mu = 0.27$.

FOR TANGENTIAL STRESS, USING EQUATION 15.50,

$$\omega = \left[\frac{4(386)(30{,}000/10)}{(0.26)\left[(3.27)10^2 + (1-0.27)2^2\right]}\right]^{1/2} = 232.4 \text{ rad/sec}$$

USING EQUATION 15.51, $\text{rpm} = \frac{60(232.4)}{2\pi} = 2219 \text{ rpm}$

FOR RADIAL STRESS, USING EQUATION 15.52,

$$\text{rpm} = \frac{30}{\pi}\left[\left(\frac{8}{3.27}\right)(3{,}000)\left(\frac{386}{(0.26)(10-2)^2}\right)\right]^{1/2} = 3940 \text{ rpm}$$

THE MAXIMUM SAFE SPEED, THEN, IS

$\boxed{2219 \text{ rpm}}$

**9**

(b)

$$\omega = \frac{120 \text{ RPM}}{60} \times 2\pi = 12.56 \text{ RAD/SEC}$$

AT $\theta = 60°$, $t = \frac{\left(\frac{60}{180}\right)(\pi)}{12.56} = 0.08338 \text{ sec}$

SINCE THE ACCELERATION IS CONSTANT, $\frac{d^2x}{dt^2} = a$, AND $x = \frac{1}{2}at^2$, SO,

$$a = \frac{(2)(0.5)}{(0.08338)^2} = \boxed{143.84 \text{ in/sec}^2}$$

(c) AT $\theta = 60°$ SINCE $a = \frac{dv}{dt}$,

$v = at = (143.84 \text{ in/sec}^2)(0.08338)$

$= 12 \text{ in/sec}$

THE CAM FOLLOWER RETURNS TO REST AT $\theta = 150°$, SO WITH CONSTANT ACCELERATION, STARTING AT $v_i = 12$ in/sec AT $t = 0$,

$v_f = 0$

$t_f = \frac{\left(\frac{90}{180}\right)(\pi)}{12.56} = 0.1251$ SEC (TIME BETWEEN $\theta = 60°$ AND $\theta = 150°$)

$v_f - v_i = a(t_f - t_i)$

$12 - 0 = a(0.1251 - 0)$

$\boxed{a = -95.92 \text{ in/sec}^2}$

(a) THE DISTANCE MOVED BY THE CAM FOLLOWER DURING $60° \le \theta \le 150°$ IS

$x = \left(\frac{1}{2}\right)at^2 = \left(\frac{1}{2}\right)(95.92)(0.1251)^2$

$= 0.75$ in

TOTAL FOLLOWER DISTANCE IS

$0.5" + 0.75" = \boxed{1.25 \text{ in}}$

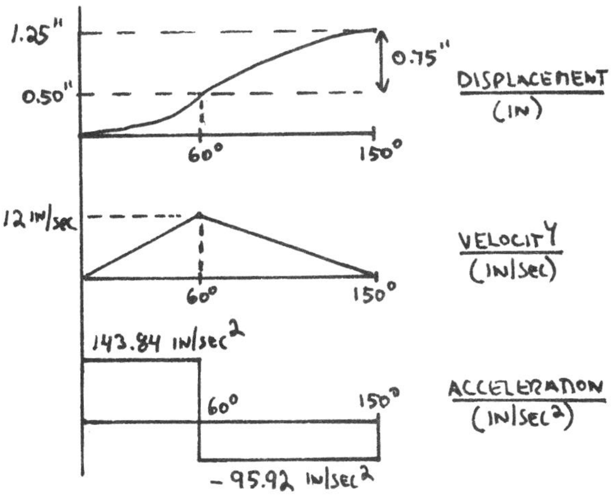

# Dynamics

**1**

$$\omega = \left(\frac{\#\text{REV}}{\text{SEC}}\right)(2\pi)$$

$$= \frac{(28)\frac{\text{MI}}{\text{HR}}(5280)\frac{\text{FT}}{\text{MI}}(12)\frac{\text{IN}}{\text{FT}}(2\pi)}{\pi(24)\text{ IN }(3600)\frac{\text{SEC}}{\text{HR}}}$$

$$= 41.07\ \frac{\text{RAD}}{\text{SEC}}$$

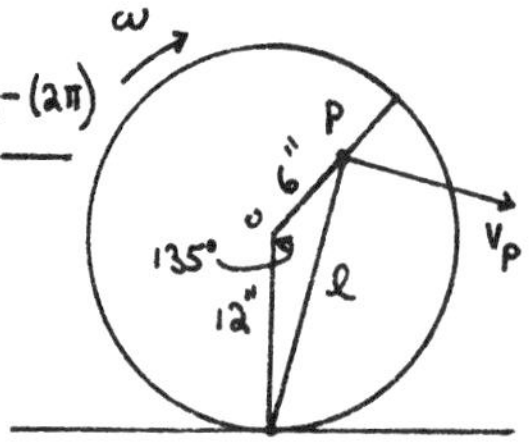


FIND DISTANCE $\ell$

$$\ell^2 = (12)^2 + (6)^2 - 2(12)(6)\cos 135°$$

$$\ell = 16.79\ \text{IN}$$

$$V_p = \omega\ell = (41.07)\frac{\text{RAD}}{\text{SEC}}(16.79)\text{ IN} = 689.6\ \frac{\text{IN}}{\text{SEC}}$$

$$= 57.46\ \text{FPS}$$

**2**

$$x = (V_0\cos\phi)t$$

$$= (900\cos\phi)t$$

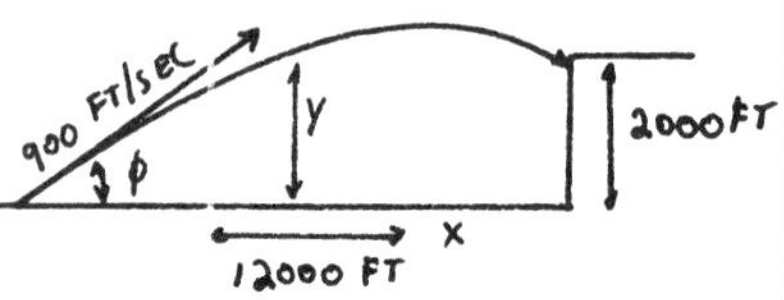


OR

$$t = \frac{12000}{900\cos\phi} = \frac{13.33}{\cos\phi}$$

$$y = (V_0\sin\phi)t - \tfrac{1}{2}gt^2 = (900\sin\phi)t - \left(\tfrac{1}{2}\right)(32.2)t^2$$

SUBSTITUTING $y = 2000$ AND $t = 13.33/\cos\phi$

$$2000 = \frac{900\sin\phi(13.33)}{\cos\phi} - (16.1)\left(\frac{13.33}{\cos\phi}\right)^2$$

$$1 = 6\tan\phi - \frac{1.43}{\cos^2\phi}$$

BUT $\dfrac{1}{\cos^2\phi} = 1 + \tan^2\phi$

SO $\tan^2\phi - 4.2\tan\phi + 1.7 = 0$

$$\tan\phi = \begin{Bmatrix} 3.747 \\ .453 \end{Bmatrix} \quad \boxed{\phi = \begin{Bmatrix} 75.06° \\ 24.37° \end{Bmatrix}}$$

## CONCENTRATES

**1** THE STATIC DEFLECTION IS THAT CAUSED BY THE 300 LBM MAGNET

$$\delta_{ST} = \frac{300\ \text{LBM}}{1000\ \frac{\text{LBF}}{\text{IN}}} = .3\ \text{IN}$$

$$f_{NAT} = \frac{1}{2\pi}\sqrt{\frac{g}{\delta_{ST}}} = \frac{1}{2\pi}\sqrt{\frac{386}{.3}} = \boxed{5.71\ \text{HZ}}$$

MINIMUM TENSION OCCURS AT THE UPPER LIMIT OF TRAVEL. THE DECREASE IN TENSION AT THAT POINT IS THE SAME AS THE INCREASE AT THE LOWER LIMIT CAUSED BY THE 200 # OF SCRAP

$$F_{MIN} = 300 - 200 = \boxed{100\ \text{LBF}}$$

**2** IF THE SPRINGS REDUCE THE FORCE FROM 25 LBF TO 3 LBF, THE TRANSMISSIVITY IS

$$TR = \frac{-3}{25} = -.12$$

{TR AND β ARE NEGATIVE ANYTIME THEY ARE < 1}

THE ANGULAR FORCING FREQUENCY IS

$$\omega_f = 2\pi f = 2\pi\ \frac{\text{RAD}}{\text{SEC}}\left(\frac{1200\ \frac{\text{REV}}{\text{MIN}}}{60\ \frac{\text{SEC}}{\text{MIN}}}\right) = 125.66\ \frac{\text{RAD}}{\text{SEC}}$$

BUT $TR = \dfrac{1}{1 - \left(\frac{\omega_f}{\omega}\right)^2}$ SO $\omega = 41.13\ \dfrac{\text{RAD}}{\text{SEC}}$

FOR THE SPRING-MOUNTED MASS,

$$\omega = \sqrt{K/M} \quad \text{SO} \quad 41.13 = \sqrt{\frac{K(386)}{800}}$$

$$K = 3506.1\ \frac{\text{LBF}}{\text{IN}}$$

$$K_{SPRING} = \frac{K}{4} = 876.5\ \frac{\text{LBF}}{\text{IN}}$$

THE STATIC DEFLECTION IS

$$\delta_{ST} = \frac{W}{K} = \frac{800\ \text{LBM}}{3506.1\ \frac{\text{LBF}}{\text{IN}}} = .228\ \text{IN}$$

THE ADDED DEFLECTION DUE TO THE OSCILLATORY FORCE IS

$$\frac{3\ \text{LBF}}{(3506.1)\frac{\text{LBF}}{\text{IN}}} = \boxed{.000856\ \text{IN}}$$

**3** THE MOMENT OF INERTIA OF THE ARM ABOUT POINT A IS

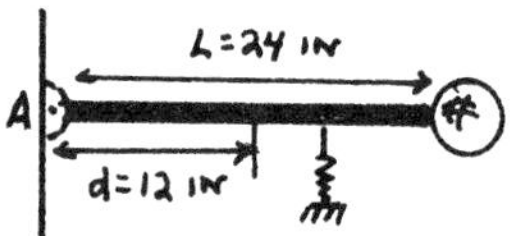


$$J_{A,ARM} = \tfrac{1}{12}ML^2 + Md^2$$

$$= \tfrac{1}{12}\left(\tfrac{5}{386}\right)(24)^2 + \left(\tfrac{5}{386}\right)(12)^2 = 2.487\ \text{LBF-IN-SEC}^2$$

THE MOMENT OF INERTIA OF THE CONCENTRATED LOAD IS

$$J_{A,LOAD} = ML^2 = \left(\tfrac{3}{386}\right)(24)^2 = 4.477\ \text{LBF-IN-SEC}^2$$

$$J_{total} = 2.487 + 4.477 = 6.964\ \text{LBF-IN-SEC}^2$$

THE MOMENT CAUSING DEFLECTION {WHICH IS RESISTED IN TOTAL BY THE SPRING MOMENT} IS FOUND BY SUMMING MOMENTS ABOUT POINT A.

$$\Sigma M_A: 3(24) + 5(12) - M_{SPRING} = 0$$

SO $M_{SPRING} = 132$ IN-LBF

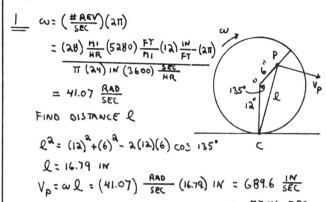

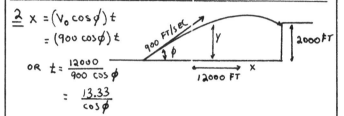

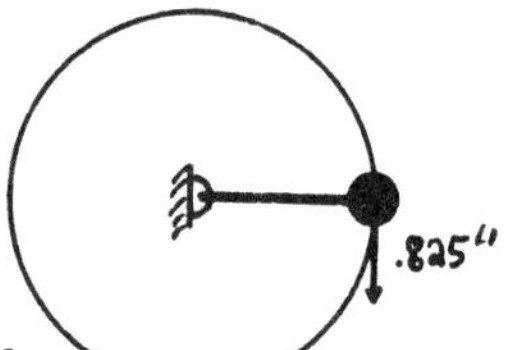


THE DEFLECTION AT THE TIP IS

$$(.55)\left(\tfrac{24}{16}\right) = .825''$$

THIS DEFLECTION IN RADIANS IS

$$\frac{(.825)\text{IN}\ (2\pi)\frac{\text{RAD}}{\text{REV}}}{(2\pi)(24)\frac{\text{IN}}{\text{REV}}} = .0344\ \text{RAD}$$

THE ANGULAR STIFFNESS OF THE SYSTEM IS THE MOMENT WHICH WOULD CAUSE A 1-RADIAN DEFLECTION

$$K_R = \frac{132\ \text{IN-LBF}}{.0344\ \text{RAD}} = 3837.2\ \frac{\text{IN-LBF}}{\text{RAD}}$$

THE NATURAL FREQUENCY IS

$$f = \frac{1}{2\pi}\sqrt{\frac{K_R}{J}} = \frac{1}{2\pi}\sqrt{\frac{3837.2}{6.964}} = \boxed{3.74\ \text{HZ}}$$

**4**

$$M = \frac{(1)\,oz}{(16)\,\frac{oz}{LBM}\,(386)\,\frac{IN}{SEC^2}} = .000162\ \frac{LBM\text{-}SEC^2}{IN}$$

$$\omega_F = \frac{(2\pi)\,\frac{RAD}{REV}\,(800)\,\frac{REV}{MIN}}{(60)\,\frac{SEC}{MIN}} = 83.78\ \frac{RAD}{SEC}$$

$r = 5$ IN

$$V_t = \omega r = (83.78)\,\frac{RAD}{SEC}\,(5)\,IN = 418.9\ IN/SEC$$

$$F_c = \text{CENTRIFUGAL FORCE} = \frac{M V_t^2}{r} = \frac{(.00162)(418.9)^2}{5} = 5.69\ LBF$$

THE NATURAL FREQUENCY IS

$$\omega = \sqrt{\frac{Kg}{W}} = \sqrt{\frac{(4)(1000)(386)}{50}} = 175.7\ \frac{RAD}{SEC}$$

$$\frac{C}{C_{CRIT}} = \frac{1}{8} \text{ AND } \beta = \frac{1}{\sqrt{\left[1-\left(\frac{\omega_f}{\omega}\right)^2\right]^2 + \left[2\left(\frac{C}{C_{CRIT}}\right)\left(\frac{\omega_f}{\omega}\right)\right]^2}}$$

SO $\beta = 1.28$

THE MAGNIFIED EXCURSION IS

$$\delta = \frac{\beta F_c}{K} = \frac{(1.28)(5.69)\,LBF}{(4)(1000)\,\frac{LBF}{IN}} = \boxed{0.00182\ IN}$$

**TIMED**

**1**

THE ANGULAR FORCING FREQUENCY IS

$$\omega_f = \frac{(1200)\,2\pi}{60} = 125.7\ RAD/SEC$$

THE CENTRIFUGAL FORCE DUE TO ROTATING THIS UNBALANCE IS GIVEN BY EQUATIONS 16.20 AND 16.21

$$F_c = m a_N = m r \omega^2$$

$$= \left(\frac{3.6}{32.2}\right)\left(\frac{3}{12}\right)(125.7)^2 = 441.6\ LBF$$

THE TRANSMISSIVITY IS GIVEN AS .05.

THE NATURAL FREQUENCY CAN BE FOUND FROM EQUATION 16.110

$$TR = \frac{1}{\sqrt{\left(1-\left(\frac{\omega_f}{\omega}\right)^2\right)^2}} =$$

$$.05 = \frac{1}{\left|1-\left(\frac{125.7}{\omega}\right)^2\right|}$$

$$\omega = 27.43\ RAD/SEC$$

FOR A SPRING-MOUNTED MASS, THE SPRING CONSTANT CAN BE FOUND FROM

$$\omega = \sqrt{K/m}$$

$$27.43 = \sqrt{\frac{K}{\left(\frac{175}{386}\right)}}$$

{386 IS g IN IN/SEC²}

$K = 341.1$ LB/IN (TOTAL)

SINCE THE 4 CORNER SPRINGS ARE IN PARALLEL,

$$K_{total} = K_1 + K_2 + K_3 + K_4$$

SO,

$$K_{EACH} = \frac{341.1}{4} = \boxed{85.28\ LB/IN}$$

THE AMPLITUDE OF VIBRATION (ZERO TO PEAK) IS GIVEN BY EQN 16.109.

$$x = (.05)\left(\frac{441.6}{341.1}\right) = \boxed{.0647''}$$

**2** THE 1ST PLATE UNDER THE LOAD IS A SIMPLE BEAM

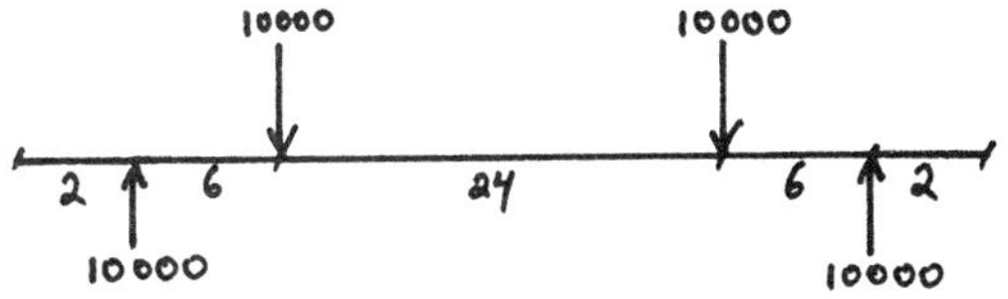


ASSUME $E = 2.9$ EE7 psi

$$I = \frac{bh^3}{12} = \frac{(30)(1/2)^3}{12} = .3125\ IN^4$$

THE DEFLECTION AT THE POINT OF LOADING IS GIVEN BY CASE 10 ON PAGE {THE OVERHANG CONTRIBUTES NOTHING TO THE RIGIDITY.}

$$y = \frac{Fx}{6EI}\left[(3a)(L-a) - x^2\right]$$

$$= \frac{(10{,}000)(6)}{(6)(2.9\,EE7)(.3125)}\left[(3)(6)(36-6) - (6)^2\right]$$

OR, $y = .556''$

THE 2ND PLATE IS LOADED EXACTLY THE SAME AS THE TOP PLATE, ONLY UPSIDE DOWN. THEREFORE ITS DEFLECTION IS ALSO .556".

THE TOTAL DEFLECTION OF 8 PLATES IS

$$y_{total} = (8)(.556) = \boxed{4.448''}$$

OF COURSE THIS RESULT ASSUMES THE YIELD POINT IS NOT EXCEEDED.

FOR CASE 10 AGAIN,

$$M_{max} = Fa = (10{,}000)(6) = 60{,}000\ IN\text{-}LB$$

$$\sigma_{max} = \frac{M_{max}\,c}{I} = \frac{(60{,}000)(.25)}{.3125} = \boxed{48{,}000\ psi}$$

THE TOTAL SPRING CONSTANT IS

$$K = \frac{F}{x} = \frac{20{,}000}{4.448} = 4496\ LB/IN$$

FROM EQUATION 16.91

$$f = \frac{1}{2\pi}\sqrt{\frac{Kg}{W}}$$

$$= \frac{1}{2\pi}\sqrt{\frac{(4496)(386)}{20{,}000}} = \boxed{1.48\ HZ}$$

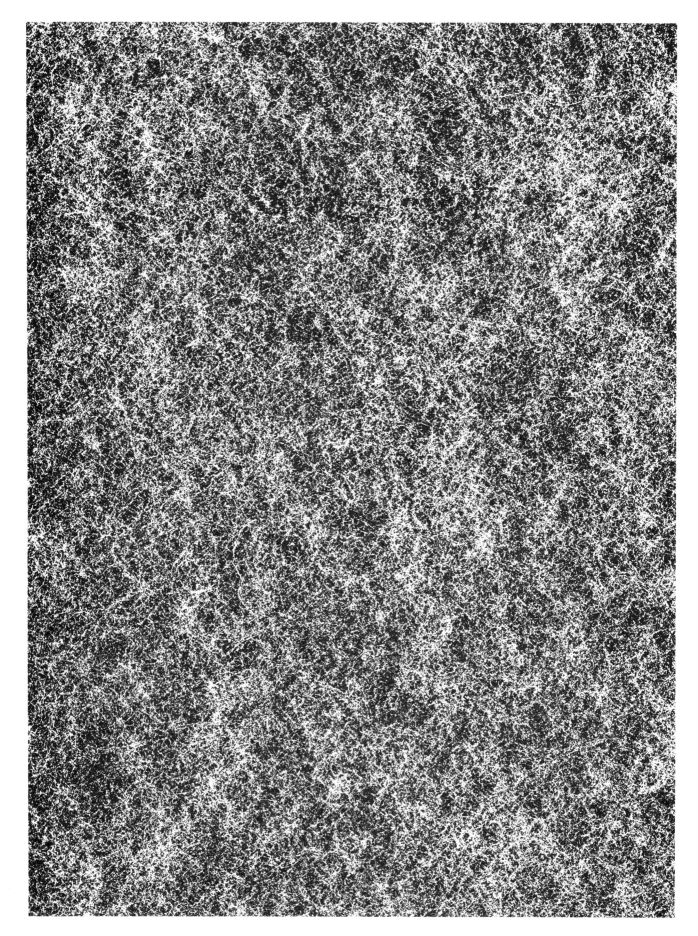

# Noise Control

WARM-UPS

1 METHOD 1 USE FIGURE 17.1

$\Delta L = 40 - 35 = 5$

FROM FIGURE 17.1, INCREMENT = 1.2

$L_{total} = 40 + 1.2 = 41.2$ dB

METHOD 2 USE EQUATION 17.6

$L_{total} = 10 \log_{10}\left[10^{\left(\frac{40}{10}\right)} + 10^{\left(\frac{35}{10}\right)}\right]$

$= 41.19$ dB

2 IF THE BACKGROUND SOUND LEVEL IS 43 dB, USE EQN 17.6

$45 = 10 \log_{10}\left[10^{\left(\frac{43}{10}\right)} + 10^{\left(\frac{L}{10}\right)}\right]$

$\text{ANTILOG}\left(\frac{45}{10}\right) = 10^{4.3} + 10^{L/10}$

$11670.2 = 10^{L/10}$

$\log_{10}(11670.2) = \frac{L}{10}$

$L = 40.67$ dB

3 BEFORE BEING ENCLOSED, THE OBSERVED SOUND LEVEL IS 100 dBA. AFTER BEING ENCLOSED, THE OBSERVED SOUND LEVEL IS 110 − 30 = 80.

THE DIFFERENCE IS 100 − 80 = 20 dB

4 FROM TABLE 17.2, 4 HOURS

5 USING EQUATION 17.10, WITH Q = 1 FOR AN ISOTROPIC SOURCE, AND ASSUMING PERFECTLY ABSORBING SURFACES ($R = \infty$),

$L_W = 92 - 10 \log\left(\frac{1}{4\pi(4)^2}\right) - 10.5 = 104.5$

THEN, AT r = 12 FT.,

$L_p = 104.5 + 10 \log\left(\frac{1}{4\pi(12)^2}\right) + 10.5$

$= \boxed{82.5 \text{ dB}}$ re $20\,\mu N/m^2$

6 THE GENERAL SOUND PRESSURE LEVEL IS

$L_{total} = 10 \log_{10}\left[10^{(85/10)} + 10^{(90/10)} + 10^{(92/10)} + 10^{(87/10)} + 10^{(82/10)} + 10^{(78/10)} + 10^{(65/10)} + 10^{(54/10)}\right]$

$= 95.6$ dB

IF THE A-WEIGHTED SOUND LEVEL IS WANTED (THE PROBLEM IS NOT SPECIFIC) THEN CORRECTIONS FROM TABLE 17.3 MUST BE ADDED TO THE MEASUREMENTS, IGNORING THE 31.5 Hz FREQUENCY,

$L_{total} = 10 \log_{10}\left[10^{\frac{85-26.2}{10}} + 10^{\frac{90-16.1}{10}} + 10^{\frac{92-8.6}{10}} + 10^{\frac{87-3.2}{10}} + 10^{\frac{82-0}{10}} + 10^{\frac{78+1.2}{10}} + 10^{\frac{65+1}{10}} + 10^{\frac{54-1}{10}}\right]$

$= 88.6$ dBA

7 $L_{500} = 87$

$L_{1000} = 82$

$L_{2000} = 78$

247

$\frac{247}{3} = 82.3$

8 $\frac{S}{A} = .5$ MEANS 50% OF THE SOUND ENERGY IS REMOVED. FROM EQUATION 17.14,

$\Delta L = 10 \log\left(\frac{.5}{1}\right) = -3.01$ (DECREASE)

CONCENTRATES

1 $A_W$ = WALL AREA $= 20(100 + 100 + 400 + 400) = 20{,}000$ FT²

$A_{CF}$ = CEILING + FLOOR AREA $= 2(100 + 400) = 80{,}000$ FT²

FROM PAGE 17-9, CHOOSE NRC = $\bar{\alpha} = .02$ FOR POURED CONCRETE.

$S_1 = (20{,}000 + 80{,}000)(.02) = 2000$

AFTER TREATMENT,

$S_2 = (.4)(.8)(20{,}000) + (.6)(.02)(20{,}000) + (.02)(80{,}000)$

$= 8240$

FROM EQN 17.14

$\Delta L = 10 \log\left(\frac{8240}{2000}\right) = 6.15$

2 REFER TO THE PROCEDURE ON PAGE 17-8

AS FOUND

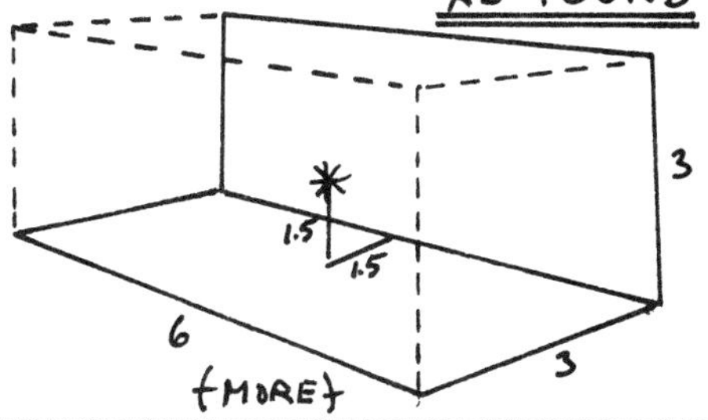


(MORE)

CONCENTRATES # 2 CONTINUED

| AREA NAME | A (AREA) | B | AB |
|---|---|---|---|
| FRONT | 18 | 1 | 18 |
| TOP | 18 | 1/3 | 6 |
| SIDES | 18 | 1/3 | 6 |
| | 54 | | 30 |

THE EXISTING SABIN RATIO IS

$$R_{s1} = \frac{30}{54} = .556$$

AFTER BEING ENCLOSED,

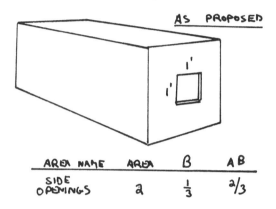

AS PROPOSED

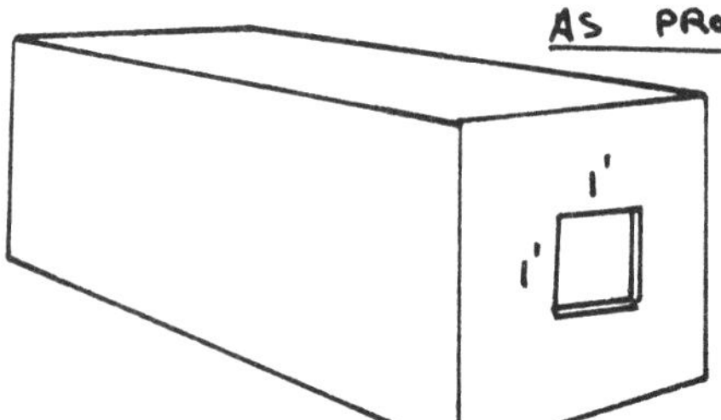


| AREA NAME | AREA | B | AB |
|---|---|---|---|
| SIDE OPENINGS | 2 | 1/3 | 2/3 |

THE NEW SABIN RATIO IS

$$R_{s2} = \frac{2/3}{2} = .333$$

SO, $\Delta L = 10 \text{ LOG}\left(\frac{.556}{.333}\right) = 2.23$

**3**

AS FOUND

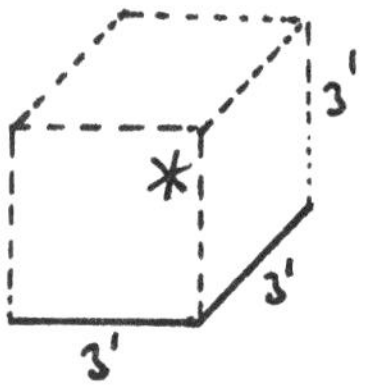


| AREA NAME | AREA | B | AB |
|---|---|---|---|
| FRONT | 9 | 1 | 9 |
| BACK | 9 | 1/6 | 1.5 |
| TOP | 9 | 1/3 | 3 |
| SIDES | 18 | 1/3 | 6 |
| | 45 | | 19.5 |

SO $R_{s1} = \frac{19.5}{45} = .433$

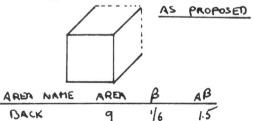

AS PROPOSED

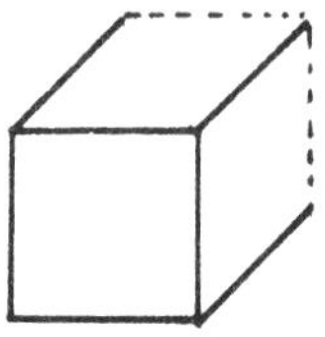

| AREA NAME | AREA | B | AB |
|---|---|---|---|
| BACK | 9 | 1/6 | 1.5 |

$$R_{s2} = \frac{1.5}{9} = .167$$

$$\Delta L = 10 \text{ LOG}\left(\frac{.433}{.167}\right) = 4.14$$

**4** THE SABINS FOR EACH ABSORBING SURFACE ARE:

FLOOR: $S_1 = (1000)(.03) = 30$

CEILING (ASSUME SMOOTH) $\alpha = .03$

$S_2 = (1000)(.03) = 30$

WALLS AND GLASS (BOTH HAVE SAME $\alpha = .03$)

$A = 10(20+20+50+50) = 1400$

$S_3 = (1400)(.03) = 42$

OCCUPANTS (FROM PAGE 17-9)

ABOUT 5 SABINS EACH

$S_4 = (15)(5) = 75$

DESKS (ABOUT $1\frac{1}{2}$ SABIN EACH)

$S_5 = (15)(1\frac{1}{2}) = 22.5$

MISCELLANEOUS

$S_6 = 5$ (GIVEN)

$$S_{total} = 30 + 30 + 42 + 75 + 22.5 + 5 = 204.5$$

AFTER TREATMENT,

$S_2 = (.7)(1000) = 700$

SO

$$S_{total} = 30 + 700 + 42 + 75 + 22.5 + 5 = 874.5$$

SO $\Delta L = 10 \text{ LOG}\left(\frac{874.5}{204.5}\right) = 6.31$

**5** ORIGINALLY

$A_{ceiling} = (20)(30) = 600$

$S_1 = (.5)(600) = 300$

AFTER PARTITIONING

$S_x = \frac{1}{2}(300) = 150$

$S_z = 150$

$A_0 = (20)(2) = 40$

USING THE PROCEDURE ON PAGE 17-8

$$PR = \left[\frac{\frac{40}{150} + \frac{150}{150}}{1 + \frac{150}{150}}\right]\left(\frac{150}{40}\right) = 2.375$$

$$L_{IL} = 10 \text{ LOG}(2.375) = 3.76$$

**6**

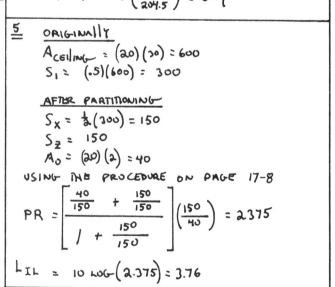

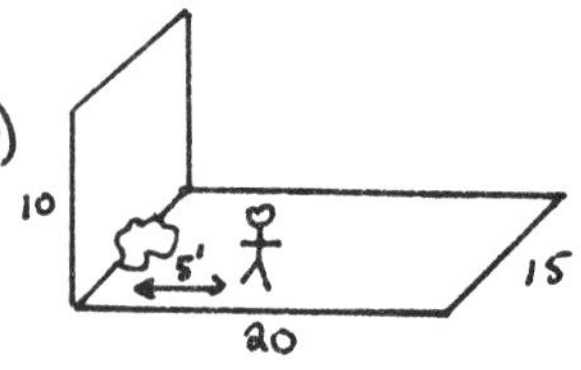


$A_{walls} = 10(15+15+20+20) = 700$

$A_{ceiling} = 20(15) = 300$

$A_{floor} = 300$

{MORE}

CONCENTRATES # 6 CONTINUED

FROM EQN 17.9

$$\bar{\alpha} = \frac{700(.06)+300(.03)+300(.5)}{700+300+300} = .155$$

FROM EQN 17.7

$$R = \frac{.155(700+300+300)}{1-.155} = 238.5$$

FROM EQN 17.10,

$$L_p = L_W + 10\log_{10}\left[\frac{Q}{4\pi r^2} + \frac{4}{R}\right] + 10.5$$

Q = 4 BECAUSE THE SOURCE IS AT THE INTERSECTION OF 2 WALLS

$$L_p = 65 + 10\log\left[\frac{4}{4\pi(5)^2} + \frac{4}{238.5}\right] + 10.5$$

$$= 60.19$$

ADDING IN THE BACKGROUND NOISE,

$$L_{total} = 10\log\left[10^{\frac{60.19}{10}} + 10^{\frac{50}{10}}\right]$$

$$= 60.58$$

7 FROM THE FAN

ROTATION: $\frac{600 \text{ RPM}}{60}$ = 10 HZ

DRIVING BLADES: $\frac{(600)(8)}{60}$ = 80 HZ

FAN BLADES: $\frac{(600)(64)}{60}$ = 640 HZ

FROM THE MOTOR

ROTATION: $\frac{1725}{60} \approx$ 29 HZ

POLES: $\frac{(1725)(4)}{60}$ = 115 HZ

ELECTRICAL HUM 60 HZ

PULLEYS

MOTOR PULLEY {SAME AS MOTOR} 29 HZ

FAN PULLEY {SAME AS FAN} 10 HZ

BELT

BELT SPEED $= \pi D(\text{RPS})$

$= \pi(4)\left(\frac{1725}{60}\right) = 361.3$ IN/SEC

FREQUENCY $= \frac{361.3}{72} =$ 5 HZ

8 $f_{FORCED} = \frac{1725}{60} = 28.75$ HZ

FROM EQN 16.92,

$$f_{NATURAL} = \frac{1}{2\pi}\sqrt{g/\delta} = \frac{1}{2\pi}\sqrt{\frac{386 \text{ IN/SEC}^2}{.02}} = 22.11 \text{ HZ}$$

FROM EQUATION 16.110

$$TR = \frac{1}{\left(\frac{28.75}{22.11}\right)^2 - 1} = 1.448$$

THIS IS AN INCREASE IN FORCE.

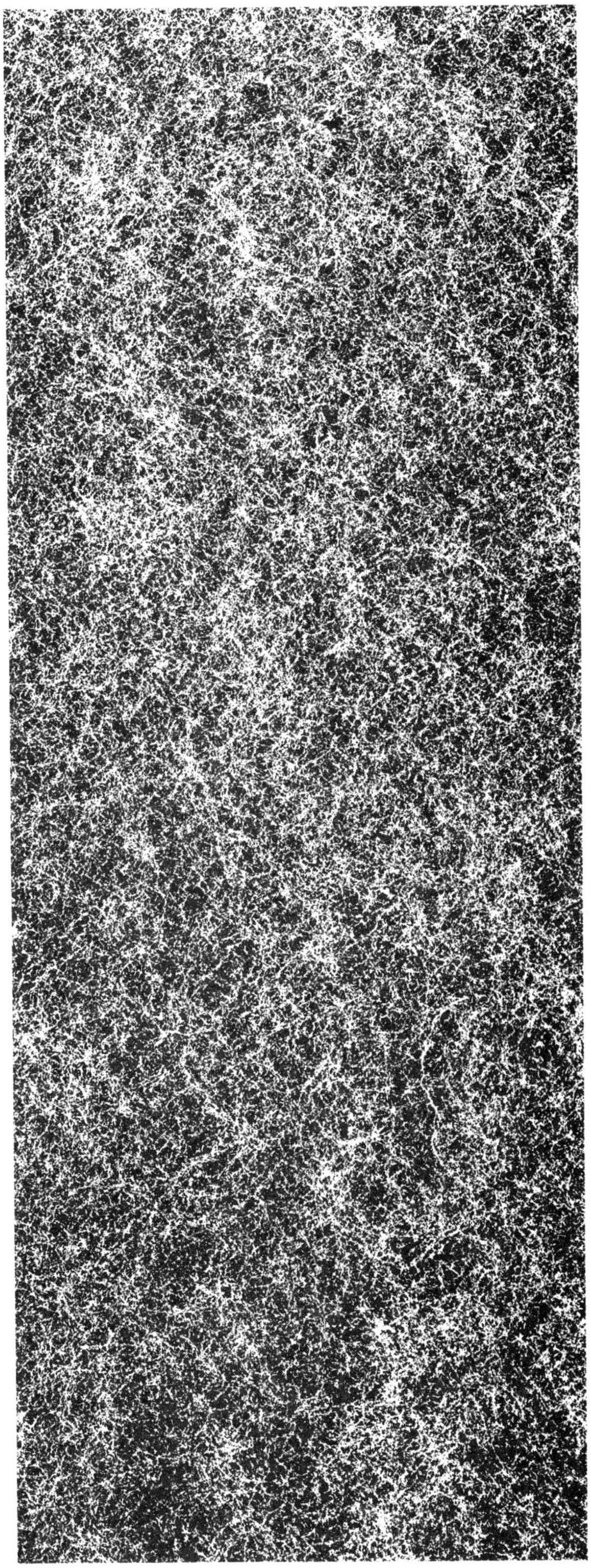

# Nuclear Engineering

WARM-UPS

**1**

USING CARBON-BASED AMU VALUES, THE MASS INCREASE IS

$(1.007825 + 1.008665 - 2.0410) = .00239$ AMU

OR $(.00239)(931.481) = 2.226$ MeV INCREASE

THUS $2.75 - 2.226 = .5238$ MeV IS SHARED BY THE NEUTRON + HYDROGEN ATOM. SINCE BOTH HAVE THE SAME MASS THE NEUTRON WILL RECEIVE HALF $= .262$ MeV

**2**

$\lambda = .693/6.47 = .1071$

$.05 = \text{EXP}(-.1071\,t)$

$t = 27.97$ DAYS

**3**

ASSUME THE SHIELD WILL BE 10 CM THICK, FOR A 2 MeV SOURCE,

$\mu_e/\rho = .0457,$

OR $\mu_e = (.0457)(11.34) = .5182$

THEN $\mu_e X = 5.182$ AND $B = 2.78$ BY INTERPOLATION. THEN

$.01 = 2.78\ \text{EXP}[-(.5182)X]$

OR $X = 10.86$

SINCE OUR INITIAL ESTIMATE WAS CLOSE, A SECOND ITERATION IS NOT NEEDED.

**4**

$\phi_u = (\text{EE}6)\,e^{-\mu_e X} = (\text{EE}6)\,e^{-5.182} = 5.63\ \text{EE}3\ \frac{\gamma}{\text{cm}^2\text{-s}}$

WHICH IS UNCOLLIDED FLUX.

THE BUILD-UP FLUX IS

$\phi_B = 2.78\,\phi_u = 1.56\ \text{EE}4$

THE DOSE IS GIVEN BY $.0659\ E_0(\phi_B)\left(\frac{\mu_m}{\rho}\right)_{\text{AIR}}$

AND SINCE $\left(\frac{\mu_m}{\rho}\right)$ FOR AIR AND 2 MeV GAMMAS IS $(.0238)$

DOSE $= .0659(2)(1.56\ \text{EE}4)(.0238)$

$= 48.9$ mR/HR

**5**

FOR 20°C GOLD, $\sigma_a = 98$ BARNES AND $\rho = 19.32$

$$\bar{\sigma}_a = \frac{98}{1.128}\sqrt{293/273} = 82.75\ \text{b}$$

$$N = \frac{(.4909)(19.32)(6.023\ \text{EE}23)}{197} = 2.9\ \text{EE}22$$

THE ACTIVATED GOLD ($Au^{198}$) HAS $t_{1/2} = 2.7$ d, SO

$\lambda = .693/2.7 = .2567$ d

$$A = \frac{(\text{EE}8)(82.75)(\text{EE-}24)(2.9\ \text{EE}22)\left[1 - e^{(-.2567)(1)}\right]}{3.7\ \text{EE}10}$$

$= 1.468\ \text{EE-}3$ CURIE

**6**

FOR AN ISOTROPIC POINT SOURCE,

$$\phi = \frac{S\,e^{-r/L}}{4\pi \bar{D} r}$$

FROM EQN 18.77

$\bar{D} = L^2 \Sigma_a$, SO

$$\bar{D} = (2.85)^2\,\frac{(.66\ \text{EE-}24)(1)(6.023\ \text{EE}23)}{18} = .18$$

{ACTUALLY, $\bar{D} = .16$}

$$\phi = \frac{(\text{EE}7)\,e^{-20/2.85}}{(4\pi)(.16)20} = 2.23\ \text{EE}2\ \frac{\text{NEUTRONS}}{\text{cm}^2\text{-s}}$$

**7**

$\sigma_f = 4.18,\ \sigma_a = 7.68$

$P\{\text{FISSION}\} = 4.18/7.68 = .544$

**8**

ASSUME A SHIELD 20 CM THICK. FOR A 1 MeV GAMMA IN IRON, $\mu_e/\rho = .0595$. FOR IRON, $\rho = 7.87$ SO $\mu_e = .4683$ AND $\mu_e X = 9.366$. $B = 14.93$ BY INTERPOLATION.

$$\phi_B = \frac{(14.93)(\text{EE}8)\,e^{-.4683X}}{4\pi X^2} = \frac{1.19\ \text{EE}8\ e^{-.4683X}}{X^2}$$

THEN, THE EXPOSURE RATE IS

$$1 = \frac{.0659\,(1)(1.19\ \text{EE}8)\,e^{-.4683X}\,(.0280)}{X^2}$$

OR $X = 14.8$ BY TRIAL + ERROR.

## CONCENTRATES

**1**

IF THE CONVERSION RATIO IS NOT ASSUMED, CALCULATE FOR FAST FISSION

U-238: $\sigma_a = .59$ $\sigma_f = .5$

Pu-239 : $\sigma_a = 1.95$ $\sigma_f = 1.8$

$$\eta = .8(2.45)\left(\frac{.5}{.59}\right) + (.2)(2.95)\left(\frac{1.8}{1.95}\right) = 2.21$$

ASSUME $\epsilon = 1.05$, SO

$$CR = (2.21)(1.05) - 1 = 1.305$$

$$\text{LINEAR } t_d = \frac{(2000)(1000)}{(1.305-1)(1.23)(1000)} = 5331 \text{ DAYS}$$

$$\text{EXPONENTIAL} = (.693)(5331) = 3695 \text{ DAYS}$$

**2**

$$P = \frac{\bar{\phi}\Sigma_f V}{3.1\text{ EE}10} = \frac{\left(\frac{4.5\text{ EE}15}{3.29}\right)(.005)\left(\frac{4}{3}\pi(40)^3\right)}{3.1\text{ EE}10}$$

$$= 5.914\text{ EE}7 \text{ WATTS}$$

**3**

$$r_1 = \sqrt{(25.4)^2/\pi} = 14.33 \text{ cm}$$

AND FROM EQNS 18.119 AND 18.120

$$E = 1 + \frac{(14.33)^2}{2(54)^2}\left[\frac{\ln\left(\frac{14.33}{1.02}\right)}{1-\left(\frac{1.02}{14.33}\right)^2} - .75 + \left(\frac{1.02}{28.66}\right)^2\right] = 1.0563$$

$$F \approx 1 + \frac{1}{2}\left(\frac{r_0}{2L}\right)^2 - \frac{1}{12}\left(\frac{r_0}{2L}\right)^4 + \frac{1}{48}\left(\frac{r_0}{2L}\right)^6$$

AND $L = 1.55$ FOR NATURAL RADIATION

$$F \approx 1.0551$$

$$\Sigma_{aF} = \frac{(18.7)(6.023\text{ EE}23)(7.68)\text{EE-}24(.984)}{(237.98)(1.128)} = .3171$$

$$\Sigma_{aM} = \frac{(1.6)(6.023\text{ EE}23)(4.65\text{ EE-}27)}{(12)(1.128)} = 3.311\text{ EE-}4$$

$$V_F \approx \pi(1.02)^2 = 3.2685$$

$$V_M = (25.4)^2 - 3.2685 = 641.8915$$

THEN,

$$\frac{\bar{\phi}_M}{\bar{\phi}_F} = 1.0551 + \left(\frac{3.2685}{641.8915}\right)\left(\frac{.3171}{3.311\text{ EE-}4}\right)(.0563) = 1.33$$

AND

$$1/f = 1 + (1.33)\left(\frac{3.311\text{ EE-}4}{.3171}\right)\left(\frac{641.8915}{3.2685}\right) = 1.2727$$

$$f = .7857$$

## TIMED

**1**

USING APPENDIX A, PAGE 22-7, $m = 4.54\text{ EE}4$ KG.

FROM EQUATION 18.104,

$$N_f = \frac{0.02(4.54\text{ EE}4\text{ kg})(1000\text{ g/kg})(6.023\text{ EE}23 \frac{\text{particles}}{\text{gmole}})}{(V\text{ cm}^3)(235\text{ g/gmole})}$$

$$= \frac{2.23\text{ EE}27}{V} \text{ particles/cm}^3$$

INSERT THIS INTO EQUATION 18.103 AND REARRANGE TO GET

$$\phi = \frac{(3.1\text{ EE}10\text{ fissions/sec/W})(5.0\text{ EE}8\text{ W})}{547\frac{\text{barns}}{\text{particle}}\left(1\text{ EE-}24\frac{\text{cm}^2}{\text{barn}}\right)(2.23\text{ EE}27\text{ particles})}$$

$$= \boxed{1.22\text{ EE}13\text{ fissions/cm}^2\text{-sec}}$$

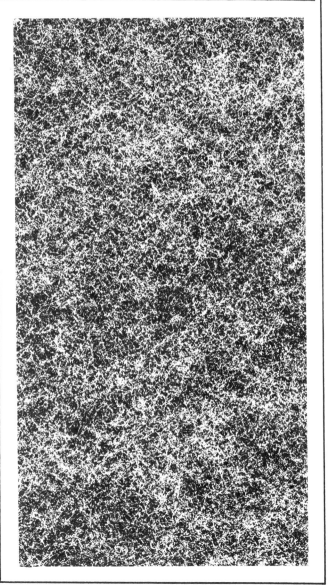

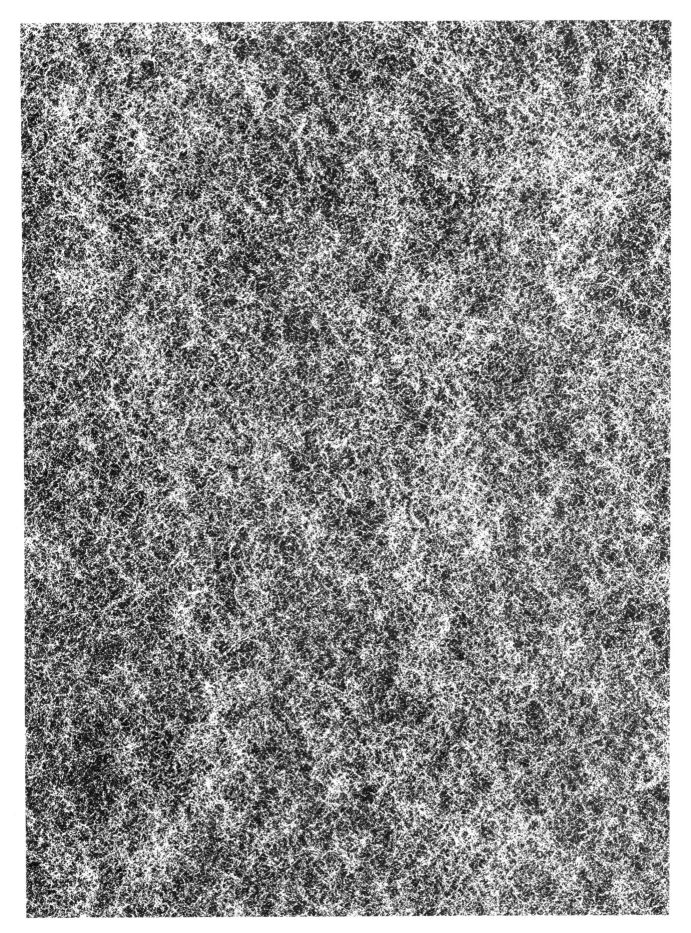

# Modeling of Engineering Systems

**1**

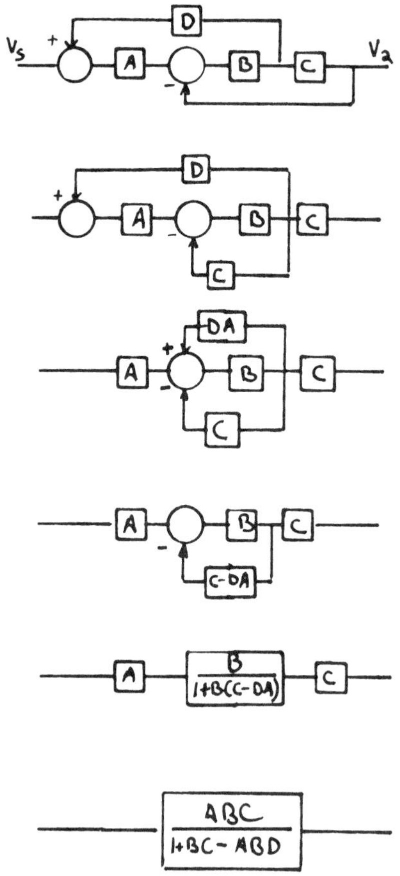

**2**

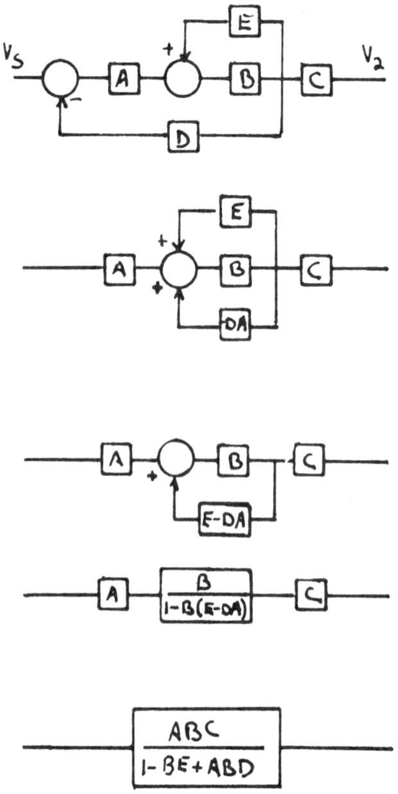

**3.1**

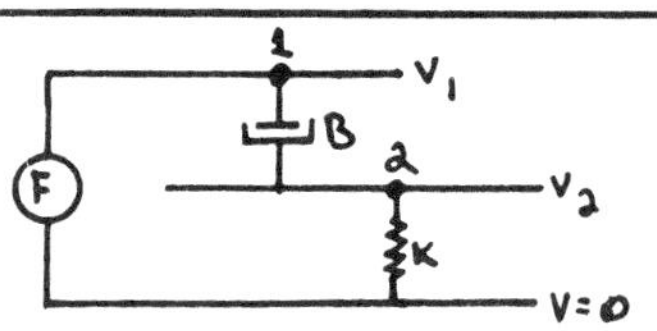

AT NODE 1

$$F = B(V_1 - V_2) \quad OR \quad F = B(x'_1 - x'_2)$$

AT NODE 2

$$F = K(x_2 - 0) \quad OR \quad 0 = kx_2 + B(x'_2 - x'_1)$$

**3.2**

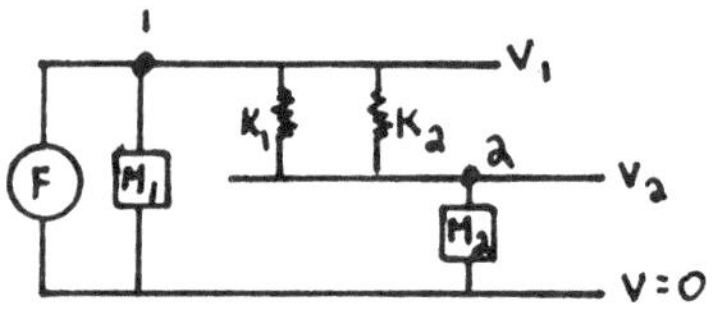

AT NODE 1

$$F = M_1 a_1 + (K_1 + K_2)(X_1 - X_2)$$
$$= M_1 x''_1 + (K_1 + K_2)(X_1 - X_2)$$

AT NODE 2

$$0 = M_2 a_2 + (K_1 + K_2)(X_2 - X_1)$$
$$= M_2 x''_2 + (K_1 + K_2)(X_2 - X_1)$$

**3.3** DESPITE ITS APPEARANCES, THIS IS A ROTATIONAL SYSTEM.

$\tau$ = APPLIED ROTATIONAL TORQUE = FL
$\Theta$ = ROTATED ANGLE
$X_1$ = ARC DISTANCE
$X_2$ = ARC DISTANCE
$I$ = MOMENT OF INERTIA OF BEAM ABOUT AN
 END (FROM P. 12-20) = $\frac{1}{3}mL^2$
$M$ = RESISTING MOMENT = $kx_2 l = kl\sin\Theta l$
 = $kl^2\Theta$ FOR small values of $\Theta$

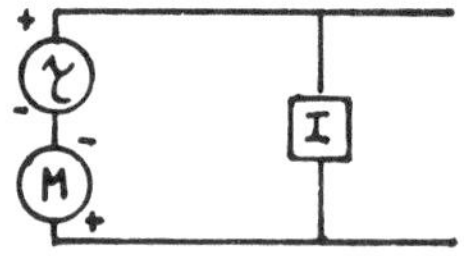

IN GENERAL, $\tau = I\alpha$, SO

$$FL - kl^2\Theta = \frac{1}{3}mL^2\alpha$$

BUT $\alpha = \Theta''$

$$\Theta'' + \left(\frac{3kl^2}{ML^2}\right)\Theta = \frac{FL}{I}$$

**3.4**

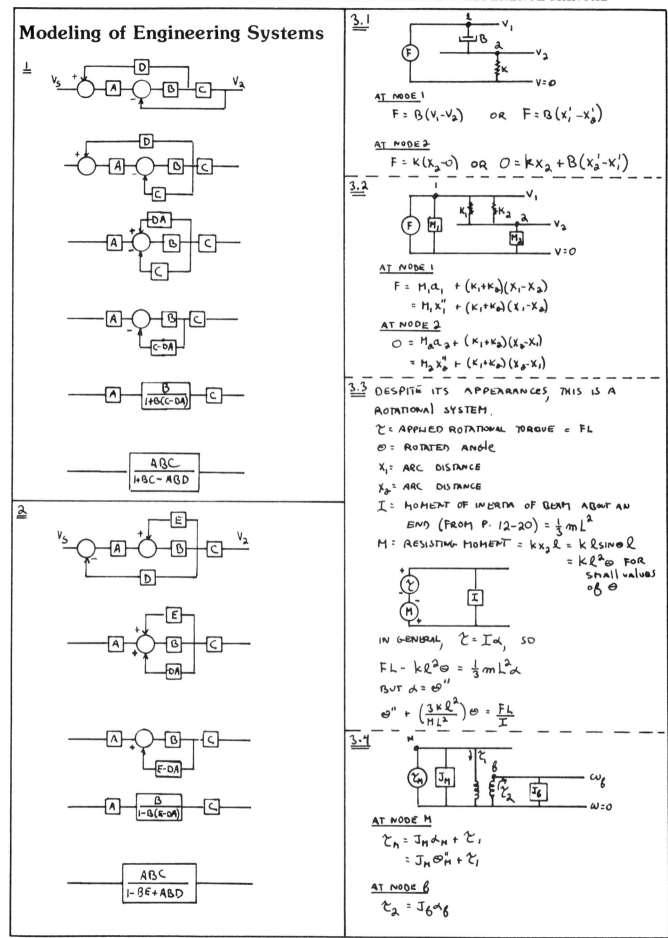

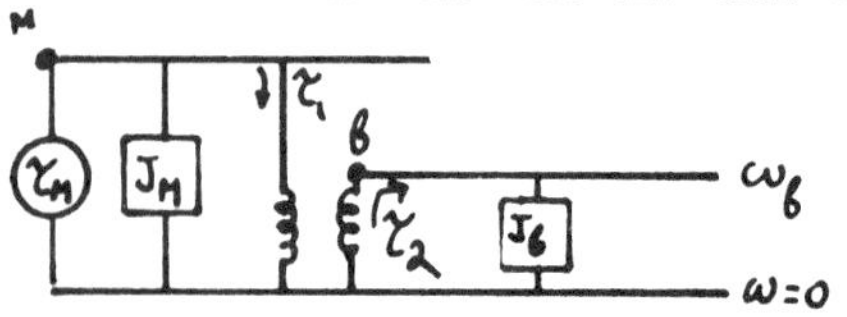

AT NODE M

$$\tau_M = J_M \alpha_M + \tau_1$$
$$= J_M \Theta''_M + \tau_1$$

AT NODE $\beta$

$$\tau_2 = J_\beta \alpha_\beta$$

But $\tau_2 = \left(\frac{N_2}{N_1}\right)\tau_1$

so $\left(\frac{N_2}{N_1}\right)\tau_1 = J_f\theta_f''$

THE THIRD EQUATION REQUIRED IS

$$\theta_M = \left(\frac{N_2}{N_1}\right)\theta_f$$

---

**3.5** THIS IS AN EXAMPLE OF AN UNTUNED VIBRATION DAMPER. CONSIDER THE FLUID TO ACT AS A DAMPER WITH COEFFICIENT $B$.

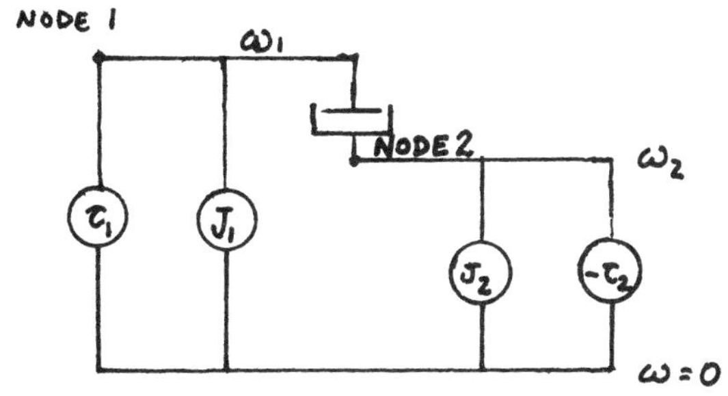


FOR NODE 1:

$$\tau_1 = J_1\alpha_1 + B(\omega_1-\omega_2) = J_1\theta_1'' + B(\theta_1'-\theta_2')$$

FOR NODE 2:

$$-\tau_2 = J_2\alpha_2 + B(\omega_2-\omega_1) = J_2\theta_2'' + B(\theta_2'-\theta_1')$$

FROM EQUATION 19.6,

$\tau = B(\omega_1-\omega_2)$, SO THE THIRD EQUATION IS

$$B = \frac{\tau}{\omega_2-\omega_1} = \frac{\tau}{\theta_2'-\theta_1'}$$

---

**3.6**

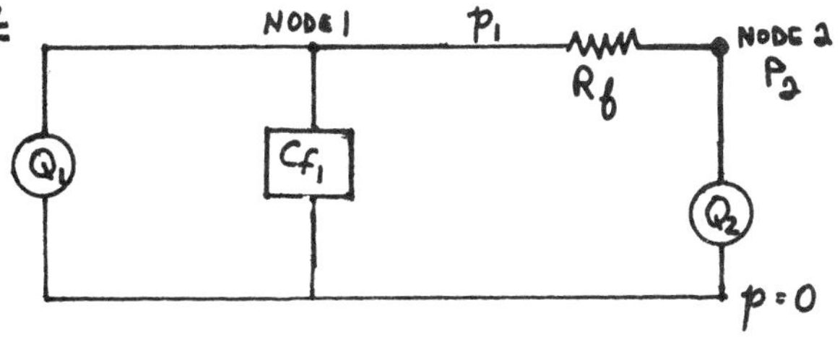


USING EQUATIONS 19.12 AND 19.13,

$$Q = A_t\frac{dh}{dt} = \frac{A_t}{\rho}\frac{dp}{dt} = C_f\frac{dp}{dt}$$

NODE 1: $Q_1 = C_{f_1}\left(\frac{dp_1}{dt}\right) + \frac{1}{R_f}(p_2-p_1)$

NODE 2: $Q_2 = \frac{1}{R_f}(p_1-p_2)$

---

3.7 A PUMP IS REQUIRED TO FILL THE TANK AGAINST THE STATIC HEAD.

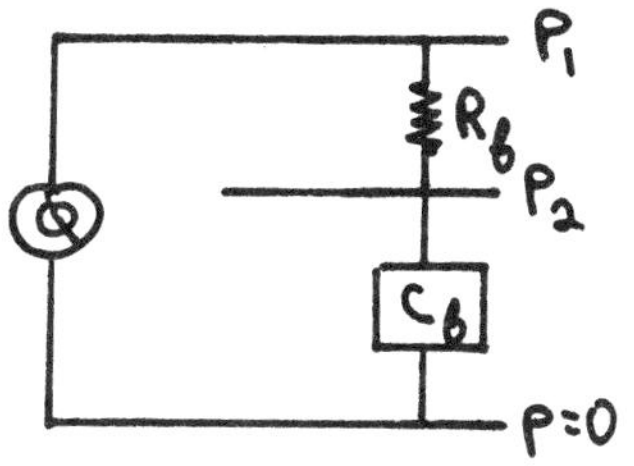


THE SYSTEM EQUATIONS ARE

$$Q = \frac{1}{R_f}(p_1-p_2) = C_f\left(\frac{dp_2}{dt}\right)$$

---

**TIMED**

1 FIRST, SIMPLIFY THE SYSTEM.

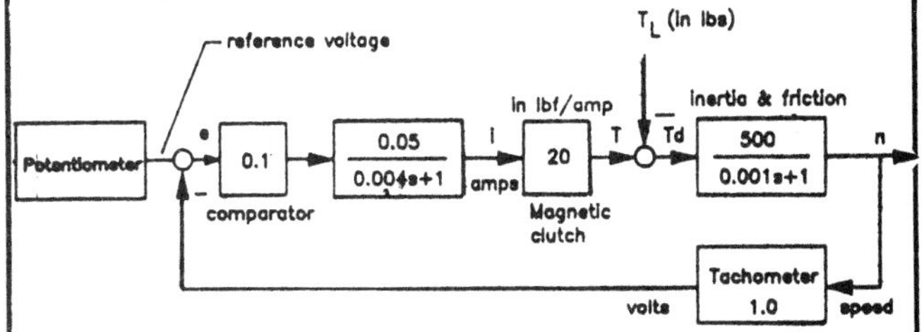


REDRAW THE SYSTEM IN MORE TRADITIONAL FORM.

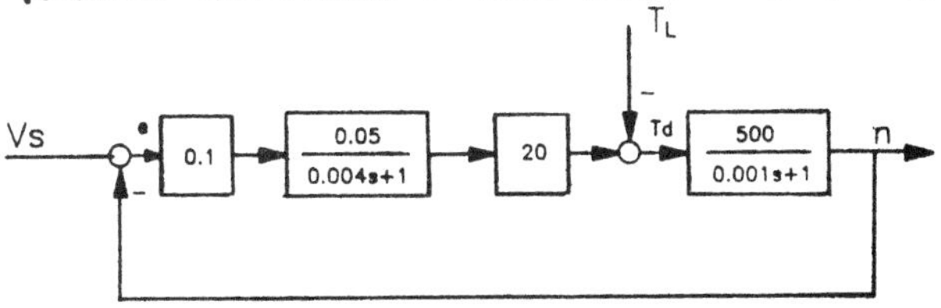


USE TABLE 19.1, CASE 1:

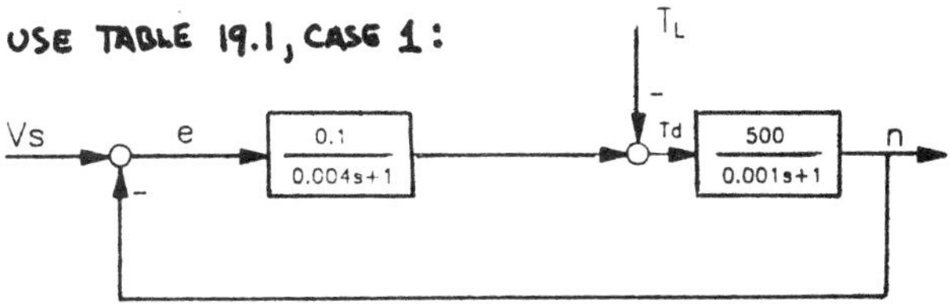


USE TABLE 19.1, CASE 5:

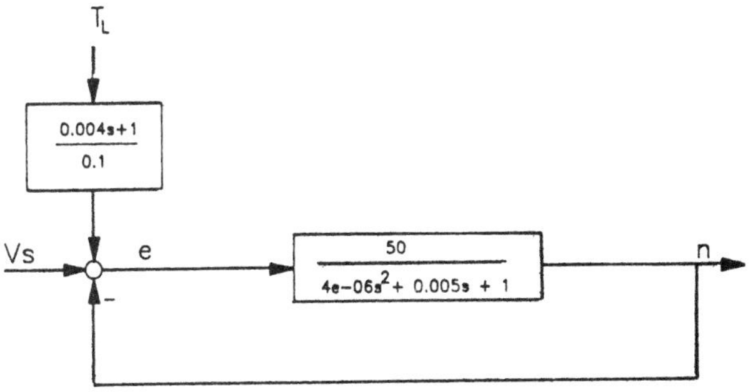


(a) IGNORING THE FEEDBACK LOOP, WE CAN OBTAIN TWO OPEN-LOOP TRANSFER FUNCTIONS: ONE EACH FROM INPUTS $V_s$ AND $T_L$

$$\frac{N}{V_s} = \frac{50}{(0.001s+1)(0.004s+1)}$$

$$\frac{N}{T_L} = \frac{500}{0.001s+1}$$

TIMED #1 CONTINUED

THE TOTAL OPEN LOOP RESPONSE IS THE SUM OF THE RESPONSE TO THE TWO INPUTS.

$$N = \frac{50 V_s}{(0.001S+1)(0.004S+1)} + \frac{500 T_L}{0.001S+1}$$

THE TWO TRANSFER FUNCTIONS HAVE BEEN PLOTTED BELOW. THE OPEN LOOP STEADY-STATE GAIN (OLSSG) CAN BE READ OFF THE PLOTS OR OBTAINED BY SETTING S=0 IN THE OPEN-LOOP TRANSFER EQUATIONS.

$$OLSSG(V_s) = 50$$

$$OLSSG(T_L) = 500$$

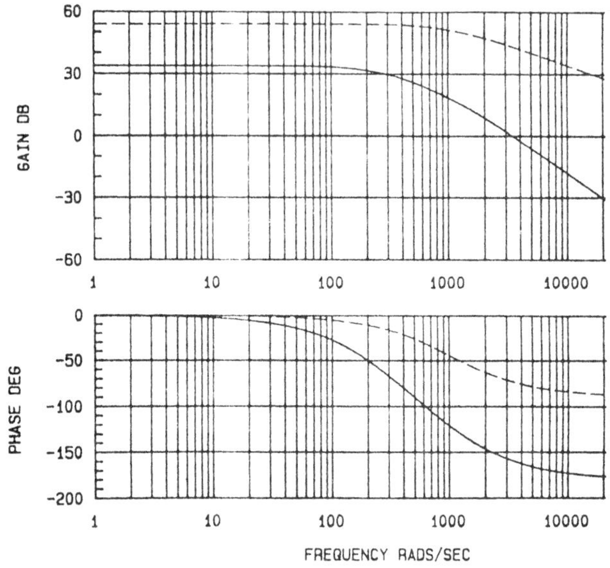


OPEN LOOP FREQUENCY RESPONSE

(b) CLOSING THE LOOP USING NEGATIVE FEEDBACK, AND USING CASE 3 ON TABLE 19.1 RESULTS IN

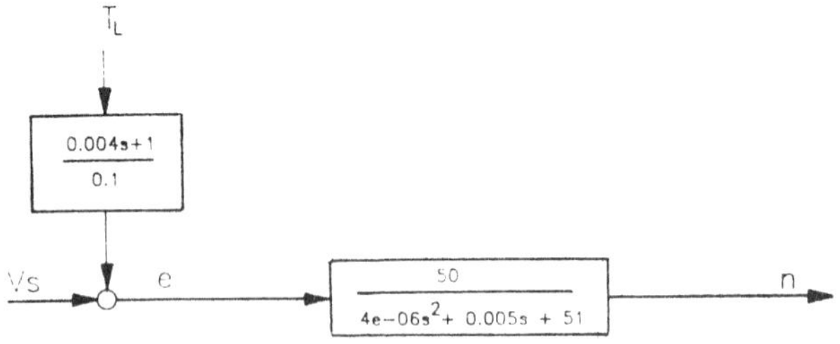


AGAIN, THERE ARE TWO TRANSFER FUNCTIONS: ONE FOR $V_s$ AND ONE FOR $T_L$. THE SUM OF THE TWO IS THE TOTAL CLOSED LOOP RESPONSE.

$$N = \frac{V_s}{\frac{(0.001)(0.004)S^2}{50} + \frac{(0.001+0.004)S}{50} + 1} + \frac{10(0.004S+1)T_L}{\frac{(0.001)(0.004)S^2}{50} + \frac{(0.001+0.004)S}{50} + 1}$$

THE CLOSED LOOP RESPONSES ARE

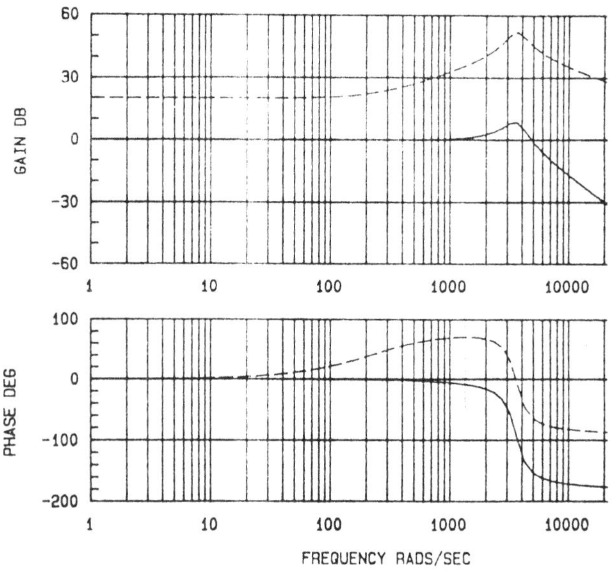


CLOSED LOOP FREQUENCY RESPONSE

THE RESPONSE OF N TO $V_s$ IS UNITY AND FLAT OUT TO 1000 RAD/sec. ALSO, NOTE THE LOW FREQUENCY RESPONSE OF N TO $T_L$ HAS BEEN REDUCED FROM 500 (54 dB) TO 10 (20 dB). BOTH OF THESE ARE DESIRABLE EFFECTS OF CLOSING THE LOOP.

WE ALSO SEE THAT ONE OF THE EFFECTS OF CLOSING THE LOOP IS TO CHANGE THE STEADY STATE RESPONSES, $V_s$ TO $N = 1.0$, AND $T_L$ TO $N = 10$.

(c) THE SENSITIVITY CAN BE OBTAINED FROM EQ. 19.34 BY RECOGNIZING THIS AS A NEGATIVE FEEDBACK SYSTEM.

$$S = \frac{1}{1+GH}$$

THE SENSITIVITY FOR THE $V_s$ TO N TRANSFER FUNCTION IS

$$S = \frac{(0.004S+1)(0.001S+1)}{(0.004S+1)(0.001S+1)+50}$$

$$= \frac{0.02(0.004S+1)(0.001S+1)}{\frac{S^2}{3536} + \frac{2(0.176)S}{3536} + 1}$$

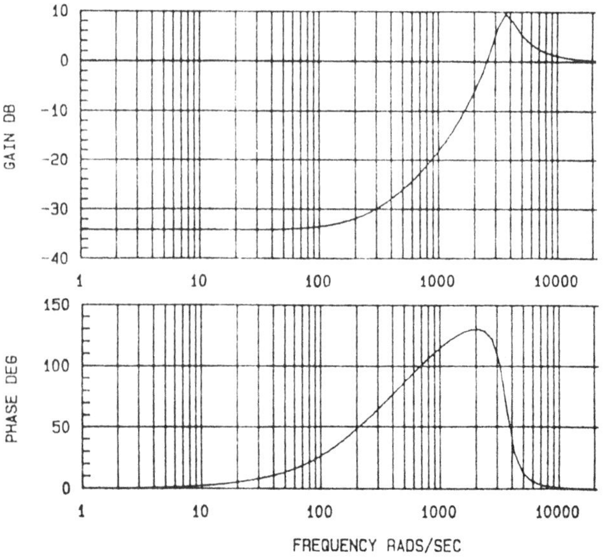


SENSITIVITY VS FREQUENCY

(d) FROM THE EQUATION FOR THE TOTAL CLOSED LOOP RESPONSE, N, THE RESPONSE DUE TO A STEP IN $V_s$ WILL BE SECOND ORDER. PUTTING THE EQUATION FOR N INTO STANDARD FORM (WITH $T_L = 0$),

$$N = \frac{V_s}{\frac{s^2}{(3536)^2} + \frac{2(0.176)s}{3536} + 1}$$

THIS SHOWS THAT THE NATURAL FREQUENCY IS $\omega = 3536$ 1/SEC, AND THE DAMPING FACTOR IS $\zeta = 0.176$.

THIS MEANS THAT THE RESPONSE TO A STEP WILL BE FAST (90% RISE TIME OF 0.4 MS), BUT WILL HAVE SIGNIFICANT OVERSHOOT (ABOUT 55%). THE RESPONSE WILL BE OSCILLATORY SINCE $\zeta < 1.0$.

AS LONG AS $T_L = 0$, THERE WONT BE ANY STEADY STATE ERROR.

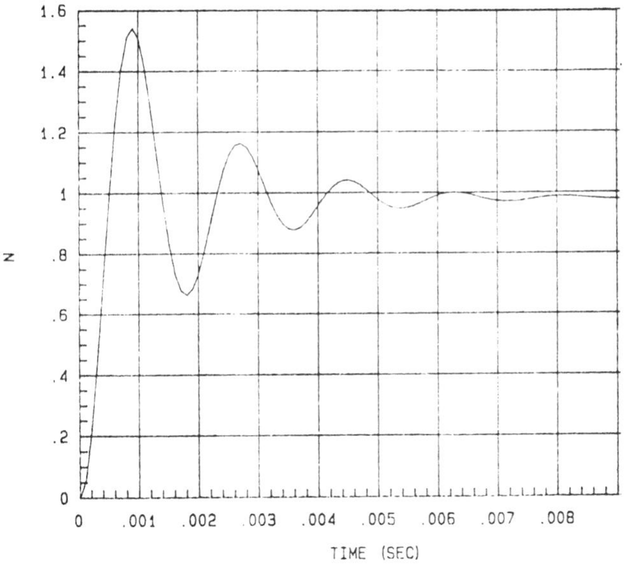


CLOSED LOOP STEP RESPONSE, VS TO N

(e) THE CLOSED LOOP RESPONSE TO A STEP LOAD CHANGE WILL BE SIMILAR TO THE STEP CHANGE IN $V_s$ EXCEPT THE NUMERATOR IN THE TRANSFER FUNCTION (N) WILL CAUSE THE RESPONSE TO DEVIATE FROM SECOND ORDER.

HOWEVER THE RESPONSE WILL STILL BE OSCILLATORY, AND THERE WILL BE OVERSHOOT. ALSO SINCE $T_L \neq 0$, THERE WILL BE AN ERROR FROM THE SETPOINT, $V_s$.

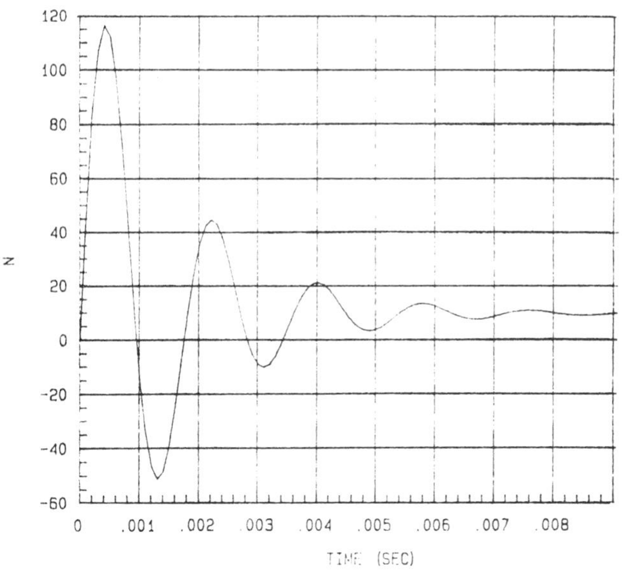


CLOSED LOOP STEP RESPONSE, TL TO N

(f) LET THE COMPARATOR GAIN EQUAL K, AND RESOLVE FOR THE OPEN LOOP TRANSFER FUNCTION FOR $V_s$.

$$\frac{N}{V_s} = \frac{500\,K}{(0.001\,s+1)(0.001\,s+1)}$$

CLOSING THE LOOP YIELDS

$$\frac{N}{V_s} = \frac{1}{1\times10^{-6}s^2 + 0.005\,s + 500K}$$

THIS IS A SIMPLE SECOND ORDER PROBLEM. THERE ARE ONLY TWO ROOTS AND THE SYSTEM WILL BE STABLE FOR ANY $K > 0$, AND WILL BE UNSTABLE FOR $K < 0$.

PRACTICALLY, ELEMENTS OF THE PHYSICAL SYSTEM WILL LIMIT THE GAIN, K, BUT SUCH LIMITATIONS ARE NOT PART OF THE LINEAR MODEL.

(g) THE SYSTEM STEADY STATE RESPONSE CAN BE IMPROVED BY ADDING INTEGRAL CONTROL. THIS WILL EFFECTIVELY COMPENSATE FOR ANY STEADY STATE DISTURBANCE DUE TO $T_L$. THIS ADDITION, HOWEVER HAS A SIDE EFFECT OF REDUCING THE STABILITY MARGIN OF THE SYSTEM. HOWEVER, IF PROPERLY DESIGNED, THE SYSTEM WILL STILL BE STABLE

**2** THIS IS A STANDARD SECOND ORDER SYSTEM.

$$F = Mx'' + Bx' + K_s x$$

SO, THE NATURAL FREQUENCY IS

$$\omega = \sqrt{K_s/M} = \sqrt{\frac{1200\ \text{LBF/FT}}{(100\ \text{LBM})\left(\frac{1}{32.2}\ \frac{\text{LBF-SEC}^2}{\text{LBM-FT}}\right)}}$$

$$= \boxed{19.6\ \text{RAD/SEC}\ \ (3.13\ \text{Hz})}$$

$$\zeta = \frac{B}{2\sqrt{K_s M}} = \frac{60}{2\sqrt{(1200)\left(\frac{100}{32.2}\right)}}$$

$$= \boxed{0.49}$$

FREQUENCY RESPONSE

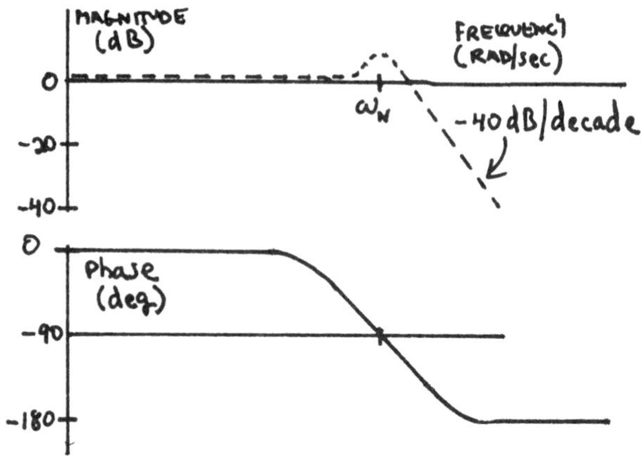


UNIT STEP RESPONSE

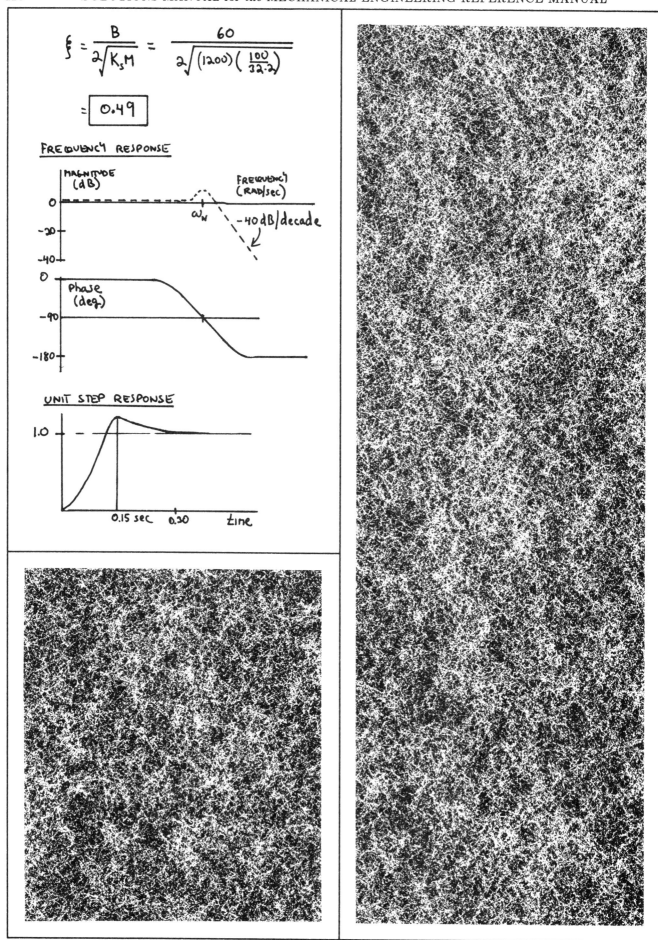

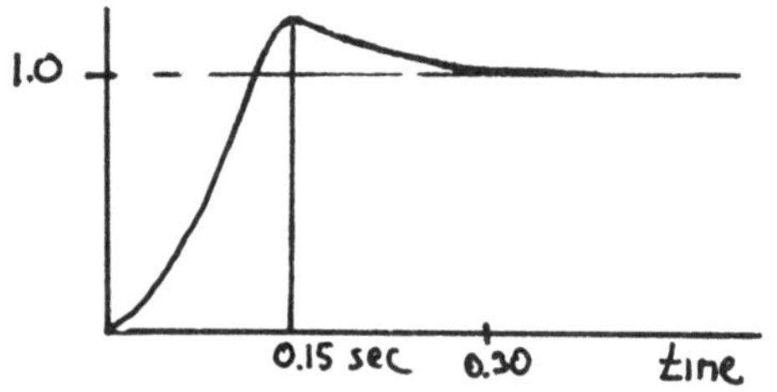

# QUICK — *I need additional study materials!*

Please send me the review materials I have checked. I understand any item may be returned for a full refund within 30 days. I have provided my bank card number as method of payment, and I authorize you to charge your current prices and shipping/handling charge against my account. (Don't forget a solutions manual for your reference manual.)

| **For the E-I-T Exam:** | **Solutions Manuals:** |
|---|---|
| ☐ Engineer-In-Training Reference Manual | ☐ |
| ☐ Engineer-In-Training Sample Examinations | |
| ☐ Engineering Fundamentals Quick Reference Cards | |
| ☐ E-I-T Mini-Exams | |
| ☐ 1001 Solved Engineering Fundamentals Problems | |
| **For the P.E. Exams:** | |
| ☐ Civil Engineering Reference Manual | ☐ |
| ☐ Civil Engineering Sample Examination | |
| ☐ Civil Engineering Quick Reference Cards | |
| ☐ Seismic Design of Building Structures | |
| ☐ Timber Design for the Civil P.E. Exam | |
| ☐ Mechanical Engineering Reference Manual | ☐ |
| ☐ Mechanical Engineering Sample Examination | |
| ☐ Mechanical Engineering Quick Reference Cards | |
| ☐ Electrical Engineering Reference Manual | ☐ |
| ☐ Electrical Engineering Sample Examination | |
| ☐ Electrical Engineering Quick Reference Cards | |
| ☐ Chemical Engineering Reference Manual | ☐ |
| ☐ Chemical Engineering Practice Exam Set | |

**Recommended for all Exams:**
- ☐ Expanded Interest Tables
- ☐ Engineering Law, Design Liability, and Professional Ethics
- ☐ Engineering Unit Conversions

For fastest service call
**415•593•9119**
Allow up to two weeks for UPS Ground shipping.

**SHIP TO:**

NAME ________________ COMPANY ________________
STREET ________________ APT ________
CITY ________________ STATE ________ ZIP ________
DAYTIME PHONE NUMBER ________________
CHARGE TO *(REQUIRED FOR IMMEDIATE PROCESSING):*
VISA/MC/AMEX NUMBER ________________ EXP. DATE ________
NAME ON CARD ________________
SIGNATURE ________________

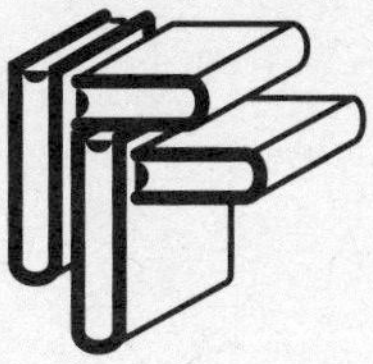

# ❖ *Send more information* ❖

Please send me descriptions and prices of all available E-I-T and P.E. review books. I understand there will be no obligation on my part.

**NAME** ________________
**ADDRESS** ________________
**CITY** ________________
**STATE** ________________ **ZIP** ________

A friend of mine is taking the exam, too. Send additional literature to:

**NAME** ________________
**ADDRESS** ________________
**CITY** ________________
**STATE** ________________ **ZIP** ________

# *I have a comment ...*

☐ I think you should add the following subject to page ________.
☐ I think there is an error on page ________. Here is the way I think it should be:

Title of this book: ________________
Edition: ________________ Printing: ________________
Contributed by (optional): ☐ Please tell me if I am correct.

NAME ________________
ADDRESS ________________
CITY ________________
STATE ________________ ZIP ________

NO POSTAGE NECESSARY IF MAILED IN THE UNITED STATES

**BUSINESS REPLY MAIL**

FIRST CLASS MAIL PERMIT NO. 33 BELMONT, CA

POSTAGE WILL BE PAID BY ADDRESSEE

PROFESSIONAL PUBLICATIONS INC
1250 FIFTH AVE
BELMONT CA 94002-9979

NO POSTAGE NECESSARY IF MAILED IN THE UNITED STATES

**BUSINESS REPLY MAIL**

FIRST CLASS MAIL PERMIT NO. 33 BELMONT, CA

POSTAGE WILL BE PAID BY ADDRESSEE

PROFESSIONAL PUBLICATIONS INC
1250 FIFTH AVE
BELMONT CA 94002-9979

NO POSTAGE NECESSARY IF MAILED IN THE UNITED STATES

**BUSINESS REPLY MAIL**

FIRST CLASS MAIL PERMIT NO. 33 BELMONT, CA

POSTAGE WILL BE PAID BY ADDRESSEE

PROFESSIONAL PUBLICATIONS INC
1250 FIFTH AVE
BELMONT CA 94002-9979